Shark Biology and Conservation

Shark
Biology and Conservation

Essentials for Educators, Students, and Enthusiasts

Daniel C. Abel
R. Dean Grubbs

With contributions from **Tristan Guttridge**

Illustrated by
Elise Pullen and Marc Dando

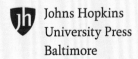
Johns Hopkins
University Press
Baltimore

Johns Hopkins University Press
2715 North Charles Street
Baltimore, Maryland 21218-4363
www.press.jhu.edu

Library of Congress Cataloging-in-Publication Data

Names: Abel, Daniel C., author. | Grubbs, R. Dean, author. | Guttridge, Tristan, author.
Title: Shark biology and conservation : essentials for educators, students,
 and enthusiasts / Daniel C. Abel, R. Dean Grubbs, with contributions from
 Tristan Guttridge ; illustrated by Elise Pullen and Marc Dando.
Description: Baltimore, Maryland : Johns Hopkins University Press, 2020. |
 Includes bibliographical references and index.
Identifiers: LCCN 2019045796 | ISBN 9781421438368 (hardcover) | ISBN
 9781421438375 (ebook)
Subjects: LCSH: Sharks. | Sharks—Conservation.
Classification: LCC QL638.9 .A245 2020 | DDC 597.3/4—dc23
LC record available at https://lccn.loc.gov/2019045796

A catalog record for this book is available from the British Library.

*Special discounts are available for bulk purchases of this book. For more information,
please contact Special Sales at specialsales@press.jhu.edu.*

Johns Hopkins University Press uses environmentally friendly book materials,
including recycled text paper that is composed of at least 30 percent post-consumer
waste, whenever possible.

CONTENTS

This is not a coffee table book, but if you leave it in plain view we suspect people will thumb through it to take a closer look at our cool photographs and artwork. Nor is it a field guide, but in a pinch it can help you identify more common species, or narrow your choices. And this book is neither a textbook nor a scholarly volume, though you would be excused for concluding the opposite due to the breadth of coverage, scientific citations, and terminology not typically found in coffee table books and field guides.

So exactly *what kind of book is this?* As the title denotes, we wrote *Shark Biology and Conservation: Essentials for Educators, Students, and Enthusiasts* for those in need of or curious about accurate information on sharks presented in what we hope is an accessible but not watered-down format. In doing so we do not disparage coffee table books, field guides, or scholarly tomes—a glimpse of our bookcases would reveal shelves of these. But no books in these categories have it both ways; that is, be popular and scholarly at the same time. So, that is in part our goal and challenge.

One of the reasons we wrote this book is an attempt to supplant fear of sharks (known as *galeophobia* and *selachophobia*) with respect, and myth with knowledge. We hope to do so by providing a comprehensive overview of sharks, including recent scientific advances that will feed your fascination, calm any fears you might harbor, and explain why sharks are critical to ocean health. We will also broaden the discussion to include the rays, or *batoids*, an overlooked group closely related to sharks, which are equally fascinating and, in a few cases, are far more endangered than the vast majority of sharks.

The idea for this book arose during Coastal Carolina University's *Biology of Sharks* course, a three-week class held in part on wondrous Winyah Bay and at the extraordinary Bimini Biological Field Station. The class has been offered annually since 1996, making it perhaps the longest running shark biology course globally.

The main drawback to the course was the lack of a suitable textbook. To be sure, there are a number of exceptional books on sharks and their relatives. However, most shark-related nonfiction books for nonspecialists are field guides, coffee table books, personal narratives, natural histories, stories of shark attacks, and so on, written mainly for a general audience.

Highly specialized books on sharks are more technical and their content and writing are more accessible to graduate students and specialists than to other students. These provide an exhaustive survey and synthesis of facts and concepts and contain complex graphs and diagrams. Both of us, along with others in the shark research community, use these texts regularly, but generally not as textbooks for our undergraduate courses. Indeed, without this assemblage of more technical books, this book would not have been possible.

The books closest to meeting our needs were *Sharks, Skates, and Rays*[1] by William Hamlett, *Biology of Sharks and Their Relatives*, edited by Jeffrey Carrier, John Musick, and Michael Heithaus,[2] and Peter Klimley's *The Biology of Sharks and Rays*.[3] We highly recommend these outstanding books but, as great and thorough as they are, they did not hit the sweet spot we were seeking.

Moreover, as we contemplated writing the book for students in our shark biology course, we realized that the shark booklist had largely overlooked a surprisingly expansive market: educators, students, advanced enthusiasts, field biologists, naturalists, and marine biologists who might not have a background in fish biology or sharks. What the market is missing for this group is a comprehensive, systematic overview of the diversity, evolution, ecology, behavior, physiology, anatomy, and conservation of sharks and their relatives written in a style that is sufficiently detailed but not too technical or intimidating.

The field of shark biology has blossomed in the last several decades. The American Elasmobranch Society, the professional organization for biologists and fishery scientists specializing in sharks, skates, and rays, boasts an estimated membership of around 400 and holds well-attended national meetings. Similar societies exist both regionally and globally.

Courses in shark biology have appeared on numerous college campuses. There is even a shark MOOC (massive open online course) called *Sharks! Global Biodiversity, Biology, and Conservation*. These all have coincided with renewed public interest in this group of organisms, whose populations in some cases are threatened by a number of human impacts and with whom there is an enormous depth of fascination.

This book draws on our combined 65+ years of experience as shark biologists and educators, as well as our extensive connections to the close-knit shark biology community (and the generous and enthusiastic contributions from many of them). We hope you have fun reading this book. Even more, we would deem our time writing it worthwhile if you use this book to expand your knowledge of these wondrous beasts and make it a springboard to educate others, stoke your enthusiasm and passion, and work so that there is always a place for sharks, and all of their relatives (including humans).

One final word about the material covered by this book. To keep the book affordable, the length of this book, including the number of illustrations, was limited. To meet this limit, we made tough choices and shortened some topics and excluded others. We hope we did this skillfully, but if you think that we omitted or devoted insufficient space to key concepts, now you know why.

NOTES

1. Hamlett, W. C. (ed.). 1999. *Sharks, Skates, and Rays: The Biology of Elasmobranch Fishes*. Johns Hopkins U. Press.

2. Carrier, J. C., Musick, J.A., and Heithaus, M.R. (eds.). 2012. *Biology of Sharks and Their Relatives*. CRC Press.

3. Klimley, A. P. 2013. *The Biology of Sharks and Rays*. U. of Chicago Press.

ACKNOWLEDGMENTS

The acknowledgment section is the last part of this book that we wrote, our last gasp. After expending every remaining ort of creativity, knowledge, energy, humor, and so on, in our arsenal writing *Shark Biology and Conservation*, we find ourselves struggling to recognize and offer our gratitude to our family, friends, colleagues, students, editors, and staff at Johns Hopkins University Press, and numerous others, in proportion to their exponential contributions to this book. Some clichés exist because they are truths, and this one is not *fake news*: We could not have written a grocery list, much less a book, without their presence in our lives and their support.

Let us begin by thanking those colleagues and friends who stepped up, sometimes with unexpected zeal for the project, when we asked, or even volunteered before we could hit them up for favors. All of these talented professional and amateur photographers below willingly offered their works. We thank, in no particular order: Jeff Carrier, Robert Johnson, George Boneillo, Emily Marcus, Marcus Drymon, Lesley Rochat, Annie Guttridge, Chelle Blais, Matt Smukall, Steven Kessel, Craig O'Connell, Nick Wegner, Lai Chin, Andrew Raak, Bryan Keller, D. Ross Robertson, Caroline Collatos, Matt Larsen, James St. John, Gavin Naylor, Charles Cotton, Steven Kajiura, Dave Itano, Jason Romine, Joshua Bruni, Ken Goldman, Shmulik Blum, Matt Potenski, Erin Burge, Bob Crimian, Eugene Kitsios, David Gandy, Kathryn Dickson, Sandra Brooke, Chugey Sepulveda, Lance Jordan, Brendan Talwar, Charles Messing, Jim Luken, Mackellar Violich, Joel Blessing, Nick Whitney, Amanda Brown, Michael Scholl, Laura Stone, and Trey Spearman. In some cases we were saved by Creative Commons photos, and the photographers are cited in figures in which these have been used. We thank Greenpeace and the Florida Fish and Wildlife Conservation Commission for photographs, as well as Shutterstock, Kelvin Aitken of Marine Themes, and Barbel Knieper of Biosphoto.

For permission to use graphics and photographs, we also thank Oxford University Press, NOAA and NOAA Fisheries, and the Bimini Biological Field Station.

Skillful and talented scientific illustrators are rare and expensive to engage. We were very fortunate to have not one, but two principal artists for this book, neither of whom will earn near what they are worth for their contributions to *Shark Biology and Conservation*. Almost all of the illustrations except those in Chapter 3 were the result of extremely fortuitous serendipity. In the fall of 2017, Elise Pullen became one of Dan's graduate students. To his surprise, she was already a professional artist. Her work is so detailed and accurate that it may surprise readers that this effort is her inaugural entry into scientific illustration, but we think you will agree that it will not be her last! And her professionalism, enthusiasm, flexibility, and attention to detail ensure a great career for her.

All of the magnificent shark watercolors for Chapter 3 were done by artist

Marc Dando, whose reputation as an exceptional wildlife illustrator is long-established. We think you will agree that *exceptional* is an understatement in Marc's case. Marc gleefully met our demanding schedule and was among the easiest people with whom to work, in spite of his numerous other impending deadlines.

We also thank other illustrators and colleagues who provided illustrations. These include Mark Grace, Matthew Kolmann, Kurt Smolen, James Gelsleichter, Roi Gurka, and Read Frost.

When we determined that shark behavior was the single topic we were entirely uncomfortable, even unqualified, writing by ourselves, we turned to colleague and friend Tristan Guttridge, a shark behaviorist at the top of his field. By the time this book is published, you may well know him more as a *Shark Week* personality, but his credentials as a behaviorist are beyond reproach, and we are very grateful for his participation here.

Even after friends and colleagues provided photos and line art for the book, we had numerous gaps to fill. We discovered that we were naïve to think that we could write a book with all of the world-class photographs and artwork we would need without a budget. Thanks to the previously acknowledged friends and colleagues listed above, we *almost* did. A generous grant from the Save Our Seas Foundation, and the active involvement and enthusiasm of CEO Michael Scholl allowed us to engage our two main illustrators and contributing author Tristan Guttridge, plus obtain some additional photographs. *Shark Biology and Conservation* would likely be destined to be a bargain bin book, wasting the gorgeous photos and illustrations from friends and colleagues, without Michael and the Save Our Seas Foundation's integral role.

We both are also grateful to the anonymous reviewers of both the prospectus for this book and the first complete manuscript. We hope that the finished book makes them proud.

This book would never have materialized were it not for the opportunities provided by the Bimini Biological Field Station. We thank the staff, past and present, for accommodating our courses and providing access to the wondrous natural world surrounding Bimini to us and more than a thousand of our students.

And a whale-shark-sized thanks and endless espressos on demand to our editor, the unflappable Tiffany Gasbarrini. When we submitted a preliminary manuscript with nearly 500 illustrations, 150 more than our contractual agreement, instead of the venomous response we deserved, she wrote that she was *somewhat discomfited* with the overage. She gave us mostly free rein throughout the process to write the book we wanted, as long as we kept in mind that it had to be *affordable*, which meant shortening the verbiage. We are making her a t-shirt with *It's the Word Count, Stupid*, on the front. We could not imagine a better editor.

Except for Tiffany Gasbarrini, we have crossed paths with almost all of the Johns Hopkins University Press production and marketing team only by email, and thus we cannot with certainty know that they are indeed people and not

Artificial Intelligence. Still we owe them our most profound gratitude for the pride they have obviously taken in publishing our book. When cloning of people becomes a thing, we urge our publisher to replicate Editorial Assistant Esther Rodriguez first. She made our lives as authors so much easier by her general excellence, professionalism, and responsiveness. Some of the Johns Hopkins University Press team that worked on this book have remained anonymous to us. We thank you, as well as managing editor Juliana McCarthy, senior production editor Kimberly F. Johnson, acquisitions associate Meredith Gaffield, production editor Hilary S. Jacqmin, publicity manager Kathryn Marguy, publicist Rebecca Rozenberg, and art director Martha Sewall.

Finally, we thank indexer Michael Tabor as well as our copy editor Liz Radojkovic, devotee of the Oxford Comma and cat lover, for her kind, strong editing hand, and her marginal notes of encouragement.

Daniel C. Abel

In addition to the above, I thank Coastal Carolina University for providing the sabbatical leave during which most of my part of this book was written, the Marine Science Department faculty, for taking over my committee assignments during my sabbatical absence, and the department's staff, especially Tammy Parker, who cheerfully protected me from my administrative loose cannon proclivities. I thank MSCI department chair Jane Guentzel for encouraging this project. I also thank our boat captains Sam Gary, Edwin Jayroe, Richard Goldberg, Jaime Phillips, and Ed Keelin; the legions of undergraduate volunteers willing to awake at ungodly hours just to be a part of the Coastal Carolina University Shark Project; and especially my MSc students past and present. My former MSc student, and Dean's current PhD candidate, Bryan Keller reviewed the entire manuscript, and Vivian Turner helped with the chapter on swimming in sharks. Numerous friends, especially Terry Munson, William Holliday, and Robert Sturgis, through their interest and words of encouragement, provided inspiration beyond which they could know. Journalist James Borton reviewed the manuscript and offered suggestions based on his considerable prowess as a writer.

I also thank my co-author Dean Grubbs. Dean and I have been friends and teaching colleagues for over 25 years. Dean is many things: consummate educator, passionate and scientifically informed conservationist, keeper of a llama and fainting goat, and perhaps the most knowledgeable and insightful shark biologist on the planet. I would not have ever envisioned writing a book on sharks if Dean had not been a part of it.

More than gratitude is also owed to the professorial giants who modeled for me what a teacher-scholar-citizen should be. There are too many to name, but five of whom I think about constantly are Reid Wiseman, Norman Chamberlain, and Chris Koenig of the College of Charleston, and Jeff Graham and Ralph Shabetai, my PhD co-advisors from Scripps Institute of Oceanography and the School of

Medicine at the University of California at San Diego, respectively. They taught me how to think and comport myself like a scientist. Chris continues to inspire me, both from the seeds of wisdom he planted nearly four decades ago about science and life and, more recently, for recommending that I use honey in my caipirinhas. If my undergraduate and graduate students profited in meaningful ways from their association with me, they should thank Chris.

Then there are my late parents, Harris and Ruth, my brothers, Billy and Alan, and my sister, Sara, all of whom made sacrifices and contributions that nurtured my love of knowledge and marine life in my childhood and beyond. And finally, my wife Mary, daughter Juliana, and son Louis. For them, words are inadequate, but I am confident they know how they have sustained me to work tirelessly for a world in which sharks always have a place to swim safely and just be sharks, this book being only one relatively small manifestation.

R. Dean Grubbs

I am eternally grateful to my academic mentors that encouraged but also challenged me throughout my career. This includes my undergraduate advisor Dan Diresta, who encouraged me to lead a scientific reading group on sharks. I thank my friend and confidant, John Morrissey, a masterful lecturer who strongly encouraged me to pursue graduate school, gave me my first opportunities to teach, and helped me push forward when my confidence in this career path wavered. None of my career would have been possible without the opportunities for research and education provided by Sonny Gruber, with whom I conducted my first shark research and developed my first shark biology courses. Dr. Gruber was my academic father and I was privileged to maintain a relationship with him for 30 years until his passing in 2019. I am also indebted to Arthur Myrberg who, in addition to being Sonny Gruber's doctoral advisor, was my favorite undergraduate professor. Whatever skill I have as an educator comes from being a very poor mimic of the two best I have known, Art Myrberg and John Morrissey. I am also incredibly grateful to my doctoral advisor, Jack Musick, a renowned scientist and naturalist cut in the mold of David Starr Jordan. In a time when naturalists have become scarce in academia, Jack encouraged those tendencies as well as my propensity to engage in disparate lines of research. He showed me that there are still places for our ilk in academic careers. I thank my post-doctoral advisor Kim Holland, who, among many things, taught me the value of balancing one's professional and personal life, a skill I continue to struggle to hone. Finally, I also thank Captains Tony Pinello and Durand Ward, friends and confidants that taught me much about life and the sea.

I thank the FSU Coastal and Marine Lab administration that has given me the freedom to pursue my research and educational interests unfettered, and the FSUCML staff, without whom none of my work would be possible. I am also thankful to the many colleagues with whom I have collaborated on many projects, some that are highlighted in this book. This list is too long to include as I have been extremely fortunate to have a large number of incredible collaborators. As the academic generations progress forward, I am also extremely grateful to the many undergraduate and graduate students that have chosen to conduct research in my lab. They teach me as much as I them and continually make me proud and make me look better than I am.

I also thank my co-author, collaborator, co-instructor, and friend Dan Abel. Dan is an incredibly talented and humorous writer and educator, and over the many years that he and I taught our shark courses we mused over the possibility of writing this book. Dan (and I) knew it would only happen if he took the initiative to push it forward and I am grateful to him for all of the hard work he put in to seeing this project through and to produce a book of which we are very proud.

Finally, I am especially grateful to my parents Audrey and Ralph Grubbs, who exposed me to marine life through fishing and snorkeling in nearly every family vacation, encouraged my crazy childhood notion to become a marine biologist who studies sharks, and sacrificed to help make it possible. I would not be where I am if not for them.

Dedication

In April 2019, our friend and mentor Samuel "Doc" Gruber died. No topic in this book escaped his fingerprints, and neither of us would be where we are without having had Doc in our lives. While our sadness at his loss is profound, we hope that dedicating this book to Doc honors his outsized legacy.

Authors' Note

Readers of this book know exactly which fish we mean when we say *Great White Shark*. There is only one such species, and it is easily distinguished from even closely related species like the Shortfin Mako and Porbeagle. Scientifically, the Great White Shark, which more accurately is called the White Shark, is known as *Carcharodon carcharias* everywhere in the world, in accordance with the rules of scientific nomenclature that assign a unique two-part name to every species.

But do you instinctively picture one specific animal called a *Sand Shark*? In the Southeast United States, the name *Sand Shark* is used by local residents for numerous species that include Blacktip, Sandbar, Dusky, Silky, Atlantic Sharpnose, and Blacknose Sharks, and even Guitarfish, which are not sharks but rather are rays, a closely related group. Or perhaps even what is pictured in the figure below. Basically, any shark that is not a Hammerhead and is found nearshore in the Southeast United States, especially near sandy beaches, has been called a Sand Shark.

And that is more than merely a trivial problem, since assigning one name to numerous different species can easily lead to cases of mistaken identity with serious consequences. In Chapter 11, for example, we discuss a case where grouping the Dusky Smoothhound and Spiny Dogfish both as *dogfish* in managing a fishery nearly led to potentially catastrophic consequences for the latter.

The problem is this: in science, a species is a precisely defined, unique entity, with its own distinct genetic makeup, distribution, life history, behavior, environmental requirements, and so on. Calling two different species by the same common name denies at least one of them its unique biological heritage.

A *species* is a real concept in nature, not a human construct. For sexually reproducing organisms, a species consists of individuals that will interbreed with each other, are reproductively isolated from (i.e., do not interbreed with) other species, and which will produce offspring that are viable (capable of surviving until adulthood) and fertile (capable of having offspring).

And every species is assigned its own unique binomial (two-part) name, by a rigorous process specified in the *International Code of Zoological Nomenclature* (ICZN). Each and every reader of this book, excluding artificial intelligence, is a *Homo sapiens* and nothing else, no matter your race or nationality. *Homo sapiens* is the species name. *Homo* is our genus, and *sapiens* is our specific epithet.

Among sharks, we have both studied *Carcharhinus plumbeus*, which is variously known as a Sandbar Shark, Brown Shark, and yes, Sand Shark. There are over 500 named species of sharks, and a total of over 1250 species of sharks and shark relatives—the skates, rays, and chimaeras (or ghost sharks).

The result of using a standard system of nomenclature is an organized classification that scientists and policymakers all over the world can reliably use when studying an organism, determining its habitat requirements, assessing the health of its populations, making policy decisions, and so on.

This takes us to our dilemma: whether to use common or scientific names, or both, in this book. Whether, on the one hand, to risk alienating readers happy with the use of a common name and perhaps intimidated or otherwise put-off by difficult-to-pronounce names derived from Latin or Greek, or on the other hand to betray our fealty to scientific accuracy.

We lied. It really is not a dilemma to us. We will use scientific names at least the first time a species is mentioned in every chapter (except Chapter 1), but not in every instance. When we do use them, we will also include widely accepted common names.

Speaking of common names, the latest convention, in accordance with the American Fisheries Society's authoritative *Common and Scientific Names of Fishes from the United States, Canada, and Mexico,* is to capitalize the first letters of all common names. Although not all publications use this convention, we do in this book, except when we refer to a group of sharks in the same genus or family. For instance, it would be hammerheads or threshers, but S̲calloped H̲ammerhead and B̲igeye T̲hresher.

There is a second major issue as well that we need to address. This is a book about sharks, but much of what is true about sharks also applies to their close relatives, the skates and rays, with whom sharks are classified as *elasmobranchs.* Rays and skates are so closely related to sharks that the former are sometimes called *pancake sharks.* Sharks, skates, and rays together are closely related to a more obscure group called *chimaeras* or *ghost sharks,* and collectively sharks, skates, rays, and chimaeras are known as *chondrichthyans.*

So, how to proceed? Chances are, you bought this book because you wanted to read about sharks (*Sharks,* after all, is in the title) so we will focus on sharks. Occasionally we will broaden the discussion by referring to elasmobranchs or sharks and their relatives. And there are times when we discuss only skates and rays, given their close relationship to sharks, and when we do this, we usually call them *rays* or *batoids.*

There, now that *that* is settled, let us focus on this superb assemblage of cartilaginous fishes, especially the sharks!

Overview

1 / Introduction

Introduction: What Is a Shark?

Scary, vicious, dangerous, ancient, primitive, threatening, and *monsters.* These are some of the descriptions typically used by the public to answer the question *What is a shark?* These responses are based more on perceptions than on fact, and they do not answer the question.

If you ask people to name a typical shark, the answer is usually an iconic species like a White, Tiger, or Bull Shark. These, however, are decidedly not *typical* sharks. Nearly two-thirds of the more than 500 species of living sharks are no larger than 1 m (3.3 ft), and over half of the species live deeper than 200 m (660 ft). A typical shark is not large, gray, fast-swimming, and found along coastline but, rather, is small, brown, and lives in the deep ocean.

And we cannot forget the rays, often referred to as *batoids,* which have a common ancestor with modern sharks and together with sharks are classified as *elasmobranchs* (see box 1.1). The batoids are a group of more than 600 species of dorsoventrally (top-to-bottom) flattened fishes, and about half of these species also live in the deep sea.

Let us start our journey into the biology of sharks (and rays) by first describing the features that distinguish them from other groups.

External Anatomy and Distinguishing Features

Examine Figure 1.1, the basic external anatomy of a shark. Note the anterior head, with snout, nostrils, mouth, and eyes, and the posterior caudal fin, the main propulsive fin.

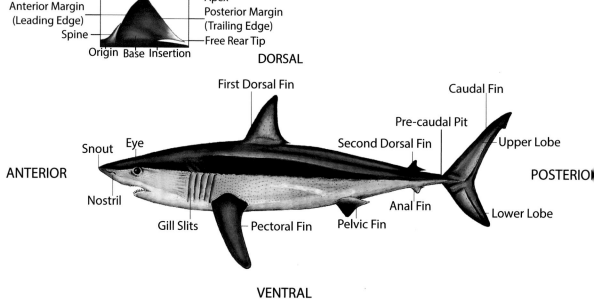

Figure 1.1. External anatomy of a shark (in this case a female Shortfin Mako), with a close-up of a dorsal fin (from a Gulper Shark) showing terminology useful in describing sharks. The Mako has no spiracles.

Sharks have both median fins (along the centerline) and bilateral fins (on both sides). Most sharks have three median fins: two dorsals and a single anal fin. Some groups have lost a dorsal fin through evolution and possess only one—for example, the cow sharks, which include the sixgill and sevengill sharks, in the order Hexanchiformes.[1] The entire 160 species of dogfish sharks, a diverse group in the superorder Squalomorphii, have lost their anal fin. No batoids possess an anal fin and many possess no defined dorsal fins.

There are two sets of paired bilateral fins: the more anterior pectorals and the posterior pelvics. These fins can also technically be called *limbs* (although no shark biologist would do so commonly), and all vertebrates have them except where they are lost or vestigial, as in snakes. In terrestrial vertebrates, we call the pectorals *arms* or *wings*, and the pelvics, *legs*.

The primary visible difference between sharks and rays is the placement of the gill slits (five in most species), which are lateral in the sharks and ventral in the rays. Of the approximately 1200 different species of sharks and rays, only nine—eight sharks and a single ray—have more than five bilateral gill slits.

Many sharks have *spiracles*, paired, bilateral openings behind the eyes that connect the mouth to the water environment. In the requiem sharks (members of the family Carcharhinidae) like the Blue, Bull, and Lemon Sharks, the spiracles are vestigial, meaning that the structure is rudimentary and has lost its core function.

In most rays, the primary source of incurrent water for the gills is through the spiracles. Exceptions are pelagic batoids (i.e., those that actively swim in open water) like the mantas. Mantas swim with their mouths open, as do most sharks, a phenomenon appropriately called *ram ventilation*, and in doing so irrigate their gills with oxygenated water. No need for spiracles if you are always swimming and merely opening your mouth serves the same purpose.

Other external features in some species include a precaudal pit (for increased tail flexibility when present), one or more ridges and keels along the body, and spines on the leading edge of one or more of the dorsal fins.

Now that you know some of the terminology of the external features, here are the distinguishing features that together define a shark. Some of these we necessarily gave away above, and some are not unique to sharks, but together these define sharks (and rays) and no other organisms.

First, a shark is a *fish* (i.e., a vertebrate with fins for locomotion and gills for breathing). Biologists distinguish between two major groups of fishes, *bony* and *cartilaginous* (box 1.1). The former group, including tuna, mullet, grouper, and so on, numbers approximately 33,000 species and is the most *speciose* (species-rich) and diverse assemblage of all vertebrates, whereas there are only about 1250 different kinds of cartilaginous fish; that is, sharks, rays, plus a relatively obscure but ancient group known as *chimaeras* or *ghost sharks*.

Twelve hundred and fifty species of cartilaginous fishes may seem like a lot of variety until you realize that there are over three times more kinds of *catfish* (3900 species), birds (> 10,000), amphibians (> 7000), and mammals (> 5000). We discuss reasons for this disparity in Chapter 3.

The essence of the difference between bony and cartilaginous fishes is an anatomical feature you cannot discern externally: the material comprising the internal skeleton. Bony fishes, which are in the class Osteichthyes (*oste* = bone; *ichthyes* = fish), have a skeleton composed of dense, rigid, *ossified* tissue, hardened by calcium and other mineral salts, with nerves and blood vessels.

Sharks, as members of the class Chondrichthyes (*chondr* = cartilage), possess a skeleton that is tough yet lighter and more flexible than bone, but which lacks nerves and blood vessels (fig. 1.2).

Mammals, including humans, also have cartilage, three different kinds in fact, in places like the nose, ears, bronchial tubes, trachea, ribs, ends of long bones, and the discs between your vertebrae. Cartilage comprises about 0.6% of a typical mammal's body weight but from 6% to 8% of a shark's.

The cartilage of both chondrichthyans and humans is not as hard as bone, because it lacks bone's high levels of calcium and phosphorous salts. However, in sharks, at certain stress points the cartilage is strengthened by incorporating calcium in the form of the mineral apatite. Because these crystals are arranged in a pattern resembling a prism, this type of cartilage is known as *prismatic calcified cartilage* (fig. 1.2A).

Chemically, cartilage is composed principally of protein and sugar molecules,

BOX 1.1

Who's Who in the World of Sharks

In the Linnaean system of nomenclature, every species is given a unique two-part name that is recognized globally and unambiguously as that, and only that, species; for example, *Carcharhinus plumbeus* is most commonly known as the Sandbar or Brown Shark.

Recall the classic taxonomic hierarchy of life: Domain—Kingdom—Phylum—Class—Order—Family—Genus—Species.* Thus *C. plumbeus*** is also, starting with the most inclusive category, a eukaryan, animal, chordate, vertebrate (or craniate), gnathostome, chondrichthyan, elasmobranch, galeomorph, carcharhiniform, and carcharhinid in the genus *Carcharhinus*.

Sandbar Shark
on a longline near
Kaneohe, Hawaii

Domain: **Eukarya**
Refers to organisms that are eukaryotic (i.e., possess membrane-bound organelles and a nucleus surrounded by a nuclear envelope with two membranes).

Kingdom: **Animalia**
Organisms that are multicellular, with mode of nutrition known as heterotrophic (i.e., not photosynthetic) by ingestion (as opposed to absorption, like Fungi).

Phylum: **Chordata**
Organisms with a notochord (skeletal supporting rod) and pharyngeal gill slits at some point during development, as well as a single, hollow, dorsal nerve chord.

Subphylum: **Vertebrata** (or Craniata)
The most prominent characteristic is a skull and vertebral column. Vertebrates also exhibit bilateral symmetry, a brain, and sensory organs.

Superclass: **Gnathostomata**
"Super" in this case is not a superlative. Gnathostomes are the jawed vertebrates and include all vertebrate classes except the jawless fishes (hagfish and lampreys).

Class: **Chondrichthyes** (or Elasmobranchiomorpha)
Chondrichthyan fish (sharks, rays, and chimaeras) possess an endoskeleton made of cartilage, employ internal fertilization using claspers, lack a swim bladder, and have a heterocercal tail and ampullae of Lorenzini.

Subclass: Elasmobranchii (also called Euselachii)

Elasmobranchs (*ee-las'-mo-branks*), the sharks and rays, are the sister-group (i.e., the closest related group) to the subclass Holocephali, or the chimaeras, and together these groups form the class Chondrichthyes.

Superorder: Galeomorphii

Galeomorphii is one of three superorders and the sister-group to the other two, Squalomorphii and Batoidea. Diverse group of small, demersal (living on the bottom) and large, pelagic predators.

Order: Carcharhiniformes

The Carcharhiniformes (*kar-kar-eye'-nɪ-form-eez*) is the largest order of sharks, with about 296 species of sharks in nine families, commonly called the *ground sharks*.

Family: Carcharhinidae

The Carcharhinidae (*kar-kar-eye'-nɪ-dee*) is the second largest family of sharks, with about 60 species. The family is commonly known as the *requiem sharks*.

Genus: Carcharhinus

Carcharhinus (*kar-kar-eye'-nɪs*) is the genus of 36 species of carcharhinid sharks, including the Sandbar, Blue, Oceanic Whitetip Sharks, and so on.

Species: *Carcharhinus plumbeus*

Carcharhinus plumbeus (*kar-kar-ey'e-nɪs* and *plum'-bee-us*), written in italics or underlined, and all in lower case except the first letter, is the unique name assigned to the shark we commonly refer to as the Sandbar Shark. This two-word appellation is the *species*. Technically, *Carcharhinus* is the genus, and *plumbeus* is the specific epithet; the term *species* is reserved for, and only for, both words together: *Carcharhinus plumbeus*. Note that the term genus-species, though commonly used, is wrong, and would be the meaningless *Carcharhinus Carcharhinus plumbeus*.

Although there are exceptions, there are standardized endings for most of the above taxonomic categories. These are:

- -morph = superclass
- -morphii = superorder
- -iformes = order
- -oidei = suborder
- -idae = family
- -inae = subfamily
- -ini = tribe
- -i or -ii = patronym (named for a male)
- -ae = matronym (named for a female)
- -ensis = location (named for a location)

----

* If needed, or for entertainment, consult the Internet for mnemonic methods of recalling the hierarchy of major taxonomic categories. On mnemonic-device.com, we found the following: <u>D</u>runken <u>K</u>angaroos <u>P</u>unch <u>C</u>hildren <u>O</u>n <u>F</u>amily <u>G</u>ame <u>S</u>hows and <u>D</u>o <u>K</u>oalas <u>P</u>refer <u>C</u>hocolate <u>O</u>r <u>F</u>ruit, <u>G</u>enerally <u>S</u>peaking? We selected these as examples so we can sell more books in Australia.

** Note that we abbreviated *Carcharhinus* here using only the letter C, followed by a period, which is permissible once you have typed out the entire scientific name previously.

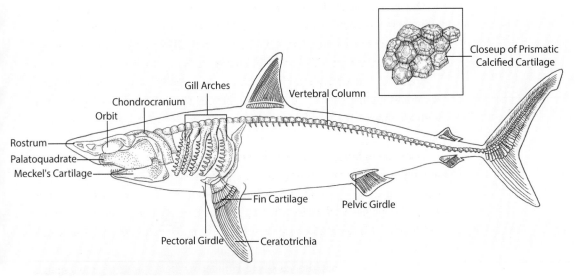

Closeup of Prismatic Calcified Cartilage

Gill Arches

Chondrocranium

Orbit

Vertebral Column

Rostrum

Palatoquadrate

Meckel's Cartilage

Fin Cartilage

Pelvic Girdle

Pectoral Girdle

Ceratotrichia

A

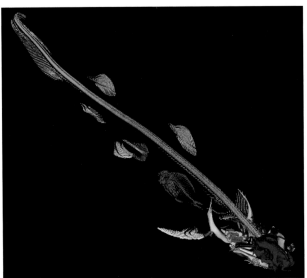

B

Figure 1.2. (A) Cartilaginous skeleton of a shark. (Adapted from Whitely 2015. http://www
.fossilsofnj.com/shark/cartilage.htm. Accessed 01/23/19) (B) Computed tomography (CT) scan
of a Lemon Shark showing the cartilaginous skeleton. The bony skeleton of a prey item
can be seen in the Lemon Shark's stomach. (Courtesy of Gavin Naylor/Chondrichthyan Tree of
Life Project/sharksrays.org)

specifically *collagen*, an abundant fibrous protein, and *proteoglycans*, protein-sugar
combinations.

In addition to being less rigid, cartilage is also lighter than bone. The density[2]
of bone in humans is about 1.85 g/ml, and in bony fishes, like mackerel and pike,
ranges from 1.3 to 1.5. The density of shark cartilage is about 1.1. For reference,
the density of pure water is 1.000, seawater is about 1.025, honey is 1.36, and concrete is 2.4. So, honey is denser than cartilage.

One benefit of cartilage to sharks is that its reduced density in part compensates for this group's lack of a swim bladder, a gas-filled internal structure found in most bony fishes that allows them to adjust their buoyancy, and in the process save precious energy and expand their range of living spaces.

Sharks are relatively muscle-bound (fig. 1.3) and are heavier than seawater, with most having whole body densities between 1.03 g/ml and 1.09 g/ml, even with an appreciably less-dense skeleton and other adaptations that we discuss later.

Does the lightness of the shark skeleton make it weaker, as one might reasonably conclude? Surprisingly, the answer is *probably not*. Because cartilage is light and flexible, it is often inferred to be inferior to bone as a structural material and incapable of tolerating the high forces that impact bone during vigorous swimming. This seems like a silly assertion, given that many sharks are powerful swimmers.

Now we know that the inference is indeed false, at least in some sharks, thanks to a 2006 study[3] in which researchers examined properties of vertebrae in seven kinds of sharks. The study found that the shark vertebrae were as *stiff* and *strong* as bone. The hardness of bone thus does not equal mechanical superiority, just a different pathway to evolutionary success.

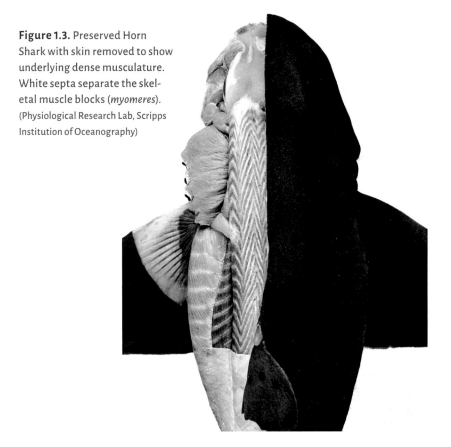

Figure 1.3. Preserved Horn Shark with skin removed to show underlying dense musculature. White septa separate the skeletal muscle blocks (*myomeres*). (Physiological Research Lab, Scripps Institution of Oceanography)

Let us examine other distinguishing characteristics of sharks, and in many cases rays. Many of these differences may be obvious to you—a shark just *looks like a shark* and a barracuda, mullet, and so on, does not. Refer to Figure 1.4 as you read about these shark features that as a group distinguish them from bony fishes:

- Five to seven bilateral gill slits
- Heterocercal caudal fin
- Internal fertilization and claspers
- Serial tooth replacement
- Placoid scales
- Simple chondrocranium (skull, braincase)
- Tribasic cartilage support of pectoral fins
- Ceratotrichia (soft, unsegmented fin rays)
- Ampullae of Lorenzini
- Absence of swim bladder
- Nictitating membrane in some taxa
- Retention of urea
- Rectal gland
- Oil-filled liver
- Intestinal valve
- Pericardio-peritoneal canal

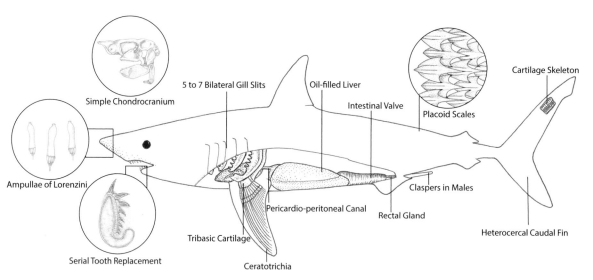

Figure 1.4. Features that distinguish a typical shark from a typical bony fish.

Five to Seven Bilateral Gill Slits

One of the quickest ways to identify a fish as a shark is to count gill slits. Most sharks have five gill slits on each side, but eight species have six or seven, whereas all bony fishes have only a single bilateral flap, the operculum, covering the gills, which makes the inch-long fish in pet stores called a red-tailed *shark* in fact a bony fish. All but one species of batoid possess five bilateral gill slits, but they are located ventrally rather than laterally.

Heterocercal Caudal Fin

Sharks have an asymmetrical caudal fin in which the upper lobe is typically longer than the lower lobe, whereas the lobes of the tail fin in most bony fishes are symmetrical and equal in length. The *heterocercal* (meaning *different tail*) caudal fin of sharks is characterized by the vertebral column extending almost to the tip of the fin's upper lobe. Some primitive bony fishes (e.g., sturgeon), have a heterocercal tail, whereas in others, like the tarpon, the vertebral column extends slightly up into the upper lobe. In more advanced bony fishes, these caudal vertebrae have fused over evolution into a *homocercal* (meaning *same tail*) caudal fin.

Internal Fertilization and Claspers

All sharks use internal fertilization. To transfer sperm to the female, male sharks all have special organs known as *claspers* that they insert into the female cloaca.[4] Among bony fishes, only a handful (e.g., the guppy), fertilize internally; the overwhelming majority fertilize externally. When you consider that internal fertilization is largely a terrestrial phenomenon (except for marine reptiles and mammals, which evolved from terrestrial ancestors), that sharks evolved that ability independently early in their evolutionary history makes claspers evolutionary marvels.

The term *claspers* dates back to Aristotle, who incorrectly inferred that these modifications of the pelvic fin were used by males exclusively to clasp onto females. Although Aristotle's interpretation of the use of the claspers was incorrect, he was accurate in concluding that some coupling mechanism was required for males to transfer semen to females while both were swimming. We think readers would acknowledge that difficulty.

Figure 1.5 shows claspers from two sharks. Note the hook and spine at the tip of the splayed claspers from the Atlantic Sharpnose Shark. Maybe *cloacal grapplers* is a more descriptively accurate term for the structures.

Claspers were in the news a few years ago when the *Underwater Times*[5] reported that a Malaysian fisher caught a shark with what looked to the world like webbed feet. Had an unsuspecting fisher serendipitously unearthed one of the proverbial missing links? Well, no. In fact, the *webbed feet* were the claspers from a male shark splayed out as they would be when attached inside a female shark's cloaca, a condition that sometimes also occurs among dead male sharks.

Sharks also practice serial tooth replacement (fig. 1.6), meaning that their teeth are not firmly embedded in bone like those of humans and other vertebrates, but rather are loosely rooted in collagen, one of the most abundant fibrous proteins used to connect and support parts of the body.

Sharks have from one to as many as 10 rows of functional teeth (e.g., White Shark and Dusky Smoothhound, respectively), with several rows of teeth in varying stages of maturity lying in wait on a very slow-moving collagen conveyor belt. In rays, there may be as many as 13–15 rows of functional teeth, which form plates for prey-handling.

A row of teeth of suitable cutting, penetrating, grasping, or crushing ability is thus always available, and if one or more teeth on the front line are lost, replace-

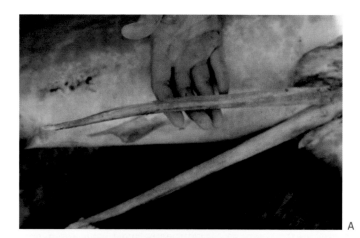
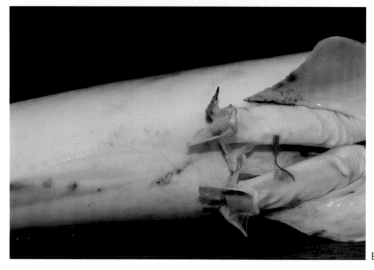

Figure 1.5. Claspers from thresher shark (A) and Atlantic Sharpnose Shark (B). The latter has splayed claspers, displaying specialized structures for securing the clasper in the female's cloaca.

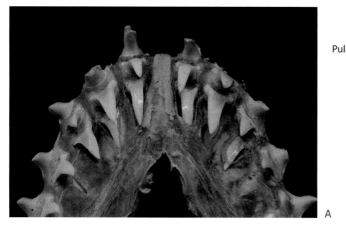
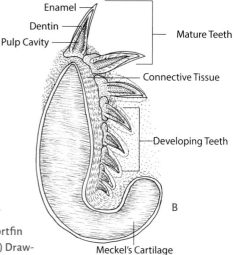

Enamel
Dentin
Pulp Cavity
Mature Teeth
Connective Tissue
Developing Teeth
Meckel's Cartilage
A
B

Figure 1.6. Examples of serial tooth replacement. (A) Jaw of Shortfin Mako showing multiple teeth rows. (Courtesy of Robert Johnson) (B) Drawing of a sagittal section through a shark jaw showing teeth in various developmental stages. (Redrawn from fig. 3, Moyer, J.K. et al. 2015. J. Morphol. 276: 797–817)

ments will soon be in transit. A shark may produce and lose thousands of teeth over its lifetime, and the replacement time is typically measured in weeks.

Surely it is wasteful to continually make and discard teeth instead of producing one permanent set, especially since one of the central tenets of evolution is that the supply of energy is often limiting and thus conserving energy is paramount to success. Evolution, however, often entails compromises, and natural selection can be thought of as a continuous experiment that tests what works and what does not. The individual with the suite of characteristics that works best, that is, that enables the bearer to eat sufficiently, avoid being eaten, and reach maturity and successfully mate and produce viable offspring, is the one that passes the evolutionary test. The individual with characteristics less successful in these areas likely does not live to pass its genes to the next generation. You may know this concept as *survival of the fittest.*

Then *yes*, making and shedding teeth may waste mineral and energy resources, but in combination with other shark characteristics, it works well. Moreover, one of the biggest threats to predators, particularly terrestrial ones but also some marine forms, is starvation. If a lion or tiger loses one of its long canine teeth, its chances of surviving are significantly diminished, whereas if a Tiger Shark loses several teeth, no problem.

Placoid Scales

The body of sharks is covered with *placoid* scales, or *dermal denticles* (fig. 1.7). Structurally, these are like miniature teeth, with an inner pulp cavity, surrounding layer of dentin (hard, calcified tissue), and a hardened outer layer of enamel.

Quick deviation: Anyone who regularly handles sharks occasionally suffers from *shark burn*, a rash that begins to sting shortly after rubbing against a shark, especially back-to-front. Think *rug burn* or *artificial turf burn*. If you are the recipient of shark burn, because scales are miniature teeth, you can legitimately (well, sort of) claim that you survived being bitten by a shark. I bet you thought only teeth in the mouth could bite?

Like their teeth, scales are shed and replaced, although more slowly. Bony fishes' scales are composed of the fibrous protein collagen and mineralized bone. The scales have different functions among sharks, including reducing drag while swimming, protecting against ectoparasites, and safeguarding females from male bites during mating season. We discuss some of these in Chapter 4.

Simple Chondrocranium

Sharks have a simple chondrocranium (also called a braincase, neurocranium, or skull), whereas the skull of bony fishes is a mosaic composed of dozens of different bones (fig. 1.8). This lack of complexity played an important role in limiting the diversity of feeding styles of sharks, as we discuss in Chapter 3.

Tribasic Cartilage Support of Pectoral Fins

In the evolution of sharks, an important advancement occurred when the inflexible, solid skeletal support of each pectoral (and pelvic) fin was replaced by three basal cartilages (named *proterygium*, *mesopterygium*, and *metapterygium*; fig. 1.9). These cartilages supported and allowed slightly higher mobility of the pectoral fins and thus increased maneuverability over the more fixed pectorals of early ancestors, an evolutionary event contributing to the success of modern sharks.

Ceratotrichia

Ceratotrichia (figs. 1.9 and 1.10) provide the support for the distal (outer) part of the fins of sharks. These elongate rays are soft, flexible, unsegmented, and are

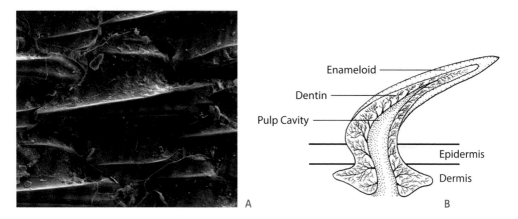

Figure 1.7. (A) Scanning electron photomicrograph of scales of the recently described shark named Genie's Dogfish. The ridges, known as *riblets*, play a role in reducing drag. (Courtesy of Charles F. Cotton) (B) Labeled cross-section of a generalized shark scale (dermal denticle). (Redrawn from fig. 5, Moyer, J.K. et al. 2015. J. Morphol. 276: 797–817)

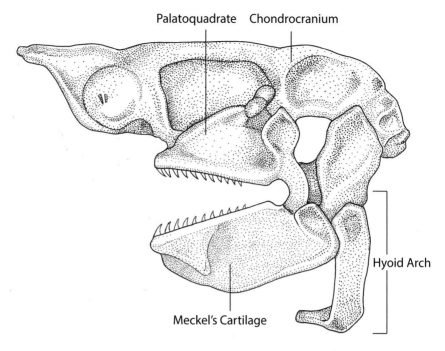

Palatoquadrate   Chondrocranium

Hyoid Arch

Meckel's Cartilage

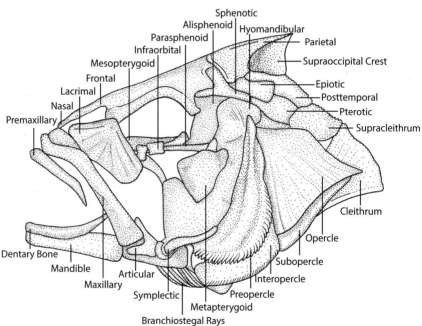

Sphenotic
Alisphenoid | Hyomandibular
Parasphenoid
Infraorbital
Mesopterygoid
Frontal
Lacrimal
Nasal
Premaxillary

Parietal
Supraoccipital Crest
Epiotic
Posttemporal
Pterotic
Supracleithrum

Cleithrum

Opercle

Dentary Bone
Mandible
Maxillary
Symplectic
Metapterygoid
Branchiostegal Rays

Articular
Preopercle
Interopercle
Subopercle

Figure 1.8. Generalized heads of a bony fish (*right*) and shark. The bony fish head is complex. Numerous bones provide the raw material for the evolution of diverse mouth and head shapes, and hence different feeding styles. In contrast, the reduced number of cartilage components in the shark head has limited their feeding styles. (Redrawn from fig. 8.3, Hyman, L. and Wake, M. 1979. *Hyman's Comparative Vertebrate Anatomy.* (3rd ed.). University of Chicago Press and Day, F. 1889. *The Fauna of British India, including Ceylon and Burma. Fishes.* London: Taylor and Francis)

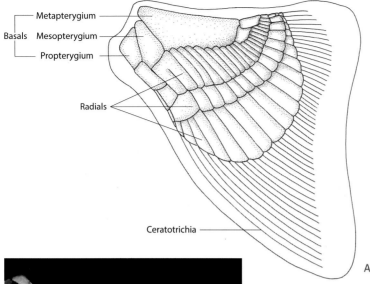

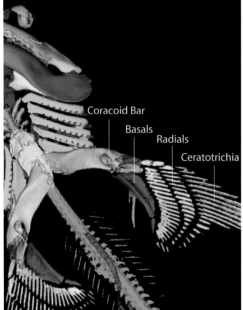

Figure 1.9. (A) Tribasic cartilage support (metapterygium, mesopterygium, and proterygium), also known as *basals*, of the shark pectoral fin. Radial cartilages and ceratotrichia are shown. (Redrawn from fig. 9, Wilga, C.D. and Lauder, G.V. 2001. J. Morphol. 249: 195–209) (B) CT scan of a Lemon Shark showing these three basal cartilages of the pectoral fin along with the radials and ceratotrichia. The coracoid bar, part of the pectoral girdle connecting the two pectoral fins, is shown. (Courtesy of Gavin Naylor / Chondrichthyan Tree of Life Project / sharksrays.org)

constructed of an elastic protein similar to the keratin of hair, nails, and claws. Unfortunately for sharks, the essential ingredient of the popular Asian delicacy, shark fin soup, is the ceratotrichia.

Ampullae of Lorenzini

Sharks also possess a means of detecting extremely minute electrical fields using a series of receptors located on the head known as *ampullae of Lorenzini* (fig. 1.11). A few primitive bony fishes (e.g., paddlefish) possess very similar organs. These structures, which we describe more fully in Chapter 5, allow sharks to detect and locate prey from distances of about 0.5 m (1.6 ft), where the sensory environment

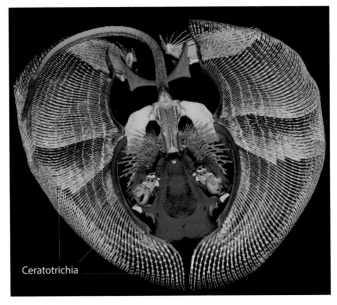

Ceratotrichia

Figure 1.10. CT scan showing ceratotrichia (light yellow) comprising the wings (i.e., pectoral fins) of a ray. (Courtesy of Gavin Naylor/Chondrichthyan Tree of Life Project/sharksrays.org)

Figure 1.11. Ampullae of Lorenzini (pores appearing as darkened dots) on the underside of the head of a Bonnethead. (Courtesy of Laura Claiborne Stone)

might be otherwise confusing because of murky water, widely diffused odors, and so on.

Absence of Swim Bladder

Sharks lack a swim bladder, and thus must either swim to maintain their position in the water column, or rest on the bottom. Refer to Chapter 3 to see the significance of a swim bladder to bony fishes, and of its absence to sharks.

Nictitating Membrane

Members of a single but large and well-known family of sharks, the carcharhinids (requiem sharks), possess a nictitating eyelid, or membrane, in the lower portion of each eye (fig. 1.12). Some birds, lizards, frogs, and even a few mammals (e.g., seals, polar bears, camels, and aardvarks) have functional nictitating membranes as well. The structure, which is covered in scales in sharks, rises to cover and protect the eye as the shark is about to eat, or when it is threatened. White and cow sharks have different methods of protecting their eyes, which we discuss later.

Retention of Urea

If you have ever prepared shark to eat (which we discourage for many species), an essential step is soaking or marinating the meat before cooking to remove some of the chemicals. Otherwise, the flesh may be unpalatable, because the liver of sharks converts nitrogen-containing wastes (e.g., those remaining after proteins and other compounds are digested), into the waste-products ammonia and urea,

Figure 1.12. Nictitating membrane covering the eye of a Tiger Shark.

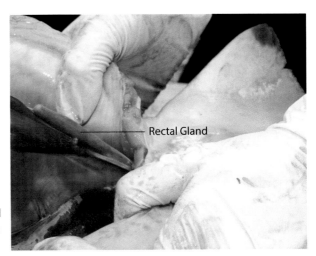

Rectal Gland

Figure 1.13. Rectal gland of a Blacktip Shark.

whose taste elicits, in scientifically accurate verbiage, the following reaction: *Blech!*

Sharks are not distasteful as a way to discourage predators but more so as part of a strategy to maintain the overall concentration of solutes (dissolved substances) in their blood and other extracellular fluids (the liquid outside of the cells) at about the same level as the seawater in which they live (see Chapter 8).

Rectal Gland

Sharks also have a small finger-like or bean-shaped structure called a rectal, or digitiform, gland (fig. 1.13). This structure also aids in eliminating the excess sodium and chloride which diffuse into the shark's body as a result of the higher concentrations of these two ions in seawater compared to the internal fluid environment of sharks. We explain this phenomenon in Chapter 8.

Oil-filled Liver

Earlier we mentioned that sharks are heavier than water, and they compensate in part by *dynamic* lift created by the caudal fins and in some sharks, the head, while swimming. The presence of an oil-filled liver (fig. 1.14), which helps decrease the overall density, or heaviness, of a shark's body, provides buoyancy, or *static* lift (i.e., lift that does not require movement). The density of oil in the shark liver is between 0.86 g/ml and 0.95 g/ml (recall the density of seawater is about 1.025).

Intestinal Valve

Sharks also have an unusual adaptation of the intestine that increases its surface area for digestion and absorption of nutrients. In most vertebrates, this same result is accomplished by increasing the length of the intestine (your small intestine is 3–5 m, or 9.8–16.4 ft, long), then convoluting, or folding, it onto itself to fit into the abdominal space. Sharks instead increase the surface area entirely within the compact length of the intestine with an intestinal valve (fig. 1.15), that has

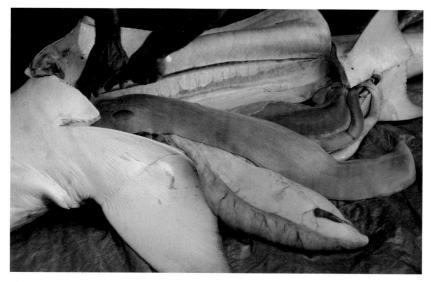

Figure 1.14. The large liver (light brown) of a Lemon Shark. Also shown are the gallbladder (green, embedded in liver) and stomach (light purple, below liver).

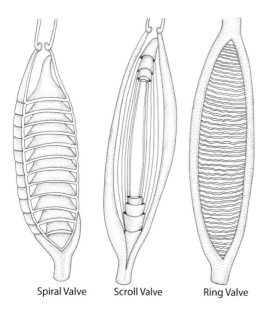

Spiral Valve   Scroll Valve   Ring Valve

Figure 1.15. Spiral, scroll, and ring intestinal valves, characteristic of Spiny Dogfish, Blacktip Sharks, and Shortfin Makos, respectively. In all three cases, the surface area for digestion and absorption is increased.

spirals, rings, or is tightly wound to resemble a scroll (depending on the species), which in each case increases the surface area for absorption in the intestine, and saves space in the process.

Pericardio-peritoneal Canal

Finally, sharks have a structure that connects the pericardial space surrounding the heart with the peritoneal, or abdominal, space surrounding the visceral organs like the stomach, pancreas, liver, spleen, and so on, with a tongue-twisting but aptly-named duct, the pericardio-peritoneal canal (fig. 1.16).

A pericardio-peritoneal canal is present in the embryological development of all vertebrates, but persists only in larval lampreys, adult elasmobranchs, hagfish, and sturgeon. This canal plays a major and often unrecognized role in heart function by protecting the heart from being dangerously compressed (e.g., by forces when the shark is eating) and by enhancing heart function in other ways (see Chapter 7). You likely will not see the pericardio-peritoneal canal listed in other references as a distinguishing feature of sharks, but it indeed is and reasons for its absence are unclear.

Now, given what you have learned about distinguishing features of sharks, examine the shark sculpture shown in Figure 1.17 and see how many mistakes you can count.[6] Then think of how you would gently inform the proprietor or artist of the errors.

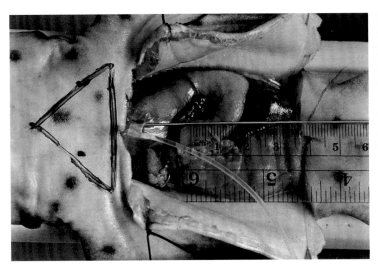

Figure 1.16. Ventral view of the open abdominal space in a Horn Shark showing the pericardio-peritoneal canal (with tube and wire inserted). The liver has been removed. The triangle drawn on the photo denotes the approximate ventral outline of the pericardial space, in which the heart resides and where the pericardio-peritoneal canal originates.

Figure 1.17. Sculpture of a shark outside a beach wares store near Myrtle Beach, South Carolina.

Where Is the Sharkiest Place on the Planet?

The location of the sharkiest place on the planet depends on what *sharkiest* means. If the term refers to the greatest biomass or number of sharks, then the winner of the title might be the NW Atlantic Ocean, specifically, the highly productive Grand Banks and adjacent waters, where the most numerically abundant shark species, the Spiny Dogfish, or Spurdog, resides. The species was heavily overfished, but overfishing is no longer occurring, the result of effective management policies (see Chapter 11).

If we interpret *sharkiest* as *biodiverse*, then we have numerous candidates. A 2011 paper[7] listed the following as global hotspots for sharks based on species-richness (the number of species): Japan (with a surprising 124 species of sharks), Taiwan, the east and west coasts of Australia (>170 species), SE Africa (about 100 species), SE Brazil, and SE USA.

In general, the richest areas were along the fringes of continents (i.e., over the continental shelves) as opposed to the open ocean, and particularly in the mid-latitudes (between 25° and 40°) in both hemispheres. Other hotspots include regions where cold, nutrient-rich deep water upwells to the surface (e.g., NW Africa) and seamounts (undersea mountains that do not breech the surface), which are like underwater island oases. Figure 1.18 shows biodiversity hotspots.

Life-history Characteristics of Sharks

For a shark (or any organism), the name of the game in evolution is surviving long enough to reproduce, thereby passing your genes, which have proven effective in getting you through life's exigencies to this point, to your offspring. If *they* survive to reproduce, then their characteristics are passed on, and so on, part of the blueprint for evolutionary success.

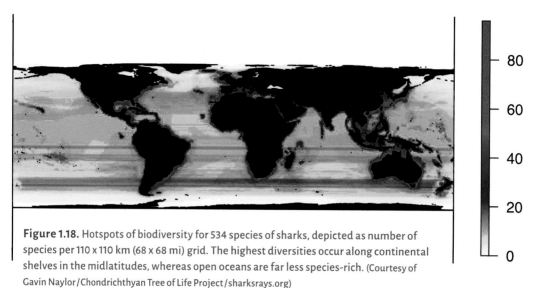

Figure 1.18. Hotspots of biodiversity for 534 species of sharks, depicted as number of species per 110 x 110 km (68 x 68 mi) grid. The highest diversities occur along continental shelves in the midlatitudes, whereas open oceans are far less species-rich. (Courtesy of Gavin Naylor/Chondrichthyan Tree of Life Project/sharksrays.org)

Organisms employ one of two main life-history strategies, with numerous intermediate stages, in this *quest* for evolutionary success. On the one hand are organisms like fruit flies, which grow quickly, become mature early, and lay lots of quickly hatching eggs, and which can prosper in a wide variety of environments.

Close to the other extreme in life-history characteristics are the sharks. Their suite of life-history characteristics has suited them well in their 400+ million-year evolutionary history. At the same time, for many sharks these same characteristics make them particularly vulnerable to pressures (e.g., overfishing, habitat destruction, and climate change impacts) that can reduce their populations and make recovery difficult.

Sharks

- are very slow growing
- are long-lived
- reach sexual maturity late in life
- have a long gestation period (the time developing embryos are retained in the female or in the egg)
- typically rely on specific mating and nursery areas, and
- have relatively low fecundity (the number of offspring produced by a female of a species; see Chapter 9).

Sharks in Our Lives

It is difficult to avoid sharks, at least in the media. Discovery Channel's *Shark Week* and National Geographic's *Sharkfest* saturate cable airways for a week each summer and in countless replays, with recent annual viewership exceeding 50 million. Feature films[8] like *Finding Nemo, Shark Tale, Open Water, 47 Meters Down, The Shallows*, the *Sharknado* franchise, and *The Meg*, include sharks as central characters. *Shark Tank* on cable TV features five *sharks*, millionaire venture capitalists, who vet ideas of erstwhile entrepreneurs for commercial success and savage them—apparently the eponymous reference of the title—if their proposals are unlikely to make money. Press coverage by TV, print, and Internet media continues unabated, especially if a shark bites a swimmer. And the ubiquity of sharks on social media? Fuggedaboutit.

In Hemingway's masterpiece *The Old Man and the Sea*, sharks represent one of nature's forces that the aged fisher Santiago battles in his epic quest first for triumph and ultimately survival. Steven Spielberg's summer blockbuster *Jaws*, based on Peter Benchley's novel, petrified beachgoers in 1975, and endures as one of the most identifiable sources of irrational and unjustified panic that has insinuated itself into the psyche of hundreds of millions of people to this day. Beachgoers are more threatened by errant surfboards, jet skis, rip currents, jellyfish, bacteria in the water, SUVs in the beach parking lot, and even toasters[9] as they prepare breakfast, than they are by sharks, but foremost in their heads as their toes sink into the hot sand is the foreboding, ominous staccato sound of the *Jaws* theme. Author Benchley was so disturbed that his book had emotionally tattooed generations with the stereotype of sharks as consummate killers, he spent the remainder of his life working for shark conservation.

Speaking of the theme from *Jaws*, a very cool study by Andrew Nosal and colleagues from Scripps Institution of Oceanography, the University of California San Diego, and Harvard, entitled *The Effect of Background Music in Shark Documentaries on Viewers' Perceptions of Sharks*,[10] concluded that attitudes toward sharks are influenced or reinforced by how ominous the background music is. Don't you just love science?

Although author Benchley denied any connection, the plot of *Jaws* is linked to a series of shark attacks in 1916 along the Jersey Shore.[11] During a 12-day period that summer, four people were killed and one was injured by a series of shark attacks.

Discerning the perpetrator or perpetrators of these attacks makes an interesting scientific detective story. At first, that a shark or sharks could be responsible for these attacks was quickly dismissed, since many ichthyologists (fish biologists) and other marine scientists were fairly certain, even in the early 20th century, that sharks did not attack swimmers at the beach, at least along the US East Coast. This seems extraordinary, even unbelievable, today, but 100 years ago there were fewer beachgoers and thus fewer interactions with sharks other than on ships at sea or at shark-bite hotspots, and, of course, no Internet to communicate and sensationalize the events.

The list of causes more likely than sharks of the Jersey Shore attacks put forth at the time by experts and others, almost hilariously if not for the seriousness of the attacks, included German U-boats, sea turtles, and even orcas. The panic that ensued from these events, and the response to it, presaged the unrelenting fear of sharks that persists today, as well as the resulting vengeful persecution of them that continues.

From a biological perspective, the most interesting question, and over 100 years later one still somewhat unsettled, is what kind of shark was responsible for the 1916 attacks. The issue was seemingly resolved at the time when a 7.5-foot, 325-pound juvenile White Shark, caught in a bay only a few miles away from the site of two of the attacks, was reported to have human remains in its digestive tract. Although the presence of human remains found in a White Shark would seem to conclusively settle the case (since the concept of *fake news* was still a century in the future), we may never know with certainty whether the Jersey Shore attacks of 1916 were caused exclusively by that species, or perhaps by two different species, the second being a Bull Shark, but we can be fairly certain that it was not an orca, sea turtle, U-boat, or toaster.

Finally, we would be remiss if we did not at least *mention* shark campiness,[12] the type in the *Sharknado* cult movie franchise, in which a waterspout sucks up toothy sharks and deposits them in various places where they consume a bevy of unsuccessful or aged actors trying to start or salvage a career.

Why so much interest in sharks? As biologists and conservationists, we wish we could state that people are interested in sharks because of the vital ecological role that sharks play in maintaining marine biodiversity, or because of the suite of fascinating adaptations they possess; for example, the streamlined shape of a Shortfin Mako that makes it one of the speediest beasts in the sea. However, it is clear that for many, the fascination with sharks owes more to fear and the perception of the danger they pose than to their ecological roles, biological attributes, or threats to their survival.

Another example of sharks in *our* lives: On May 9, 2017, both of us received this e-mail from a staff reporter for the *Palm Beach Post* that started:

> Hi Dr. Abel / Dr. Grubbs,
> *I'm sure you've seen the video of the adult film star allegedly getting bit by the shark . . .*

After we quickly recovered from the shock of a reporter we barely knew expressing certainty of our familiarity with adult film stars, we, along with other shark specialists, immediately deduced that the wound was human-inflicted as a very successful quest for web fame. The bite of one of the sharks in the video would likely have either removed a chunk of flesh or left a series of tooth punctures in an arc corresponding to the arched shape of a shark jaw, but not a clean laceration.

Why We Fear Sharks

If you are a predator, especially an apex predator, you begin life saddled with an image problem: the *perceived* threat you pose invokes a fear disproportionate to the *real* threat, and may replace the awe we should have for you with irrational terror. We thus are numbed to the very real challenges to your survival. We root for the underdog (e.g., the old, young, or otherwise infirm) wildebeest that lions have separated from its herd or the impala being chased by the stealthy, speedy leopard. And when you step into the water at the shore, your sympathy for the plight of sharks is supplanted by the comfort of knowing that, because of overfishing, habitat destruction, and so on, there are fewer of these toothy beasts lurking nearby in the shallows awaiting your entry. Yes, we love Shark Week, but for most of us, watching sharks from the safety of our screens or from a boat is just fine, if not preferred.

Dolphins are predators as well, but cute and cuddly ones according to our perceptions. In the 1980s, when people in the US saw graphic images of dolphins killed in the capture of tuna for the canned seafood industry in the Eastern Tropical Pacific Ocean, consumers united to demand action, and the *Dolphin-Safe* movement arose. No movement of equivalent scale exists for sharks. Moreover, the net effect (no pun intended) of the Dolphin-Safe campaign was taking the lives of orders of magnitude more sharks, sea turtles, small fish, and so on, as well as causing the Mexican tuna industry to crash. We discuss this further in Chapter 11.

Why do we care less about sharks than we should? Ignorance of their ecological role or threats to them may explain some of our indifference to their plight, but not all. The other factor is fear, even though many know that a shark bite constitutes a low risk. Humans fear the dark, spiders, snakes, heights, public speaking, a visit to the dentist, and so on.

Our brains conjure up reasons for our fears: What evil could be lurking in the dark? What if I stepped on a venomous snake that I did not see? What if I fell from this 9th story balcony? And the more we contemplate these unlikely occurrences, the more we nurture our fear. Fear evolved in humans as a survival mechanism, and there are times when it is rational to be fearful, but much of our fear borders on panic and is not based on high likelihoods of the occurrence of harmful things.

Fear of sharks is most like fear of the dark, which really is fear of the unknown. For all typical beachgoers know, there could be hundreds of hungry sharks anticipating their entry into the surf, especially in murky water. And layered on top of this fear are the gruesome descriptions and images of shark attack victims. That death by shark attack would be a particularly horrifying way to die, or to lose a limb, is the icing on the cake for fearing sharks.

Shark Bites

In the new millennium, the years 2001 and 2015 have stood apart from the other years, at least in the eyes of the American media, as *Years of the Shark*, because of the perception of a significant increase in the number of shark bites and attacks.

Prior to the horrific events of 9/11, the summer of 2001 was a slow news period. Beginning on that summer's 4th of July weekend, several gruesome human-shark interactions occurred, including serious injuries to an 8-year-old boy and others, and fatalities in North Carolina and Virginia. News footage of migrating Blacktip Sharks off the coast of Florida, a common annual occurrence, fueled the hysteria (fig. 1.19). Although in terms of bites the summer shark season started out furiously in 2001, by its end that year was not a particularly unusual one. Although regrettably there were fatalities, what really differed about the year 2001 was the increased focus on even minor shark bites.

The year 2015 started out similarly to 2001 along the US East Coast, with numerous bites, although no fatalities, and ended as a slightly atypical year. Explanations for the uptick include more beach visitors and thus more opportunity for interactions, and two events that could be associated with global climate change: elevated water temperatures, and wind and current patterns that may have pushed shark prey, and thus sharks as well, closer to shore.

Hand-in hand with the widely held fear of sharks, remember from the Preface that we call this particular kind of fear *galeophobia* or *selachophobia*, is the wrongheaded idea that we would face fewer risks in a world without sharks. This is bad reasoning because of the low likelihood of being attacked by a shark; the number of documented, unprovoked shark attacks hovers around 100 annually, according to the Florida Museum of Natural History's International Shark Attack File. In 2018, 66 unprovoked attacks were reported. Explanations for the reduction include declines in shark populations and public education. Along the US East Coast, where Blacktip Sharks have been responsible for numerous bites in Florida, recent increases in seawater temperatures associated with climate change are thought to have delayed or reduced migrations of these sharks from the north, and thus resulted in substantially fewer bites, although the number of bites varies annually.

Being attacked by a shark is thus a statistical possibility, but the number of shark-related fatalities and serious but non-life-threatening injuries is negligibly low compared to the threats we face every day and consider routine. We shorten our lives by smoking, drinking sugary sodas, eating too many carbs, and texting while driving. In the US, air pollution causes as many as 200,000 premature

Figure 1.19. Aerial photographs of Blacktip Shark aggregations off the east coast of Florida, which are annual occurrences. (Courtesy of Stephen Kajiura)

deaths annually. Breathing the dust emitted by our tires and running shoes as they wear is a bigger risk to our health than are sharks.

Published estimates of risks are readily available; for example, 0.00125% chance of being killed by a lightning strike, 0.00003% chance of being killed by a shark. We know far more people who have experienced direct lightning strikes than people who have been attacked by a shark, and we are in the shark business.

One of our students wrote in an essay that you are more likely to be struck by *lighting* than bitten by a shark. We are pretty sure she meant to write *lightning*, but she was correct as the word was written: More people are injured by falling lighting fixtures than by sharks annually. A few years ago, a meme circulated on the Internet that you were more likely to be killed in the tropics by strikes from random coconut falls. Being killed by an errant coconut may be one of the few risks less likely than shark attack, although no International Falling Coconut Database collects data on coconut strikes.

On a somewhat serious note, as field biologists who work at sea often in bad weather, we are aware that the odds of being struck by lightning are increased from those while ashore (BoatsUS estimates one of every 1000 boats is struck by lightning each year). So, the authors try to avoid thunderstorms while we work on the water, in part because we fear the headline that might result: *It's True! Shark Biologists Prove Adage: Struck by Lightning!*

So, it boils down to these two conclusions: First, the most dangerous activity you engage in daily is waking up and conducting your daily affairs. Second, as we stated above, given enough time, even extremely unlikely events will happen. That is why people buy lottery tickets.

Note that we have used the word *bite* in describing these shark-human interactions. Some shark biologists distinguish between *bites* and *attacks*, the former typically causing only minor injuries after *bite-and-release* behavior associated with mistaken identity.[13] A spate of these often occur in the summer months along the US southeast coast, particularly in naturally murky waters, where Blacktip Sharks, streamlined, swift-swimming eaters of small schooling fish, are thought to confuse human hands and feet with these fish and may bite. However, within fractions of a second, they apparently recognize the mistake and release, often leaving only a series of shallow puncture marks.

The term *attack* is reserved for cases involving more serious injury and may involve repeated bites after the first. In the case of White Sharks, bites may be fatal, even if they are considered mistaken identify, because the damage of even a single bite may be extensive.

Ways to Avoid Shark Bites

When we are in the water and see sharks, our instinct is to swim *toward* them. However, do not try this yourself (our liability lawyers made us write that). Seriously, the sensible action to take if you are in the water and see sharks, except perhaps small harmless species like horn and bamboo sharks, is to calmly but deliberatively return to your boat or the shore and leave the water. And do not swim into a school of small fish (sometimes called a *bait ball*), or other areas where sharks might be attracted, like a fishing pier.

Are there repellents (sometimes also referred to as *deterrents*, although technically the terms have different definitions) effective against sharks? The answer is yes, but none are foolproof, and their efficacy depends on the application. Below is a listing of some of the repellents used historically and/or currently. For a comprehensive summary, we refer you to Dudley and Cliff's *Shark Control: Methods, Efficacy, and Ecological Impact*[14] and the 2015 report by McPhee, Blount, and Lincoln-Smith,[15] whose categories of deterrents we use below.

Large Scale

Shark Surveillance

The use of lifeguards or shark spotters on the beach or in boats to locate sharks and warn bathers has had mixed success. Major drawbacks of surveillance methods involving humans to detect sharks include the needs for an appropriate vantage point, relatively clear and calm water, sharks to stay near the surface, spotters to be alert for long durations, communities to support such efforts, and appropriate temporal and spatial coverage.

One very successful effort is the *Shark Spotters* program, which employs people equipped with binoculars to detect White Sharks in False Bay and other areas along the South Peninsula of Cape Town, South Africa. Since 2004, they have reported over 2000 shark sightings. Warnings are conveyed via sirens, a system of colored flags, and social media.

More recent innovations include using aerial drones, which have been called *game changers* among shark early warning systems, high-resolution sonar (e.g., *Cleverbuoy*), and acoustic or satellite tags on sharks.

Shark Barriers

Barriers to exclude sharks date back as far as 1907, when a steel structure was erected at a beach in Durban, South Africa. A variety of other materials have been tested but the mainstay has been cheap and effective mesh beach nets. Unfortunately, these nets entrap and kill numerous species of sharks. One long-term study[16] in South Africa reported that 44 km (27 miles) of nets caught an average of 1470 sharks annually (in addition to 536 other large marine animals). Newer nets are constructed of more rigid plastics and thus ensnare fewer or even no sharks and other marine life, are more durable, and have reduced operating costs.

Researchers have also explored alternative barrier systems, including bubbles (largely ineffective), electricity, magnets, and other physical barrier systems. Electrical/magnetic fields in the range of 3–7 volts per meter have shown some efficacy as deterrents, but the response varies by species.

Another device proposed as an alternative to shark nets is the *Sharksafe Bar-*

rier, which consists of both a visual and magnetic barrier. Powerful rare earth magnets, which have been demonstrated to repel sharks, are used along with a visual barrier of PVC pipes anchored to the bottom such that they can move with waves and currents. Peer-reviewed tests[17] of the system have confirmed the barrier's efficacy in repelling White and Bull Sharks.

Culling

The last large-scale method we discuss is culling, which is a category of shark persecution and a sanitary way to say *slaughtering*. Culling involves catching and killing of sharks on both small and large scales, either as a preventive measure or in response to shark bites.

While it is undeniable that culling removes sharks from an area, at least temporarily, the long-term success as a shark deterrent is not known. What is known is that any practice that removes predators, particularly top predators, which are typically the least abundant animals in any ecosystem, from their environment likely will have reverberations on populations of other organisms throughout that system.

Personal Deterrents

In recent years, numerous personal deterrents employing different repellent techniques (physical screen, chemical, acoustic, visual/camouflage, electric, magnetic, chainmail wetsuits, and cages) have become available in the marketplace. Prior to this, personal repellents in the US were developed mainly by the Navy for use in shipwreck/plane crash scenarios. Currently, most personal deterrents are proprietary and many (but not all; see below) have not undergone meaningful, robust scientific scrutiny that assesses both their efficacy and, ideally, their underlying physiological basis.

Websites for commercial personal deterrents may feature impressive in-house tests, often with equally impressive videos, demonstrating that sharks avoid their repellent. There may also be testimonials in support of the products. Our experience with the companies that produce and/or market these devices is that they are earnest, and they encourage independent tests of their products. And many, but certainly not all, of their repellents show some promise as repellents in some situations.

Finally, we remind you that shark bites are extremely low-likelihood events, so in the overwhelming majority of situations, personal repellents are redundant; that is, you are already protected by virtue of not being among the regular prey of any sharks. If you *still* want to purchase a personal shark repellent, we suggest you ask a resident of Western Australia (WA) to make the purchase for you, since the WA government has offered a shark deterrent rebate on selected repellents.

We conclude this section with the following breaking news, which should ease the anxiety of readers in the UK:

You may rest assured that the British Government is entirely opposed to sharks.
Prime Minister Winston Churchill,
February 20, 1945, in the House of Commons

Biomedical and Other Non-food Uses of Sharks

Before we discuss the ways in which humans use sharks, we remind you that the single best *use* of sharks is allowing sharks the unfettered ability to perform their ecological roles by preserving their biodiversity. This serves sharks and humans both, as well as the entire planet.

That said, sharks have played and will continue to have roles in human health. You may have heard that *sharks don't get cancer*. Cancer refers to a disorder in which cells undergo genetic changes and lose their growth constraints and thus reproduce unchecked and invade other tissues, often leading to death. It may surprise you to learn that virtually all kinds of multicellular animals, from insects to whales, experience cancer, although the prevalence varies considerably.

Is the claim true? Unfortunately, no. In 2004, cancer researcher Gary Ostrander and colleagues published a thorough analysis[18] of this claim. We quote from the paper:

(a) sharks do get cancer;

(b) the rate of shark cancer is not known from present data; and

(c) even if the incidence of shark cancer were low, cancer incidence is irrelevant to the use of crude extracts for cancer treatment.

Irrespective of the degree to which sharks are protected from cancer, there is *something* special about their primitive but powerful immune system. One characteristic that has particular clinical promise is the presence of a class of heavy antibodies, proteins produced by the immune response system to fight invasive foreign substances or pathogens (called *antigens*). Typical vertebrate antibodies possess two light and two heavy protein chains, and sharks are no exception. However, sharks, and inexplicably camelids (camels and their relatives), have antibodies composed only of heavy chains which, despite the name, are very small and are in a class known as *nanobodies*. Clinically, nanobodies from sharks and other sources may prove extremely useful in the field known as *immunotherapeutics* (fighting viruses, tumors, and diseases), as well as in developing diagnostic technology. Australian scientists are testing a drug based on the shark antibodies to treat the disease idiopathic pulmonary fibrosis (IPF). Additionally, researchers have found that sharks and their relatives possess more genes related to wound healing than are found in bony fishes.

Examples of other biomedical uses of sharks include:

- **Skin grafts/wound dressing**—Diabetic foot ulcers are treated with a mixture of silicone, cow collagen, and shark cartilage.

- **Antibacterial coverings**—A commercial surface coating/plastic sheet designed to prevent bacteria from sticking to a surface uses a *ridge* and *ravine* micropattern inspired by the skin of sharks.

- **Antibiotic/Antiviral agents**—The compound squalamine, which is found in the liver and stomach of some sharks, may have broad antibacterial and antiviral properties.

- **Bone regeneration**—Apatite bioceramics from sharks may be useful in repairing, replacing, and/or treating bone defects or in other maxillofacial surgical applications.

- **Therapy for neurodegenerative disorders** (e.g., Parkinson's disease)—The compound squalamine also appears to prevent the buildup of a lethal protein associated with Parkinson's disease and dementia.

- **Therapy for arthritis**—The compound chondroitin sulfate, derived from both shark and cow cartilage, in combination with the compound glucoseamine, may slow the progression of osteoarthritis. In combination with other compounds, chondroitin may also have applications in cataract surgery and reduction of urinary tract infections in women.

- **Medical creams**—Shark liver oil is one of the ingredients in the anti-hemorrhoidal cream *Preparation H.*

- **Use as models for biomedical research**—Marine biomedicine is the discipline that uses marine organisms, sharks among them, to study biological problems relevant to human health.

Concluding Comments: Why Sharks Matter?

We conclude this chapter by answering the question *why do sharks matter?* Arguments in favor of caring about the plight of sharks include ecological, economic, spiritual, and ethical reasons.

Let us examine the ecological argument first. Sharks, even filter-feeding species, are predators. Some (but not most) are top, or apex, predators. Although an oversimplification, apex predators are defined as the animals at the top of the food chain, with no or extremely few natural predators as adults. In addition to some sharks like Whites, Tigers, Bulls, Oceanic Whitetips, Shortfin Makos, and Reef Sharks, marine apex predators include Orcas, Polar Bears, Saltwater Crocodiles, and Swordfish. Smaller sharks (e.g., Atlantic Sharpnose, Blacknose, and Bonnethead) and some rays (e.g., Cownose) that are one or two feeding levels below apex predators are known as *meso-predators.*

Some sharks may be *keystone species* in their ecosystems (i.e., they are of disproportionately large importance). Removing the keystone species from an ecosystem is analogous to removing the keystone from an architectural arch, which causes the entire structure to collapse. The impact of the former, however, is often less apparent and not always easy to quantify. Still, the normal function, stability, and resilience of marine ecosystems are threatened if apex predators, particularly if they are keystone species, are removed or decreased in numbers. Tiger Sharks are both top predators and possibly a keystone species in Australian seagrass ecosystems, where they prey on sea turtles and dugongs. In doing so, they play a role in preventing these species from overconsuming seagrass.

Predatory sharks, whether apex or meso-predators, help preserve an ecosystem's biodiversity. At the same time, apex predators are particularly vulnerable to threats like overfishing, habitat destruction, and pollution because they are the least abundant organisms in an ecosystem.

The next argument for shark conservation is economic, namely preserving *ecosystem services* (goods and services that are required for human well-being and that ecosystems provide for free), for which there are not sufficient technological replacements.

The list of ecosystem services provided by the world's biodiversity is extensive and includes food, clean air, clean water, flood and erosion control, building materials, pollination, medicines, and so on. We depend on these ecosystem services, and global health and the global economy would both most assuredly collapse in their absence.

In addition to the biomedical uses and ecotourism we listed above, the kinds of economic services provided by sharks include maintaining biodiversity of marine ecosystems such that populations of commercially and recreationally important fishes continue to exist. Products of economic value from sharks include shark meat, fins for the Asian delicacy shark fin soup, leather from shark skin, cartilage for the alternative health care industry, livers for omega-3 fatty acids and other products, and teeth and jaws.

Similarly, the recreational fisher, SCUBA diver, snorkeler, photographer, or ecotourist contributes to local and national economies by buying gear and permits, traveling, room and board, and other purchases.

That brings us to the final reasons, both intangibles, as to why we need sharks. First, sharks provide spiritual and aesthetic refreshment and add an ineffable value to our lives. Ecologist E.O. Wilson identified an innate human need known as *biophilia* (in his eponymous book) as *the urge to affiliate with other forms of life*. If you have ever marveled at seeing a shark in nature, or even in an aquarium, then you understand the value that a shark adds to your life.

Second, sharks, like every other species on the planet, have intrinsic value; that is, they have a right to exist based on their heritage, and this right is irrevocable.

Given all of the benefits, tangible or not, that sharks and the ecosystems in which they live provide, we ignore their plight at our own risk. Sharks are simply essential for our survival. It thus is incomprehensible that anyone would want to live in a world without sharks.

1. For a refresher on the hierarchy of zoological nomenclature, refer to Box 1.1.

2. Density is a measure of heaviness and it is expressed as mass per unit volume; for example, grams per ml (sometimes written as $g \cdot ml^{-1}$).

3. Porter, M.E. et al. 2006. J. Exp. Biol. 209: 2920–2928.

4. Cloaca (pronounced *klo-aý kah*) refers to the common and only opening for the reproductive, digestive, and urinary tracts in chondrichthyans, bony fishes, amphibians, reptiles, birds, and some mammals.

5. http://www.underwatertimes.com/news.php?article_id=73210864950. (Accessed 9/9/19).

6. Here are the mistakes: four gill slits, a single row of widely spaced teeth, wimpy pectoral fins, mouth open but no nictitating membrane deployed, and unrealistic combinations of colors. And yes, no shark even remotely resembling this one grows to this size. You will learn later that the upper jaw should also be protruded.

7. Lucifora, L.O. et al. 2011. PLoS One 6: e19356.

8. See https://www.ranker.com/list/the-best-shark-movies/all-genre-movies-lists for a pretty thorough list of 49 films featuring sharks. (Accessed 9/9/19).

9. Yes, toasters. Do not think so? Watch this: https://www.youtube.com/watch?v=htpx GXOlh3U. (Accessed 8/9/19).

10. http://journals.plos.org/plosone/article/file?id=10.1371/journal.pone.0159279&type= printable. (Accessed 9/9/19).

11. The events' history is recounted in the books *Close to Shore* by Michael Capuzzo (2002; Broadway Books) and *Twelve Days of Terror: Inside the Shocking 1916 New Jersey Shark Attacks* by Richard Fernicola (2016; Lyons Press).

12. "Consciously artificial, exaggerated, vulgar, or mannered; self-parodying, esp. when in dubious taste," according to Dictionary.com.

13. Neff, C. and Hueter, R. 2013. J. Envir. Stud. Sci. 3: 65–73.

14. Dudley, S.F.J. and Cliff, G. 2010. In: Carrier, J.C. et al. Sharks and their relatives II: Biodiversity, adaptive physiology, and conservation. CRC Marine Biology Series. CRC Press. 713 pp.

15. McPhee, D. et al. 2015. NSW Department of Primary Industries, 59916026.

16. Cliff, G. and Dudley, S.F. 1992. Mar. Freshwater Res. 43: 263–272.

17. O'Connell, C.P. et al. 2014. J. Exp. Mar. Biol. Ecol. 460: 37–46.

18. Ostrander, G.K. et al. 2004. Cancer Res. 64: 8485–8491.

2 / Evolution of Sharks

Introduction: Are Sharks Living Fossils?

Sharks, it is said, are *living fossils*, organisms whose anatomy appears little changed from their early ancestors. Think horseshoe crabs, crocodiles, coelacanths (lobe-finned fish thought to be extinct for 65 million years until rediscovered in 1938), or even ginkgo trees or ferns. To the extent that sharks and their relatives settled on a body form and predatory lifestyle that was successful relatively early in their evolutionary history (during the adaptive radiation of modern sharks, about 200 million years ago), and some aspects of these have persisted to the present, then *yes*, sharks are living fossils.

But other characteristics, for example, the loosening of the jaws and increased mobility of the pectoral fins, are very different from those of early ancestors, so that modern sharks are not primitive creatures but are advanced, and well-adapted to their current environments. Bony fishes, indeed, all lower vertebrates, trace their origins back to the same geological period as sharks, the Devonian, so if a Shortfin Mako is a living fossil, so is a mullet and a turtle.

How Do Paleontologists Study the Evolution of Sharks?

Although cartilage is not heavily mineralized and does not preserve well, and thus the fossil record of sharks is sparse compared to that of bony fishes, it is not without useful information. That is because imprints of some shark and shark-like fossils have been preserved in anoxic (devoid of oxygen) fine sediments, where

A

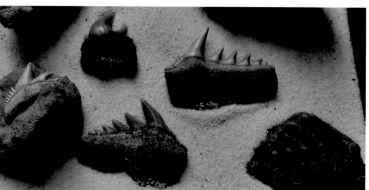

B

Figure 2.1. (A) Bluntnose Sixgill Shark, a representative cow shark, which is among the most primitive known extant sharks. Note the distinct cockscomb-shaped teeth.
(B) Teeth from fossil sixgill sharks. (From the Natural History Museum, Bonn University; Courtesy of Ghedoghedo)

there is little bacterial decomposition. Two such sites are the Cleveland Shale, in Ohio, and the Bear Gulch Limestone, in Montana and North Dakota.

Moreover, the information from fossil shark teeth (one of the most numerous vertebrate fossils), spines, and scales is not insignificant. Paleontologists thus know much less about the evolutionary history of sharks than we would like, but they know a lot about shark dentition, and have made inferences from this knowledge.

Paleontologists know, for example, that among those sharks swimming today, the most *primitive*, that is, the ones whose lineage we can trace back the farthest, are the cow sharks (order Hexanchiformes; fig. 2.1A), whose morphologically distinct teeth have been found in sediment 180 to 200 million years old (fig. 2.1B).

Additionally, similarities and differences in the anatomy of extant (living) sharks (e.g., the presence or absence of anal fins, or the type of jaw suspension), can be useful in constructing family trees of sharks.

Finally, the newest and one of the most powerful methods for understanding relationships of sharks, indeed all organisms, is using the structure of their genetic material, specifically their mitochondrial and/or nuclear DNA or RNA, to work out a group's phylogeny, or evolutionary history and relationships. This molecular biology technique has revolutionized the fields of systematics and taxonomy.

How do molecular systematics and phylogenetics work? Recall that DNA is composed of nucleotide building blocks. The more closely related organisms are, the higher percentage of common nucleotide sequences they possess. The more geological time that has elapsed since species diverged from each other, the more some nucleotides have been deleted and others have been inserted, and the longer natural selection has acted on the new combinations, selecting for some and eliminating others. Typically, meaningful information can be achieved by sequencing around 1000 pairs of nucleotides. Once the nucleotide sequence has been obtained, the analysis and evaluation take place (fig. 2.2).

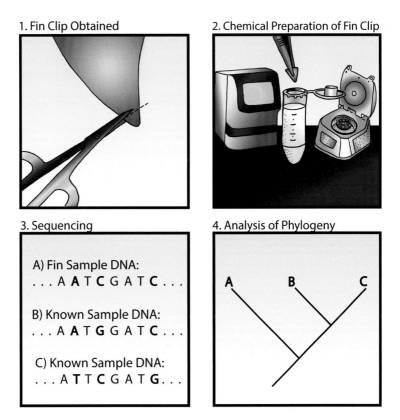

Figure 2.2. Simplified overview of how scientists construct phylogenies based on DNA sequences.

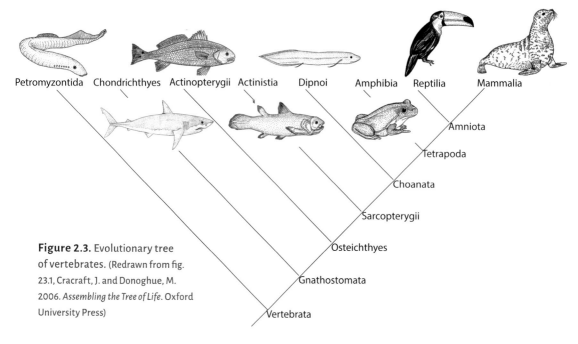

Figure 2.3. Evolutionary tree of vertebrates. (Redrawn from fig. 23.1, Cracraft, J. and Donoghue, M. 2006. *Assembling the Tree of Life*. Oxford University Press)

Using all of these methods, it is known that sharks and their relatives diverged very early from the vertebrate evolutionary tree, right after the jawless fishes (*Petromyzontida* in fig. 2.3), which are considered the stem vertebrates.

Relationships Among Extant (Living) Forms

Let us start with a quick activity that will introduce you to one of the methods, although oversimplified here, utilized by systematists and taxonomists—scientists who name and classify organisms based on their phylogeny; that is, their evolutionary history. Look at Figure 2.4, which is a composite of outlines of 15 different kinds of sharks and rays. Duplicate this page, then roughly cut out each outline, and do the following:

1 / Examine each of the outlines, which are not drawn to scale. Look for one major characteristic that differs among some of the forms and which will allow you to divide the 15 elasmobranchs into two groups, not necessarily of the same number. Do not try to find one or two unique or odd-looking organisms and separate them out first.

Write down the characteristic you used for this division. Try to be as anatomically accurate as you can as you name the characteristic, but if you do not know the correct term, then use one as descriptive as you can.

2 / Now divide each of your groups into progressively smaller groups, as you did above in step 1, recording the characteristic, or anatomical basis, for separating each group. Continue until you have separated the original stack into 15 individuals.

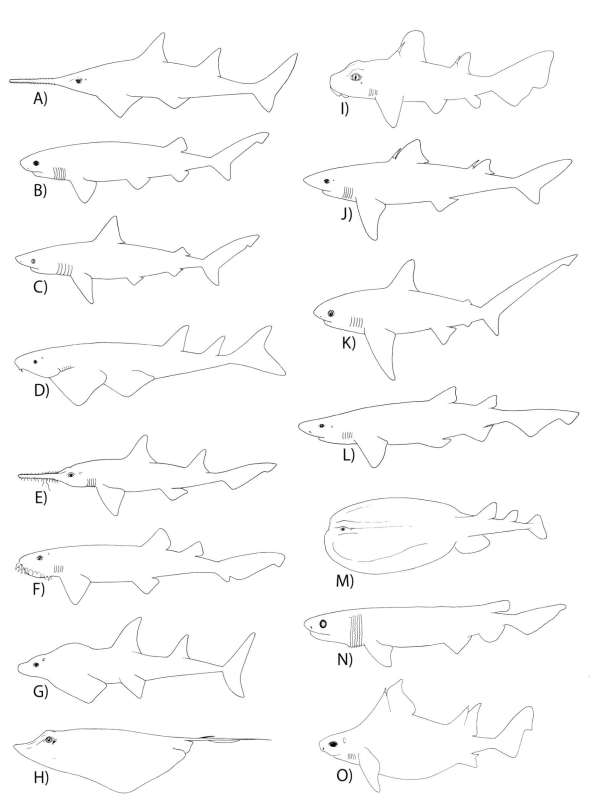

Figure 2.4. Outlines of 15 different kinds of sharks and rays (note that M is portrayed as slightly tilted toward you).

Let us see how accurate you were determining relationships among these elasmobranchs. Our prediction, based on using this activity in courses for over 20 years, is that you first attempted to divide the group broadly into what you consider *sharks* on the one hand and *rays* on the other, likely based on general body shape. Specifically, the group you might call *sharks* was *compressed* (pushed in from side-to-side) and the latter *depressed* (flattened).

Your rationale would be a good one and generally accurate, except that some sharks have evolved to live ecologically like rays, and thus to look more like a ray, and vice versa. Accordingly, you may have placed two batoids, letter G, the Bowmouth Guitarfish (*Rhina ancylostoma*), more commonly called the Shark Ray, and letter J, the Smalltooth Sawfish (*Pristis pectinata*), into the *shark* group, and letter D, the Atlantic Angel Shark (*Squatina dumeril*), with the *batoid* group.

This begs the question: *What is the difference between sharks and rays?* Evolutionarily, batoids diverged from sharks nearly 270 million years ago (mya). Until recently this divergence was known as the *Great Jurassic Split* (which sounds more like an extraordinary if painful gymnastic move than a pivotal point in the evolution of elasmobranchs) because it was formerly thought to have happened during the Jurassic, about 200 mya.

Structurally, it is tempting to state that the main difference between the two groups is that batoids are depressed and sharks are laterally compressed and more spindle-shaped. This difference is generally true, but it is secondary to the location of the gill slits. In sharks, the 5–7 bilateral gill slits are *in all cases* located above, or dorsal to, the well-defined pectoral fins. In rays, the five gill slits (six in one species) are located on the underside of the body, below where the broadly expanded pectoral fins, which we call the wings, are attached to the body.

If you divided the organisms as we suspected, then you incorrectly grouped the Bowmouth Guitarfish and the Smalltooth Sawfish with the sharks on the basis of their more *derived* body plan, and the Angel Shark with the batoids because it is flattened.

From this point, there are several different ways that you could have proceeded. Here are some of the anatomical characteristics that you may have used in creating your taxonomic tree:

1 / one or two dorsal fins

2 / presence or absence of an anal fin

3 / terminal (at the end of the snout) or subterminal (underslung) mouth

4 / head shape (tapered or hammerhead)

The identification of each of the organisms is in this note.[1] If you want to check the accuracy of your attempt to classify the elasmobranchs in this activity, jump to Chapter 3 or visit the very cool and informative *Chondrichthyan Tree of Life* website.[2]

Cladistics

As we hope you learned from the preceding activity, grouping organisms according to external anatomy is only a first step, and is fraught with potential wrong turns that may not group organisms together according to their *phylogeny*, or common ancestry, but rather on their superficial similarities.

The taxonomic system that focuses on evolutionary relationships is known as *cladistics*, or *phylogenetic systematics*.[3] Cladistics distinguishes between characteristics that are more evolutionarily older, that is, ancestral (or original), and those that are more recently evolved, or changed (or derived).

The former are known as *plesiomorphic* characteristics and the latter as *apomorphic*. (The easiest way to recall these names is that the first *p* in *plesiomorphic* stands for *primitive* and that the *a* in *apormorphic* stands for *advanced*. You should be cautioned that this does not imply that an advanced state is necessarily more adaptive or of higher value than a primitive one.)

You may be more familiar with the terms *homology/homolog* and *analogy/analog*. Traditionally, *homologous* traits are those traits due to common ancestry, and *analogous* traits are those as a result of convergent evolution and thus not inherited from a common ancestor.

Here is one example. Shortfin Makos (*Isurus oxyrinchus*) and Albacore Tuna share numerous adaptations that make them high-performance predators (e.g., fusiform, or torpedo-shaped, body; keeled caudal fin; elevated body temperature), but since the former is a chondrichthyan and the latter an osteichthyan, these common characteristics are the result of convergent evolution and are thus *analogous* traits. Similarly, the wings of flying fish, birds, butterflies, and flying squirrels are also analogous with each other, whereas the wings of flying fish are homologous with the pectoral fins of other bony fishes, since they share the same underlying structure, the result of being closely related.

Within the elasmobranchs, plesiomorphies include a simple jaw structure, tribasic fins (with three rigid cartilages at the base), and claspers on the inner margins of the pelvic fins of males. These characteristics, in other words, have been retained from ancestors.

One example of an apomorphy from the elasmobranchs is something that is not present: the loss of an anal fin in the shark order Squaliformes, a group of seven families that includes the dogfish and bramble sharks (fig. 2.5). Another elasmobranch apomorphy, from the batoids, is the presence of a tail spine, which is a modified scale, in members of the order Myliobatiformes, or stingrays, a group of 11 families.

When constructing phylogenies, it is critical to know what traits are shared by members of groups. Apomorphies that are shared by two or more taxa and recent ancestors (but not older ancestors) are called *synapomorphies* (*syn* = together, with). Thus, synapomorphies are derived traits shared by organisms in the same

Figure 2.5. An apomorphic (more recently evolved, or derived) characteristic of elasmo-branchs loss of the anal fin in the Little Gulper Shark (order Squaliformes).

taxonomic group. Synapomorphies, therefore, are very useful in establishing phylogenies and relationships among extant organisms.

Synapomorphies of chondrichthyan fishes (elasmobranchs + chimaeras) include (see fig. 1.4):

- Skeleton composed of cartilage, some of which is prismatic and calcified (also known as *tesserated*[4] cartilage). According to a 2012 review paper[5] on the subject, this characteristic is **the** (emphasis theirs) *critical defining character of this group.*
- Pelvic claspers as intromittent (copulatory) organs
- Teeth not firmly affixed to jaws, with serial replacement
- Chondrocranium (braincase) without sutures (seams)
- Soft fin rays (ceratotrichia) that are unsegmented

Taxa that have synapomorphies form what is known as *monophyletic* groups, since they all have the same characteristics as their ancestors. In other words, a monophyletic group consists of a single ancestral species and *all* its descendants. Another name for a monophyletic group is a *clade*, and the diagram that depicts relationships based on cladistics is known as a *cladogram.*

The overarching goal of cladistics is to create monophyletic groups (fig. 2.6); that is, to discover the true taxonomic relationships among organisms.

For a group (most commonly a family or order) to be monophyletic, all of the species *belonging* in that group are *classified* within that group and are not mistakenly classified in another group. Additionally, no species from other groups are spuriously included within the group in question.

If one or more of the descendants of an ancestor are missing from the group, say, because we simply do not know all of the details yet, the group is not considered monophyletic but rather is *paraphyletic* (*para* = near; fig. 2.6). Reptiles are considered a paraphyletic group if birds are erroneously excluded (as they were until recently), because birds and reptiles share a common ancestor and should have been classified together.

If a group is *polyphyletic* (*poly* = many; fig. 2.6), it means that it wrongly does not include the *common* ancestor but rather ancestors common to *some* of the members of the group. An example of polyphyly is classifying mammals and birds in the same group separate from the reptiles, as was also done in early taxonomies, because both are endothermic (warm-bodied).

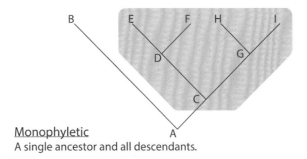

Monophyletic
A single ancestor and all descendants.

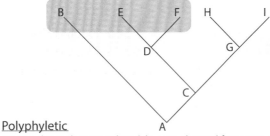

Polyphyletic
Descendants that are related, but are derived from two or more ancestors.

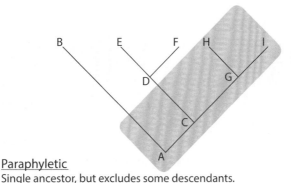

Paraphyletic
Single ancestor, but excludes some descendants.

Figure 2.6. Cladograms showing monophyletic, polyphyletic, and paraphyletic groups.

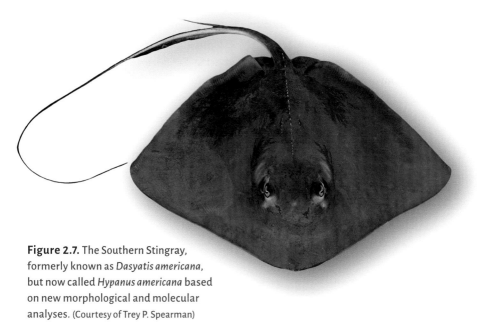

Figure 2.7. The Southern Stingray, formerly known as *Dasyatis americana*, but now called *Hypanus americana* based on new morphological and molecular analyses. (Courtesy of Trey P. Spearman)

Thus, paraphyletic and polyphyletic taxonomies are simply wrong. They usually arise as educated best guesses based on incomplete information.

While these terms may seem confusing, once you begin observing and applying the concepts, you will become more facile with them.

Here is an example. As long as we can remember, the scientific name for the most common coastal stingray of the SE Atlantic Coast and Caribbean Sea, the Southern Stingray (fig. 2.7), has been *Dasyatis americana.* The 97 known species of whip-tail stingrays, including the Southern Stingray, are in the family Dasyatidae, which has been thought to be a monophyletic group. That means that these 97 species would be more closely related to each other than to other rays, and that is because they all have the same common ancestor.

However, a recent study[6] employed a combination of morphology and molecular analyses and concluded that most of the genera were *not* monophyletic, and that the genus *Dasyatis* was polyphyletic, requiring the revision. In other words, these species were all related but not as closely as once thought and were derived from more than a single common ancestor. As a result, an entirely new classification was created, and the Southern Stingray was renamed *Hypanus americana* to distinguish it from other rays that we now know are not as closely related as previously thought.

Gnathostome Phylogeny: The Evolution of Jaws

Before we describe the evolution of the chondrichthyan fishes, we want to emphasize a preceding evolutionary development of singular significance: the evolution of jaws.

The first true fishes were the *ostracoderms* (meaning *shelled skin*; fig. 2.8), a poly- or paraphyletic group of bony-plated, small, jawless animals that arose about 460 million years ago, in the Ordovician Period, and diversified around 443–416 mya, in the Silurian Period (fig. 2.9).

Jawless does not mean that they lacked a mouth, but still, without jaws how could a fish survive? Ostracoderms, in fact, prospered. How? By emulating vacuum cleaners, likely living near the bottom of the water column and sucking up small, soft-bodied invertebrates. They are thought to have survived for about 100 million years and given rise to today's jawless fishes, the hagfish and lampreys

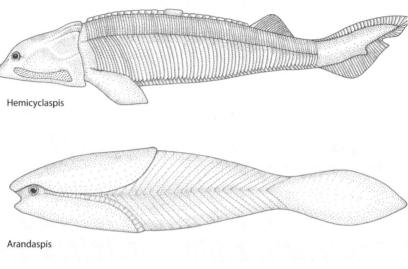

Hemicyclaspis

Arandaspis

A

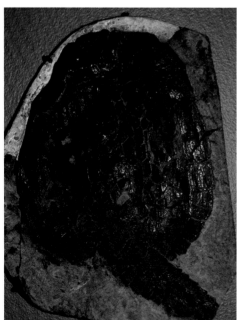

B

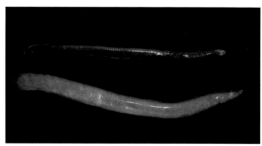

C

Figure 2.8. (A) Ostracoderms, the first true fishes. (Redrawn from Maisey, J. 1996. *Discovering Fossil Fishes.* Holt) (B) *Cardipeltis bryanti*, a fossil ostracoderm from the Devonian of Wyoming. (Courtesy of James St. John) (C) Hagfish, one of two groups of extant jawless fishes.

Millions of Years Ago	Geological Period	Event
457	Silurian	Ostracoderms arise; Divergence of Osteichthyes and Chondrichthyes
440	Silurian	Oldest known jawed vertebrates, Acanthodians, lived; Placoderms arose around the same time
420 – 360	Devonian	The Age of Fishes
421	Devonian	Divergence of Elasmobranchii and Holocephali
409	Early Devonian	Oldest articulated skeleton from a true shark (that is, a member of the Class Chondrichthyes) (*Doliodus problematicus*) discovered in New Brunswick, Canada
395 – 345	Devonian – Carboniferous	All known Gnathostome clades established. These include the classes: Acanthodii, Placodermi, Osteichthyes, and Chondrichthyes
370	Devonian	Cladoselache, a Cladodont shark (the first radiation of sharks), discovered from Cleveland Shale
360 – 300	Carboniferous	Golden Age of Sharks
250	Permian	Permian Extinction
213 (Range 281 – 364)		Batoid and shark lineages split; Post-Permian recovery, diversification of squalomorph and galeomorph sharks
225		Heterodontidae/Triakidae split
246–216		All families of modern chondrichthyans established

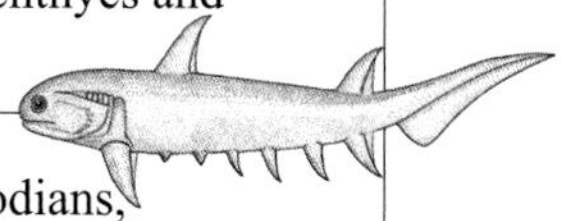

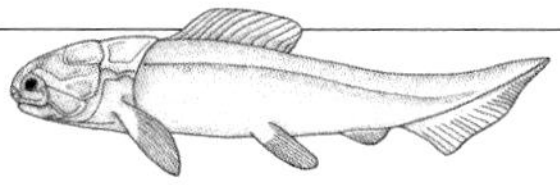

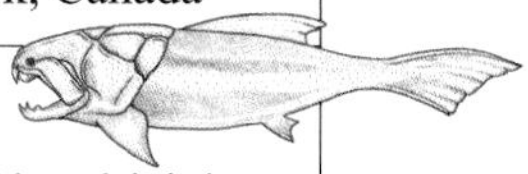

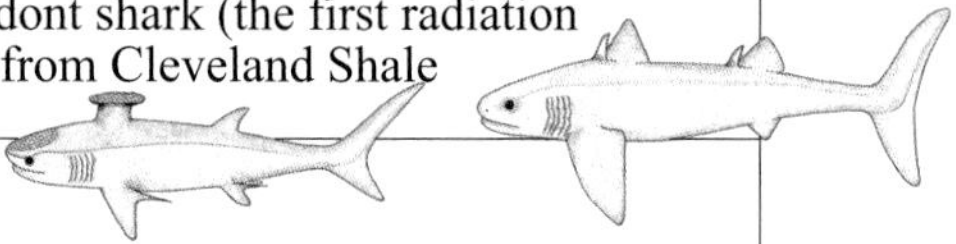

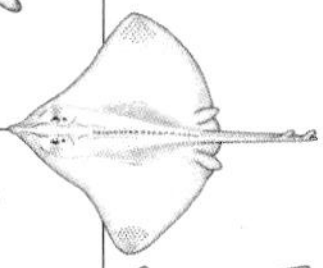

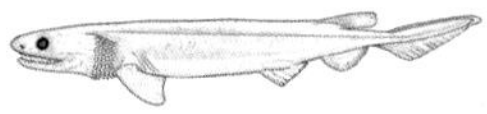

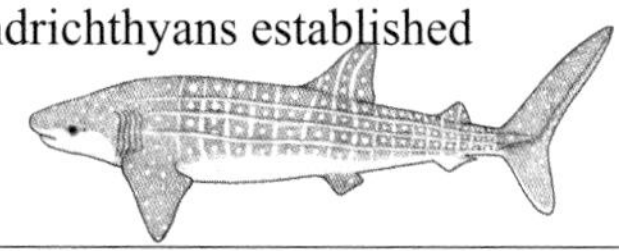

Figure 2.9. Geological Time Scale, with references to major events.

(fig. 2.8C). What led to the extinction of all but the stem group (i.e., the common ancestor that gave rise to the hagfish and lampreys)? Likely the rise of the *jawed fishes*.

There are about 122 extant species of lampreys and hagfish. They have persisted for about 500 million years, but they are not very successful in terms of diversity compared to the jawed fishes, the approximate 33,000 bony fishes and 1250 chondrichthyans.

The diversity of bony and cartilaginous fish and the evolutionary success of tetrapods (amphibians, reptiles, birds, and mammals) owe in large part to the replacement of a suctorial mouth with jaws in the Silurian Period. Jaws arose through natural selection from the first set of gill arches, which evolved into the upper and lower jaw (the *mandibular arches*), and these were structurally supported by modifications of the second set of gill arches (the *hyoid arches*; fig. 2.10). Fish with jaws, the first of which were the acanthodians and placoderms, are known as *gnathostomes* (*gnath* = jaws; *stome* = opening; see figs. 2.11 and 2.12).

How could modification of the mouth become such a meaningful evolutionary development? What are the advantages of jaws? First, grasping became possi-

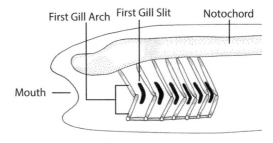

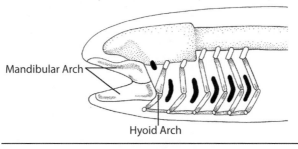

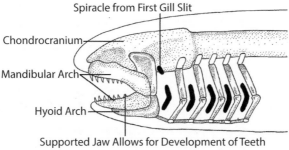

Figure 2.10. The evolution of jaws from the first set of gill arches. (Redrawn from fig. 2, Mallatt, J. 2008. Zool. Sci. 25: 990–999)

ble. Second, once grasped, prey or parts of prey could be handled or manipulated, cut, and further processed. Finally, once jaws evolved, they became raw material for natural selection to modify into the myriad forms that bony fishes possess, and the evolution of the tetrapods. These may not sound like earth-shattering advances, but in the evolution of vertebrates, the rise of jaws arguably is the key event that led to their success, and left the jawless fishes treading water, so to speak.

Overview of the Evolution of Chondrichthyans

While jawed fishes, including the species that gave rise to sharks and their relatives, existed before the Devonian Period (360 to 420 mya), it was during this 60 million-year period that bony fishes diversified such that the Devonian is known as the *Age of Fishes*. It is thought that the increased abundance of warm, shallow seas led to a profusion of biogenic reefs (formed by animals such as sponges and extinct coral relatives), which in turn led to diversification of bony fishes by providing innumerable living spaces and abundant food, which are ecological roles that current coral reefs also perform. All major clades of fishes arose during this period.

Shark ancestors also swam in the Devonian, and the earliest fossils of true sharks were found in Devonian deposits, but they were not nearly as diverse or numerous a group as the Osteichthyes. Two other clades of fishes that arose during the Devonian, the placoderms and acanthodians, are considered below as possible ancestors to the chondrichthyans. The Devonian ended with a mass extinction event (or, more likely, a string of extinction events over 20–25 million years) that killed between 70–80% of marine species, especially those in shallow, tropical waters, likely due to extensive hypoxia and anoxia (low or no dissolved oxygen).

It was not until the Carboniferous Period (300–360 mya) that chondrichthyans experienced their first major radiation (large increase in diversity), and thus this period is known as the *Golden Age of Sharks*. During this period, there were more sharks and rays than any other fishes, and they were the dominant predators in oceans, rivers, and lakes.

Perhaps the most important site contributing to our understanding of the diversification of sharks is the Bear Gulch Limestone of Montana and North Dakota, deposits dating back to about 323 million years ago. The former bay, some of whose inhabitants are preserved as fossils in the limestone, existed for only about 1000 years. Within the approximately 27 m (90 ft) thick system of sediments, over 65 species of sharks were discovered. By the end of the Carboniferous, 45 families of sharks are believed to have existed.

The next major event affecting chondrichthyan fish was one that occurred about 250 mya, and its effects were far-ranging. This event was the *Great Permian Extinction*, during which 90–95% of all marine species (including chondrichthyan fishes) were wiped out, as well as 70% of terrestrial life. Some chondrich-thyan lineages survived, most likely those living in deeper waters.

Several explanations exist for what has also been called the *Great Dying*, including a meteor strike, volcanic event, an enhanced greenhouse effect due to sea floor methane emission, the formation of the supercontinent *Pangaea* (which may have had large scale atmospheric and oceanic impacts), or some combination of events.

Modern sharks, known as *neoselachians*, or *euselachians*, first arose during the late Triassic / early Jurassic Periods, around 200 mya, and species similar to many living today arose in the Cretaceous period, 145 mya. Another major extinction event occurred about 66 mya, at the end of the Cretaceous Period. The most likely cause was a comet or asteroid strike, although volcanism, climate change, and other causes have been propounded.

Although the end-Cretaceous extinction is considered only the fifth worst extinction, over 40% of elasmobranch genera and 45% of shark species became extinct (and 75% of all marine species), and the ecological impact was severe. After this extinction, movement of the Earth's crustal plates created new continental margins in relatively warm climates in temperate and tropical latitudes. These shallow, warm waters provided conditions ideal for the development of seagrass, salt marsh, mangrove, and coral reef communities. Coincident with this was a similar proliferation of bony fishes, which took over marine systems.

We are currently in the Earth's sixth and, we reluctantly assert, *last* extinction. Unlike some of the previous extinction events, there is no ambiguity about the cause—humans, specifically—too many people using too many resources, most notably fossil fuels.

Now that we have given you a broad overview of the evolution of chondrich-thyan fishes, let us focus in on some of the details.

Who Are the Immediate Ancestors of Sharks?

To be called a shark, a fossil would be a fishlike vertebrate with a primitive jaw suspension (*jaw suspension* refers to how the upper and lower jaws connect to the skull and other supporting structures), a cartilaginous skeleton, naked gill slits (i.e., lacking the operculum found in bony fishes), claspers, various specialized hard parts (scales, spines), moderately-mineralized vertebra, a skeletal element connecting the pectoral fins, and specially-shaped teeth not firmly embedded in the jaw. Thus, to be the immediate ancestor of sharks, an organism should have one or more of these characteristics and a plausible pathway for the evolution of these characteristics if they lack them.

The oldest known jawed vertebrates, dating back at least 440 million years ago, are members of the Class Acanthodii (*acanth* = spine; fig. 2.11). Acanthodians have been referred to as *spiny sharks* because of their superficial resemblance to sharks but calling them *sharks* may have been premature until recently.

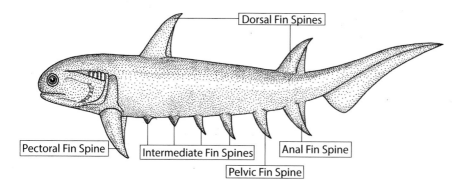

Figure 2.11. Acanthodians, or Spiny Sharks, the oldest known jawed vertebrates, which are considered the earliest known shark ancestor. Note the three sets of paired fins and the huge fin spines characteristic of the group. (Redrawn from fig. 2, Watson, D.M.S. 1937. Philos. Trans. R. Soc. Lond. B. Biol. Sci. 228: 49–146)

Acanthodians possessed a cartilaginous skeleton, but no other defining characteristics that would qualify them as sharks. In fact, they possessed more characteristics of bony than cartilaginous fishes. Acanthodians, representatives of which survived until about 250 mya, were small and had jaws with true teeth, large eyes, streamlined bodies with multiple paired fins, spines preceding the fins (hence, the name *spiny*), and small bony plates covering the body. Until recently, they were not considered a monophyletic group.

Were acanthodians ancestors to sharks? A 2018 paper,[7] using histology and advanced imaging techniques on the 385-million-year-old acanthodian shark *Gladbachus* concluded that acanthodians are indeed the true stem chondrichthyans.

The other group which at one time was thought to have given rise to sharks was the placoderms (*plac* = plate; *derm* = skin), a highly successful assemblage that arose 440 mya and lived through the Devonian (420–360 mya), or even into the Carboniferous (360–300 mya; fig. 2.12). Among the first jawed fish, placoderms were well-adapted, dominant predators with a bony endoskeleton and a hinged, ossified, helmeted head.

The best-known placoderm is the > 6 m (20 ft) long *Dunkleosteus* (fig. 2.12B), as formidable a predator as ever swam, though with its heavy armor, likely a sluggish one. *Dunkleosteus* is often pictured superimposed over a school bus, a particularly effective if unimaginable metaphor for how fearsome this beast was.

What shark-like characteristics did placoderms possess? Some had heterocercal tails. A group known as the *ptyctodonts* (*folded teeth*) had claspers, but anatomically they differed from the shark model and thus represent an example of convergent evolution. Ptyctodonts had an unarmored head and reduced bone, and superficially resembled the chimaeras. Could ptyctodonts have been the mother (stem) group for sharks? Probably not. Most of the similarities have been shown to be analogous structures or only superficially resembling those of shark relatives.

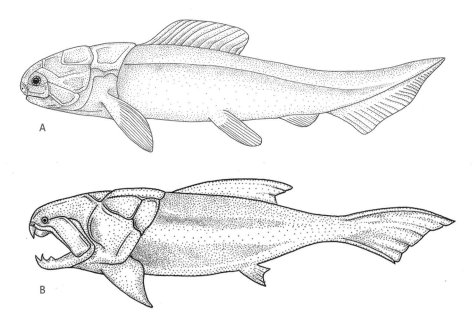

Figure 2.12 (A) A placoderm. (B) The placoderm *Dunkleosteus*. (Redrawn from Maisey, J. 1996. *Discovering Fossil Fishes*. Holt)

Placoderms survived only about 50 million years, likely succumbing to the late Devonian extinction (explanations for which include sea level changes, widespread anoxia in oceans, asteroid impacts, climatic change, and others), as well as competition from early sharks and bony fishes, the major group thought to have arisen from the placoderms.

The Problem of *Doliodus problematicus*

In 1997, an entire 23 cm (9 in) fossil fish was unearthed from mudstone in Canada, and it turned the chondrichthyan paleontology world upside down. The story is beautifully related in a 2005 article.[8] The fish was in fact a shark in the genus *Doliodus,* a taxon known previously exclusively from its teeth. What distinguished this specimen from congeners (members of the same genus) and indeed all other chondrichthyan fossils, was that this individual fossil was a complete, articulated specimen. This means that all or most of the hard anatomical parts were present and in or close to their proper orientation, an extremely unusual and lucky find.

Why was *Doliodus*, which resembled a modern angel shark, considered a shark? For one, it had a jaw-full of teeth, significantly and surprisingly in rows. It had calcified cartilage and fin spines. Other characteristics were more problematic in assigning a lineage (e.g., the presence of a pectoral spine, which is not known in modern sharks), as well as some dentition and braincase differences, and hence the specific epithet *problematicus*. *Doliodus problematicus* may represent the most primitive condition among sharks, with characteristics that evolved into those of more recent sharks. Alternatively, this group may be intermediate between acanthodians and sharks.[9]

Let us stop here and review a few of the salient points thus far:

- The fossil record for sharks is poor.
- A major evolutionary advancement was the development of jaws. The evolutionary possibilities exploded once jaws came into existence.
- The common ancestor of chondrichthyans in general and sharks in particular remains to be discovered, but evidence points to the Class Acanthodii.
- All four of the Gnathostome clades (Acanthodii, Placodermi, Osteichthyes, and Chondrichthyes) arose over an approximately 50-million-year span during the Devonian Period.

Now, we will turn to the next stages in the evolution of chondrichthyans, which occurred in three stages: *Cladodont, Hybodont,* and *Modern.*

Cladodont Sharks

The *cladodonts* (*clade* = branch; *dont* = tooth), a group named for its multiple-cusped teeth, are the earliest known assemblage of what can be called with certainty *sharks* (fig. 2.13). Cladodont sharks arose in the early Devonian, 150 million years distant from their jawless ancestors, and had a type of jaw suspension (termed *amphistylic*; see Chapter 4) characteristic of extant primitive sharks, a homocercal tail (i.e., with equal lobes), and big eyes.

The earliest known cladodont shark, which clearly resembled a modern shark and which swam in shallow Devonian Seas, was *Cladoselache* (fig. 2.13B). *Cladoselache* was most likely a fast-swimming, large (2 m, or 6.6 ft) pelagic or epipelagic top marine predator that lived contemporaneously with the aforementioned school bus-sized placoderm *Dunkleosteus*.

Cladoselache had additional shark characteristics: pectoral fins reinforced with basal and radial cartilages, five gill slits, dorsal spines, lateral keels, and notochord extending into the upper lobe of the caudal fin (even though it was homocercal from the exterior). It had dentition typical of cladodonts: multi-cusped with enamel-covered dentin (homologous with scales). No extant sharks have similar teeth.

However, one large deficiency in the *Cladoselache* fossil specimens is troublesome: none were found with claspers, which, you recall, are the *sine qua non*[10] of being an elasmobranch (a male). Claspers were present on many sharks contemporaneous with *Cladoselache*.

The absence of claspers may or may not be meaningful in determining the taxonomic position of *Cladoselache*. On the one hand, the genus may have simply not possessed them and thus either lacked internal fertilization or used a more innovative but unknown mechanism of internal fertilization.

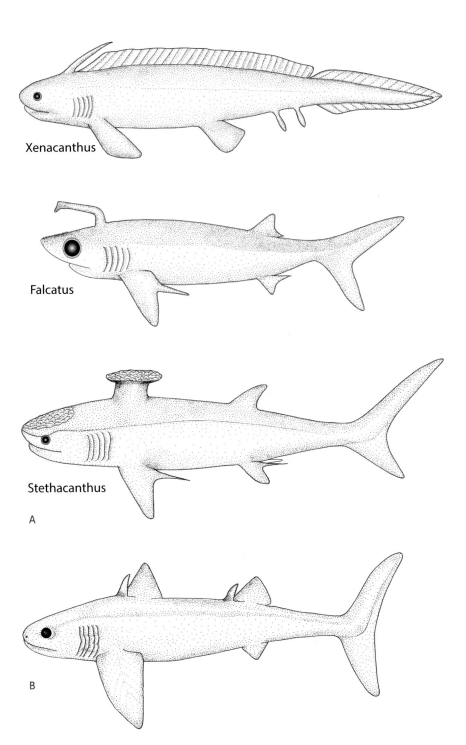

Figure 2.13. (A) Cladodont Sharks (*from top to bottom*): *Xenacanthus*, *Falcatus*, and *Stethacanthus*. (B) The earliest known cladodont shark, *Cladoselache*. (Redrawn from Maisey, J. 1996. *Discovering Fossil Fishes*. Holt)

More likely, *Cladoselache* may have practiced an activity common to modern sharks, *sexual segregation*, which is seen in numerous adults of many species (e.g., Scalloped Hammerheads [*Sphyrna lewini*] and Blacktip Sharks [*Carcharhinus limbatus*]). Thus, it could be possible that all specimens unearthed from the relatively shallow marine environment preserved in the Cleveland Shale were females which had congregated there. Other sharks found in the same deposits as *Cladoselache* possessed claspers, but this fact is not inconsistent with the sexual segregation hypothesis, since not all species of shark segregate by sex.

Also, *Cladoselache* had a terminal mouth. Most sharks have a *subterminal*, or *underslung* mouth and only a handful (e.g., the Frilled Sharks [*Chlamydoselachus*] and Megamouth [*Megachasma pelagios*]) possess terminal mouths. In addition, *Cladoselache* did not have a vertebral column but instead only an unsegmented (sometimes referred to as *unrestricted* or *unconstricted*) notochord, which in modern sharks is *constricted* (or divided) into vertebral centra.

In spite of some of the inconsistencies, *Cladoselache* is considered the first true shark in the fossil record.

Another highly successful cladodont shark was *Xenacanthus* (fig. 2.13A), which dominated freshwater lakes and streams for 200 million years throughout the late Devonian and Carboniferous Periods, until the Permian Extinction. *Xenacanthus* also resembled the Frilled Shark, including having a terminal mouth. The body shape of *Xenacanthus* suggests a flexible shark capable of navigating through trees and other obstructions in rivers and lakes.

Another successful cladodont group was the *stethacanthids* (figs. 2.13 and 2.14), 5.5 m- (18 ft-) long sharks that prospered at the end of the Devonian. Male specimens had an unusual feature in front of the dorsal fin, a structure resembling a wire scrub brush. Females lacked this, which may in males have been ornamental and involved in courtship. No modern sharks possess similar structures.

One strange stethacanthid shark was *Falcatus* (fig. 2.14), which has been called the Unicorn Shark because of its large, curved, scaled spine just posterior to the head on the dorsal surface of males of this group. *Falcatus* was a very small (0.15–0.25 m, or 6–9 in) schooling shark that preyed on small invertebrates, like shrimp. It was first described in 1883 from the St. Louis Limestone and later discovered by paleontologist Dick Lund in Montana's Bear Gulch, a limestone laid down in tropical fresh and brackish waters about 320 mya. Lund also made a serendipitous[11] discovery in the Bear Gulch of a pair of *Falcatus* in which the female appears to be grasping the head appendage, or cephalic clasper, of its male partner. Among living chondrichthyans, one group, the chimaeras possess head claspers on males, but no sharks or rays do, and in chimaeras the clasper is used to grasp on to the female.

Two aspects of the anatomy of *Falcatus* provide a glimpse of its lifestyle, ecology, and physiology. First, the genus had a slightly elongated snout, which implies the presence of scaffolding for sensory structures like *ampullae of Lorenzini*, which are electroreceptors, and perhaps an increased olfactory (smell) sense. Sec-

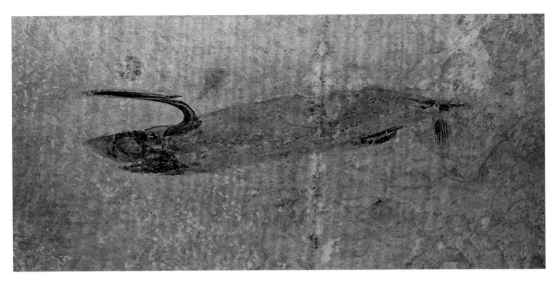

Figure 2.14. The cladodont stethacanthid shark *Falcatus falcatus*, from the Lower Carboniferous. (Courtesy of H. Zell. https://commons.wikimedia.org/wiki/File:Falcatus_falcatus_01 .JPG. Accessed 6/29/19)

ond, the tail was internally *heterocercal* (with vertebrae ending in the upper lobe) but externally *homocercal* and even *isocercal*, with equal lobes that generate high thrust, like the tail of a Shortfin Mako or tuna.

Another enigmatic cladodont group are the *ctenacanths* (meaning *comb-spine*; fig. 2.15), which appeared at about the same time as *Cladoselache*, about 380 mya. The group lasted through the Permian, and some even persisted into the Triassic, about 250 mya. The genera *Ctenacanthus* and *Goodrichthyes* are best known, based almost exclusively on preserved fin spines and some body impressions.

Ctenacanths had numerous characteristics in common with *Cladoselache* (e.g., cladodont teeth, unconstricted notochord, terminal mouth) but deviated in possessing the tribasic cartilaginous support of the pectoral fins. They also had enamel-coated spines along the leading edge of the first dorsal fin, similar to those of modern sharks like dogfish (family Squalidae) and horn sharks (Heterodontidae). The recent discovery of a well-preserved specimen of the ctenacanth *Gladbachus*, found in 385 million-year-old deposits, further supports that ctenacanths are stem chondrichthyans that may have given rise to modern sharks.

The Carboniferous period lived up to its reputation as the *Golden Age of Sharks*, with the evolution of 45 different families. Numerous forms, some very bizarre (fig. 2.16), existed then, including *Helicoprion* (*helico* = spiral; *prion* = tooth), the beast with tusk-like tooth whorls, and *Edestus*, in which scissor-like toothed structures protruded from its jaws. One explanation for these structures is that they were adaptations for feeding on the then-common Chambered Nautilus, and they used these protrusions to pry the soft-bodied nautilus out of its shell. At best this is an educated guess. Depictions of both *Helicoprion* and *Edestus* often portray them with protrusible jaws, which had not yet evolved in sharks.

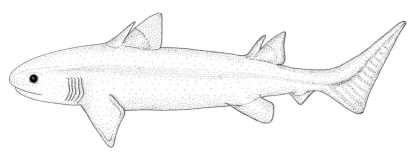

Figure 2.15. A cladodont ctenacanth. (Redrawn from Long, J. 1995. *The Rise of Fishes*. Johns Hopkins University Press)

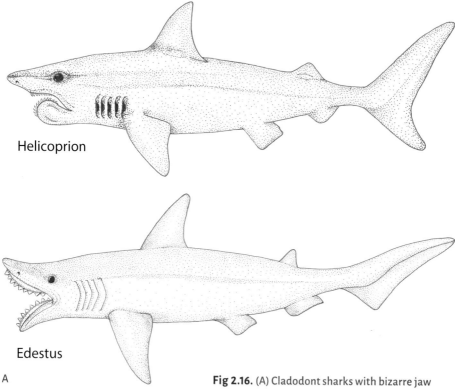

Helicoprion

Edestus

A

B

Fig 2.16. (A) Cladodont sharks with bizarre jaw structure, *Helicoprion* and *Edestus*. Artists' depictions, including ours, of these and other fossil sharks often must involve best guesses about their appearance, since they are based on incomplete fossils, often merely a few isolated parts. (Redrawn from Maisey, J. 1996. *Discovering Fossil Fishes*. Holt) (B) Fossil tooth whorl of *Helicoprion* from the Permian of Idaho. (Courtesy of James St. John)

Hybodont Sharks

The Great Permian extinction killed off most Carboniferous sharks, including the cladodonts, along with > 90% of all marine organisms. Survivors included only two groups of sharks, the *neoselachians* (modern sharks) and what is most likely their closest sister group, the *hybodonts* (*humped tooth*). Hybodonts (fig. 2.17) first appeared around 360–320 mya, in the late Carboniferous. Their success allowed them to be represented as the next major level of shark evolution. Hybodonts were the dominant sharks for about 100 million years, from the late Jurassic to the early Cretaceous Periods.

Hybodonts are perhaps the closest ancestors of the modern sharks, although it is not clear whether they or an offshoot gave rise to neoselachians. Current thought is that hybodonts represent an extinct sister-group to neoselachians (i.e., they share a common ancestor).

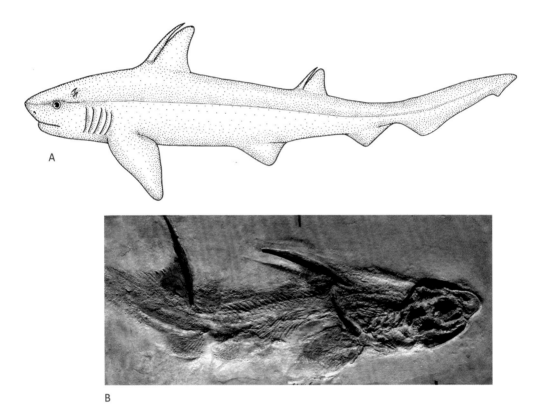

A

B

Figure 2.17. (A) Hybodont sharks. Illustrations of hybodont sharks frequently bear a striking resemblance to modern horn sharks. Why is that? Since hybodont sharks had the same kind of heterodont dentition, that is, with two different tooth morphologies, as modern horn sharks, artists interpreted that they must also look like them. (Redrawn from Maisey, J. 1996. Discovering Fossil Fishes. Holt) (B) The hybodont shark Hybodus fraasi from Eichstätt, Germany. (Courtesy of Haplochromis. https://commons.wikimedia.org/wiki/File:Hybodus_fraasi_(fossil).jpg. Accessed 6/29/19)

Superficially, hybodonts may have resembled the bullhead, or horn, sharks, a group of nine extant species in the family Heterodontidae, whose first fossils were found in the early Jurassic. This putative resemblance is based on the discovery that hybodonts possessed the heterodont dentition, with crushing teeth on the rear of the jaw and small piercing teeth in the front, for which the family Heterodontidae is named (*hetero* = different; *dont* = teeth). Heterodont dentition offers a different specialization for prey type than homodont dentition. Hybodont sharks still possessed a type of primitive but slightly loosened jaw suspension (termed *amphistylic*, which provided slightly more mobility). Also, like cladodonts, hybodont sharks had an open vertebral column with an unrestricted notochord. Additionally, they had tribasic cartilage anchoring the pectoral fins and fin spines preceding each dorsal fin. Unlike cladodonts, they had claspers, more elaborate fins, and less specialized teeth. Hybodont sharks did not survive beyond the early Cretaceous.

The Rise of Modern Sharks

The most recent radiation of sharks gave rise to the *modern sharks*, or *neoselachians*, a monophyletic group which arose in the Tertiary and diversified in the Jurassic.[12] The rise of modern sharks coincides with the breakup of the supercontinent Pangea into Laurasia in the north and Gondwanaland in the south. By the late Jurassic, about 150 mya, the outlines of North America and South America/Africa were becoming evident, separated by a young Atlantic Ocean and the tropical Tethys Sea, parts of which were destined to become the Indian Ocean. The breakup of the continents provided shallow water habitat over the now-increased continental shelves—an ideal habitat for elasmobranch fishes.

What makes a modern shark *modern* is a suite of advancements over ancestors that allowed for more mobility when swimming and more efficient feeding. These advances (see fig. 1.4) included:

- **An elongated snout with subterminal (ventral) mouth.** The elongated snout, known as a rostrum, allows sensory organs to be moved forward. The underslung jaw can accommodate heavier jaw musculature and allows the evolution of greater jaw mobility (i.e., protrusibility). Moreover, the coracoid bar, the skeletal element that connects and supports the pectoral fins, also serves as an anchor for some jaw musculature (see fig. 4.19). Relocation of the lower jaw also allows the braincase to expand to accommodate the enlarged brain associated with increased sensory capability.

- **Pectoral fins fused along the coracoid bar.** In ancestral sharks, a skeletal element connecting the left and right pectoral fins was not always present. Fusion of the pectoral fins provides more stability as well as anchoring for muscles. In modern sharks the pelvic girdle connects both pelvic fins, which ancestors also lacked.

- **Differently shaped teeth.** The variation in tooth morphologies from the cladodont and hybodont forms opens up new feeding opportunities.

- **Looser jaw suspension (hyostylic).** The loosening of the palatoquadrate (upper jaw) from the braincase along with support from the hyomandibular (part of the hyoid arch, which developed from the second gill arch evolutionarily) allows for greater jaw protrusibility and mobility, and a wider range of prey items. The significance of this larger gape—being able to open the jaws wider than ancestors—to the success of modern sharks cannot be overstated.

- **Vertebrae hardened by minerals (firm support for heavy musculature).** The notochord becomes constricted into defined vertebral centra that are calcified and noncompressible, and serve as points of attachment for stronger swimming musculature.

- **Streamlined body with placoid scales.** Moving through a relatively viscous (i.e., resistant to flow) and dense medium-like water requires significant amounts of energy. Streamlining reduces drag, thus reducing the energetic cost of swimming, allowing faster swimming, or both. Placoid scales in some species also play a role in reducing drag.

- **Fin spines disappear (with exceptions).** Spines on the leading edge of fins operated for defense or in some cases, as cutwaters, to reduce drag. As sharks became streamlined, in most cases they were no longer necessary.

- **More flexible fins.** The stiff radial cartilages, which disallowed any fin mobility, are now shortened and the supporting distal elements, the ceratotrichia, are more elongate. Increased fin flexibility translates into greater maneuverability.

- **Enamel of the teeth in three layers.** Hybodonts had a single enameloid layer. Enamel is one of the hardest substances in toothed organisms. More layers mean stronger teeth.

The upshot of these modifications of neoselachians is improved feeding and swimming over their ancestors, the hybodonts, thus expending less energy or using it more efficiently in their daily activities.

The first modern shark from the fossil record is *Paleospinax* (fig. 2.18), which appeared in the early Jurassic. Stem neoselachians likely were present in the early Triassic, but the group diversified in the Jurassic. *Paleospinax* superficially resembled a modern carpet shark (family Orectolobidae), although it possessed a spine on both dorsal fins and carpet sharks do not, leading some to conclude that it more closely resembled a modern dogfish (family Squalidae). Anatomical features of *Paleospinax* shared with extant sharks include a subterminal mouth; shortened, loosely suspended jaws; a true vertebral column (as opposed to unsegmented notochord); enamel-coated teeth; and a more elongate snout. Other early modern chondrichthyans include *Protospinax* (a bottom-dwelling ray; fig. 2.18),

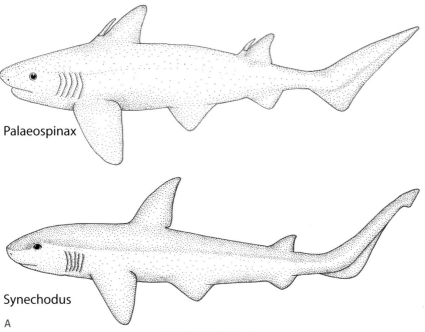

Palaeospinax

Synechodus

A

Figure 2.18. (A) Early modern sharks *Paleospinax* and *Synechodus*. (Redrawn from Maisey, J. 1996. *Discovering Fossil Fishes*. Holt) (B) Fossil *Protospinax*, which has a flattened body. (Courtesy of High Contrast)

B

Synechodus (a shark related to *Paleospinax*), and *Crossorhinus* (progenitor of the Nurse Shark).

The neoselachians comprise three major monophyletic groups, *squalomorphs*, *galeomorphs*, and *batoids*, whose extant (living) taxa will be described in Chapter 3.

Recall that batoids diverged from the chondrichthyan line nearly 270 mya.[13] Among those sharks swimming today, the most primitive (the one whose lineage we can trace back the farthest) is the group of sharks known as the cow sharks (see fig. 2.1).

The divergence times between cow sharks and the other squaliform sharks, based on mitochondrial genomic sequences from two studies, is about 115–130 mya. A candidate for the oldest known extant species of shark is either the

Broadnose Sevengill Shark (*Notorynchus cepedianus*; fig. 2.19), or one of its confamilials, which date back 78–88 mya. The next oldest groups are the angel sharks (order Squatiniformes) and dogfish (Squaliformes).

The superorder Galeomorpha is a more recent taxon than the Squalomorpha. Among galeomorph sharks, the horn sharks (order Heterodontiformes) are the oldest in the fossil record. They split off from the galeomorph lineage somewhere between 240–310 mya. Extant horn sharks date back as far as 47 mya. The next oldest galeomorph group are the carpet sharks (order Orectolobiformes; 166 mya), followed by the mackerel sharks (Lamniformes; 162 mya), and the requiem sharks (Carcharhiniformes; 179 mya).

Megalodon? Really?

We conclude this chapter with an introduction of the evolution of the White Shark and the extinct beast known as *Megalodon* (fig. 2.20). The story is no more compelling than others we have told, and indeed sheds little light on the major questions of the evolution of sharks. Still, it is a topic that shark enthusiasts care about, and thus it merits our attention here.

Figure 2.19. The Broadnose Sevengill Shark *Notorynchus cepedianus*, in a kelp forest near Cape Town, South Africa. (Courtesy of Lesley Rochat)

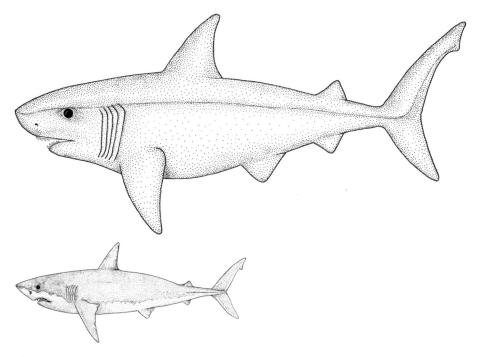

Figure 2.20. Megalodon (*top*) compared to an extant White Shark drawn roughly to scale.

White Sharks (*Carcharodon carcharias*) have been recorded as large as 6 m (20 ft) and 1900 kg (4200 lb). If you have ever seen a mature White Shark, you cannot help but be impressed by its immensity.

Of course, a full-grown White Shark would be dwarfed by the largest predatory fish to ever swim the world's oceans, *Megalodon* (meaning *giant tooth*), which refers to the extinct shark *Carcharocles megalodon* (formerly *Carcharodon megalodon*), whose estimated maximum size is conservatively > 15 m (> 50 ft) to as large as 18 m (59 ft).

Megalodon is mistakenly thought of as an ancestor of the White Shark. Based on dates of its fossil teeth and vertebrae, the genus *Carcharodon* dates back 60–65 mya. Until recently, however, there was uncertainty as to the lineage of the White Shark. The phylogeny was clarified in a 2012 paper,[14] and is based on the discovery of vertebrae and a complete set of intact teeth from a new shark fossil (*Carcharodon hubbelli*). It turns out that *Megalodon* was not a direct ancestor of today's White Shark, as nice as this story would be, but rather led to an evolutionary dead end.

Concluding Comments: Are Sharks Still Evolving?

This chapter described sharks of the past, and the next chapter describes sharks of the present. Are sharks still evolving? What will sharks of the future look like?

Yes, sharks are still evolving—evolution is not static, and all species on the planet are evolving. But no one knows what the future looks like for sharks (and other inhabitants of the planet) or what sharks of the future will look like, since humans are remaking the planet at an unprecedented rate.

NOTES

1. (A) Smalltooth Sawfish (B) Bluntnose Sixgill Shark (C) Atlantic Sharpnose Shark (D) Angel Shark (E) Longnose Sawshark (F) Ornate Wobbegong (G) Bowmouth Guitarfish (H) Southern Stingray (I) Horn Shark (J) Spiny Dogfish (K) Bigeye Thresher (L) Mouse Catshark (M) Common Torpedo (N) Frilled Shark (O) Prickly Dogfish.

2. https://sharksrays.org. (Accessed 8/9/19).

3. *Taxonomy* is the theory and practice of naming things. *Systematics* is the process of classifying organisms according to their phylogeny (i.e., based on common ancestry). The words are typically used incorrectly as synonyms. Most current taxonomies of organisms are systematic taxonomies.

4. A *tessera* (pl. *tesserae*) is a tile, usually in the shape of a cube, that comprises a mosaic.

5. Grogan, E.D. et al. 2012. In: Carrier, J.C. *et al.* Biology of sharks and their relatives. CRC Marine Biology Series. CRC Press. 3–31 pp.

6. Last, P.R. et al. 2016. Zootaxa 4139: 345–368.

7. Coates, M.I. et al. 2018. Proc. Royal Soc. B 285, No. 1870: 20172418.

8. Turner, S. and Miller, R. 2005. Am. Sci. 93: 244–252.

9. Maisey, J.G. et al. 2017. Am. Mus. Novit. 3875: 1–16.

10. *That without which there is none* (Latin). In other words, an indispensable or essential part of. Now, don't you wish you had taken classic languages in high school?

11. Serendipity, meaning *a fortunate surprise*, often is associated with hours, days, week, years, or even decades of meticulous hard work, which is clearly the case here.

12. This is an oversimplification. Three periods of diversification are recognized: the Upper Triassic, Jurassic, and Cretaceous.

13. Estimates of times of divergence between groups, which are based on a variety of methods that frequently differ from each other, should not be taken as absolutes.

14. Ehret, D.J. et al. 2012. Palaeon. 55: 1139–1153.

3 / Diversity of Sharks

Introduction: Why So Few Kinds of Sharks?

Exactly how many different species of sharks are there? The truthful answer is that we do not know, in part because the use of molecular techniques combined with increased sampling in the deep-sea and remote coastal regions make this a moving target as new species are discovered. Over the past decade, on average 400 new species of fishes are described each year, including approximately 20 chondrichthyans.

In 1984, the authoritative book on sharks, Compagno's *Sharks of the World*[1] listed 342 species of sharks. As of this writing, an annually updated checklist[2] as well as the *Chondrichthyan Tree of Life* website,[3] documented 536 kinds of sharks among 1239 species of chondrichthyans, which also included 648 batoids and 55 holocephalans. This breaks down to 14 orders, 64 families, and about 199 genera. We have used a figure of about 500 described species of sharks and 1250 total species of sharks, skates, rays, and chimaeras. This number will certainly increase in the future, assuming deep-sea research continues.

In this chapter, we survey the diversity of chondrichthyan fishes, with greater detail provided for sharks. We cover all of the orders and most families within them, as well as species that are iconic, ecologically important, commercially valuable, overfished, and/or are otherwise interesting. If the diversity of shapes, sizes, and lifestyles of sharks and other chondrichthyan fishes fascinates you, this chapter will be most interesting.

Let us consider why there are so few kinds of sharks compared to the dominant aquatic vertebrates, the bony fishes, of which there are approximately 33,000 kinds.

Bony Fishes Have More Offspring

First, all sharks use internal fertilization, and the overwhelming majority of bony fishes fertilize externally. Bony fishes typically invest a small amount of resources into a large number of eggs (e.g., several million in the case of cod). These eggs or resulting hatchlings have a high degree of *vagility*, or ability to disperse, if they are pelagic and are entrained in surface currents that carry them perhaps thousands of miles. The wider their dispersal, the greater the odds of diversifying into new species.

Sharks use an entirely different life-history strategy of investing resources into a limited number of offspring. Early development of the embryo occurs inside the female for those sharks that lay eggs, like horn and bamboo sharks, and complete development occurs internally for sharks like Whites (*Carcharodon carcharias*) or Blacktips (*Carcharhinus limbatus*). The advantage for sharks is that each individual has a greater chance of survival than a single bony fish hatchling. The disadvantage is decreased vagility (large numbers of offspring dispersing). Thus, there is a far greater opportunity for speciation in bony fishes.

Most Sharks Are Small

In addition to eggs and neonates (newborns), adults are also capable of dispersal. However, sharks are also handicapped by their size. Surprisingly, most species of shark (over 300 species) are small (< 100 cm, or 3.3 ft). Migrating across the distances required to achieve the geographic isolation required for speciation is very risky for small sharks. Moreover, about 80% of shark species are bottom-associated, and bottom-associated species are less likely to disperse great distances. Most bony fishes are also not large, but recall that they lay eggs, and these, or the hatchlings, can readily disperse.

Bony Fishes Have a Swim Bladder

Swim bladders are gas-filled spaces (think *internal balloon)* in the body cavity of most bony fishes, which act as flotation devices and allow fish that possess them to adjust their buoyancy, and hence their vertical position in the water column. Sharks are heavier than water and, without a swim bladder, must swim continuously or live on the seafloor.

The evolution of swim bladders provided two critical advantages. First, bony fishes could save energy by not needing to continuously move, and saving energy is a major driving force for an organism's success, enabling it to devote energy to other survival tasks. Second, fish with swim bladders could expand their *niches*; that is, occupy myriad habitats with more or better food, fewer predators, or appropriate environmental conditions, by simply adjusting their buoyancy and using their fins for fine scale movement.

A swim bladder enables bony fishes to exploit the many tight spaces on coral reefs, which account for most of the living space in these ecosystems. In other words, a swim bladder facilitates being small. No sharks on coral reefs are as small as the smallest and/or most common bony fish species.

Elasmobranchs had about 300 million years during which they could have evolved small forms before teleosts did; however, they did not. The teleosts, with their fancy swim bladders, came along and literally took over the marine and freshwater aquatic world, aided by the evolution of complex habitats like reef-building corals, seagrasses, and mangroves.

Sharks Possess Relatively Rigid Pectoral Fins with a Small Range of Motion

A major evolutionary advancement for sharks was replacing a rigid pectoral fin base with three cartilages (see fig. 1.9), endowing the pectoral fin with greater flexibility; however, the pectoral fins of sharks are still less flexible than those of most bony fishes. Absent the restrictions imposed by these three cartilages, many bony fishes (e.g., damselfish) have acquired the ability for fine-scale movement. Thus, no sharks capable of darting in and out of crevices exist, due largely to their inflexible pectoral fins.

Bony Fishes Have More Complex Heads and Jaws

Finally, bony fishes have evolved seemingly unlimited types of head shapes and mouths and thus different feeding styles, whereas sharks have stuck with the same basic plan with a few exceptions.

What allows this variation in bony fishes is the complexity of the head (see fig. 1.8). Each part of the head of bony fishes, especially mobile maxillae and pre-maxillae, can be molded through evolution to create jigsaw patterns culminating in a variety of mouth and head shapes. With every different mouth position and shape in bony fishes came the opportunity to exploit a new habitat or food source, and these structural and functional variations led to the evolution of new species.

In contrast, the skull of sharks is fairly simple. There is a suture-less (one solid piece) *chondrocranium*, and a single upper jaw (the *palatoquadrate*) and a single lower jaw (*Meckel's cartilage*), each composed of two cartilages (see fig. 1.8). There is little raw material for evolution to work on. To be sure, the basic shark head plan has been an unqualified success, but along with the other features described, it deprived them of the ability to diversify to the same extent as bony fishes. There will never be a shark with a mouth like a seahorse, wrasse, hound-fish, or butterflyfish.

Overview of Shark Taxonomy

To start, if you need a taxonomic refresher, consult Box 1.1, which provides an overview of the Linnaean system of classification along with some helpful conventions used in naming organisms. Box 3.1 contains definitions for some of the terms that appear frequently as distinguishing features.

BOX 3.1

Frequently encountered terminology used in descriptions of distinguishing features of chondrichthyans

Jaw Suspension—refers to how the upper jaw (palatoquadrate) and lower jaw (Meckel's cartilage) connect to the skull (chondrocranium) and other supporting structures, particularly the hyomandibular cartilage, part of the hyoid arch (the second gill arch that moved forward in evolution to support the jaws to some degree; see fig. 2.10). Consult Chapter 4 for more details. The following types are in descending order from primitive to most derived, and from least to most mobile.

Amphistylic: The upper jaw is attached by ligaments (strong, fibrous material that connects skeletal elements) at two points to the skullcase. Posteriorly, the jaw is provided with very limited support by the hyomandibular cartilage.

Orbitostylic: A projection from the upper jaw called the *orbital process* articulates with the skull, and the hyomandibular arch, while still buttressing (supporting) the jaw, is shorter.

Hyostylic: There is a ligamentous connection (the *ethmopalatine* ligament) between the palatoquadrate and chondrocranium anterior to the orbit, and there is buttressing (support) by the hyomandibular cartilage.

Euhyostylic: There are no connections (no ligaments or articulations) between the upper jaw and skullcase, and the jaws are supported exclusively by the hyomandibular cartilage at the corners of the jaws.

Modes of Embryological Development—refers to whether there is live birth or hatching from an egg, or some intermediate form, and includes the mode of nutrition for the developing embryo. This topic is discussed in detail in Chapter 6. The three main categories of embryonic development in elasmobranchs are:

Oviparity: egg-laying

Yolk-sac Viviparity (formerly Aplacental Viviparity): live birth without a placenta, a structure formed from maternal and embryonic tissues that connects the embryo to the wall of the uterus and provides nutrition to the developing embryo

Placental Viviparity: live birth with a placenta, although different from the mammalian placenta

Chondricthyans and humans are both in the phylum Chordata. So are 3000 species of tunicates (basically sea squirts and their relatives—Did you know that you are more closely related to blobs known as sea porks or sea livers than you are to, say, a squid?), 30 species of lancelets (small, obscure, sediment-dwelling fishlike creatures), and about 60,000 other vertebrates (also known as *craniates*).

We will start our taxonomic journey with the superclass *Gnathostomata*. One widely accepted scheme has the gnathostomes as a mostly monophyletic[4] group of five classes: the extinct classes *Placodermi* and *Acanthodii*, and three extant classes, the *Actinopterygii* (ray-finned fishes), *Sarcopterygii* (lobe-finned fishes as well as the tetrapods; e.g., amphibian, reptiles, birds, and mammals), and *Chondrichthyes* (sharks, skates, rays, and chimaeras).

The class Chondrichthyes includes two subclasses, the Holocephali and the Elasmobranchii, the latter of which is the dominant chondrichthyan group. Consult Figure 3.1, a cladogram of the class Chondrichthyes, for an overview of relationships among the 14 orders.

Subclass Holocephali

Superorder Holocephalomorpha

ORDER CHIMAERIFORMES

Extant members of the Holocephali (fig. 3.2) are a monotypic subclass (i.e., there is but a single order, *Chimaeriformes* [*chimaer* = monster]), and they are commonly known as the *chimaeras* (or *ghost sharks* in Australia). The > 50 described species of Chimaeriformes comprise three families: Chimaeridae (44 sp.), Callorhinchidae (3 sp.), and Rhinochimaeridae (8 sp.). Holocephalans represent a monophyletic sister-group to the Elasmobranchii.

Most chimaeras live in deep water but the best-known species, often displayed in public aquaria, is the spotted ratfish, *Hydrolagus colliei*, which is found along the US West Coast. Chimaeras are the only living chondrichthyans with holostylic jaw suspension (the upper jaw is fused to the skull) and thus have no jaw protrusibility. Other notable features include large eyes, a large spine on the first dorsal fin, and in addition to the pelvic claspers typical of all male chondrichthyans, male chimaeras possess a clasper on their heads.

Subclass Elasmobranchii

The most speciose and diverse members of the Chondrichthyes are the elasmobranchs. Most possess dermal denticles (placoid scales), protrusible jaws of varying extents that vary with the superorder, serially replaced teeth not embedded firmly in the jaws, and tribasic pectoral and pelvic fins. Recall that the group is named for their gills, which are described as naked, strapped, or plate gills. These all refer to their gills lacking the protection afforded by the operculum of bony fishes.

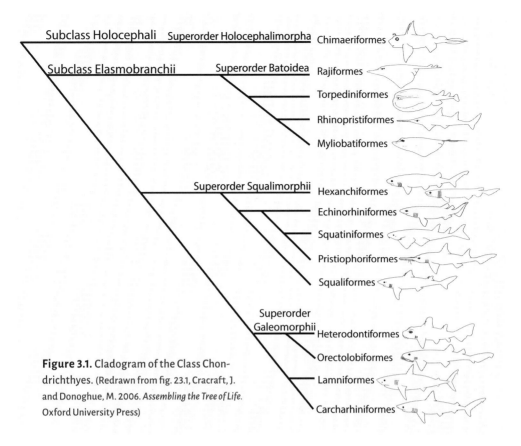

Figure 3.1. Cladogram of the Class Chondrichthyes. (Redrawn from fig. 23.1, Cracraft, J. and Donoghue, M. 2006. *Assembling the Tree of Life*. Oxford University Press)

Superorder Batoidea (Skates and Rays)

Batoids are the most speciose chondrichthyans and elasmobranchs, with nearly 650 species in 26 families identified,[5] with species remaining to be described. One explanation for this high diversity of the group is the very protrusible jaws, which allow for a variety of feeding modes.

Batoids are a monophyletic group united by many *apomorphic* (derived, or evolutionarily advanced) characters that include[6] (fig. 3.3):

- Pectoral fins fused to the head above the gill openings.
- No orbital articulation of the jaw (upper jaw does not articulate with chondrocranium), which allows for *extremely* protrusible jaws in most taxa (in some, there is less protrusibility because the jaw muscles have become hypertrophied and are used for crushing). This is known as *euhyostylic* jaw suspension.
- A synarcual cartilage, essentially a series of fused cervical vertebrae and vertebral elements. Interestingly, placoderms possessed a similar synarcual.

In addition to these derived characteristics, batoids also have a flattened (depressed) body, a reduced or absent anal fin, five gill slits on the ventral surface, and a pair of enlarged spiracles.

A

B

Figure 3.2. Holocephalans. (A) Juvenile chimaera. (Courtesy of NOAA) (B) The chimaera *Rhinochimaera*. (Courtesy of NOAA)

Less-enlightened shark biologists often describe batoids as *pancake sharks*. Underlying this condescension is the outdated idea that batoids are merely highly derived sharks. In fact, batoids diverged from sharks around 266 mya and evolved independently of sharks over that period. Recently, the tables have been turned, with emboldened batoid specialists referring to sharks as *sausage rays*.

Ecologically, most batoids are benthic, but there are pelagic forms as well, and whereas most rays you are familiar with are coastal, nearly half (mostly skates) live in the deep-sea. Many batoids feed on invertebrates and some are *duropha-gous* (eating hard-bodied prey), but some are *piscivorous* (fish-eaters) and others *planktivorous* (plankton-eaters). Many have small teeth that are used for grasping or crushing. Members of three families, Pristidae, Dasyatidae, and Potamotrygonidae, inhabit low-salinity brackish waters or even fresh water.

Batoids are threatened by habitat destruction/alteration (especially freshwater forms) as well as growing fisheries. Like sharks, rays have life-history patterns that make them extremely vulnerable to commercial fisheries. Recently, highly publicized misinformation about Cownose Rays (*Rhinoptera bonasus*) in Chesapeake Bay led to the development of a fishery that may critically threaten their population along the US East Coast (see Chapter 11).

The superorder Batoidea comprises four orders: skates (order Rajiformes), electric rays (Torpediniformes), guitarfish and sawfish (Rhinopristiformes, also known as Rhyncobatiformes), and stingrays (Myliobatiformes) (fig. 3.4). Batoids have also been called *hypotremates* (*hypo* = under; *tremo* = opening), a term dating back to the early 1900s that refers to placement of the gill openings on the ventral surface beneath the pectoral fins. Sharks have been called *pleurotremates* (gills on the side). An abbreviated overview of batoids follows. For more information, we recommend the 2016 exquisitely illustrated reference *Rays of the World*.[7]

ORDER RAJIFORMES (SKATES)

The rajiforms are the skates (fig. 3.4), and at 293 species in four families, they represent the largest order of elasmobranchs and nearly half of the diversity of batoids. Skates are benthic organisms, and occur in all oceans, from Arctic to

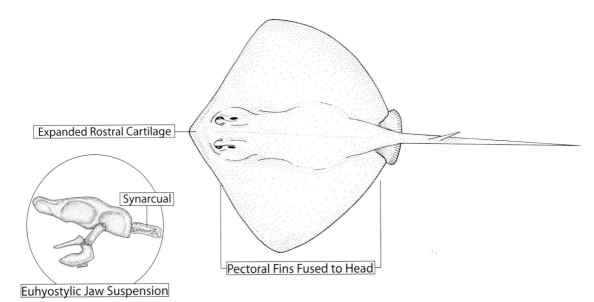

Figure 3.3. Apomorphic characters of batoids. (Redrawn from fig. 41, Yokota, L. and De Carvalho, M.R. 2017. Zootaxa 4332: 1–74)

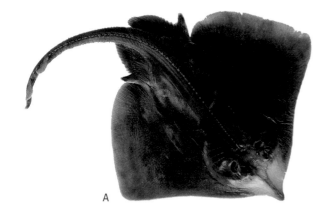

A

B

C

D

Figure 3.4. Representatives of each of the four orders of batoids. (A) Rajiformes: Clearnose Skate. (Courtesy of Trey Spearman) (B) Rhinopristiformes: Smalltooth Sawfish. (C) Torpediniformes: Marbled Electric Ray. (Aquapix/Shutterstock.com) (D) Myliobatiformes: Spotted Eagle Ray. (Courtesy of Annie Guttridge)

Antarctic, in shallow coastal shelves to abyssal regions. Skates are most common in temperate cold or deep waters, where they are the dominant batoid. Deep-sea skates are all brownish in color.

Skates are characterized by a prominent rostral cartilage that is more rigid than that in rays, a quadrangular or rhomboidal body disc, two reduced dorsal fins (or one or both are absent) placed posteriorly on a stocky but spineless tail, and no caudal fin. Skates do not possess venomous spines.

Skates have a peculiar modification of the inner margin of their pelvic fins, an additional lobe (fig. 3.5). This bi-lobed pelvic fin creates a mobile pelvic girdle that is used to crawl or scull on the bottom, a phenomenon known as *punting*. Each fin can move independently of the other. Most elasmobranchs are capable of only limited pelvic mobility, so this represents a big deviation from the norm, and a very cool adaptation.

Skates possess small cusped or flat, plate-like teeth. Many have enlarged, thorn-like scales called *bucklers* along the midline of the back and tail. Males possess

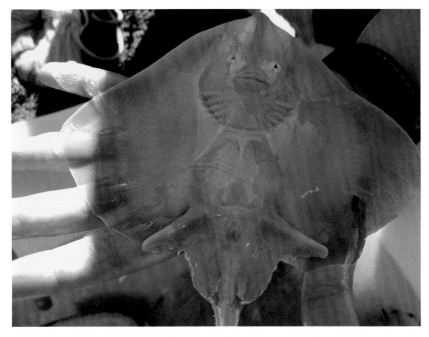

Figure 3.5. The unique bi-lobed pelvic fin of a skate. The anterior lobes rotate and are used for movement on the bottom.

rows of enlarged scales on the cheeks near the eyes and also on the wingtips (termed *malar* and *alar* spines, respectively) that are used to facilitate attachment to females during copulation (see Chapter 6). Skates are *oviparous*, with their *mermaid's purses* frequently seen by beachgoers.

Finally, some skates have a weak caudal electric organ (discharges of ~1 v) whose function is enigmatic, but which may be used for communicating with each other.

Rajiform families include Anacanthobatidae (Smooth Skates; 13 sp.), Arhynchobatidae (Softnose Skates; 101 sp.), Gurgesiellidae (Pygmy Skates; 19 sp.), and Rajidae (Skates; 161 sp.; the most speciose family of skates).

ORDER TORPEDINIFORMES (ELECTRIC RAYS)

The order Torpediniformes (*Torp* = numb, numbness), the electric rays, are a group of about 67 species in five families (fig. 3.4). They are small to moderately large (total length to about 1.8 m, or 5.9 ft, but most species are less than 1 m, or 3.3 ft, long) with no scales (naked) and with large dorsal spiracles. Torpediniform rays have two well-developed dorsal fins, a well-developed caudal fin, and small mouths with rounded to slightly cusped teeth. Their large caudal fin is atypical of batoids. Their caudal fin is used for propulsion more so than undulations of the pectorals.

The group is named for its huge pectoral-cephalic (anterior) electric organ derived from honeycomb-shaped muscle cells aligned like batteries in a series. The electric organ in some species is capable of 200-volt, high amperage discharges capable of rendering victims temporarily paralyzed or even dead.

Although the order is highly derived, having lost its rostral nasal cartilages and thus it possesses a fleshy snout, it may be the most primitive of the batoids.

Torpediniform families include Torpedinidae (Torpedo Rays; 18 sp.), Narcinidae (Numbfishes; 30 sp.), Narkidae (Sleeper Rays; 11 sp.), Hypnidae (Coffin Rays; 1 sp.), and Platyrhinidae (Fanrays; 5 sp.).

ORDER RHINOPRISTIFORMES (GUITARFISH AND SAWFISH)

The order Rhinopristiformes (sometimes called the Rhinobatiformes), also known as the shovelnose rays, was formerly two separate orders, the Rhinobatiformes (guitarfish) and the Pristiformes (sawfish).

This order consists of about 62 species of guitarfish and sawfish (fig. 3.4). All of these species superficially resemble sharks, at least posteriorly, more so than rays. Like the torpedinids, members of this order have a well-developed caudal fin, which they use for propulsion. They possess an elongated, stiffened rostrum (like the skates), two well-developed spineless dorsal fins (which also aid in propulsion), and 65–75 rows of small, pavement-like teeth. They are bottom-feeders, preferring small crustaceans and fish. This group was formerly considered the most ancestral of batoids before the Torpediniformes assumed that honorific. They are yolk-sac viviparous. This group includes the most endangered of all chondrichthyan fishes, the wedgefishes, giant guitarfishes, and sawfishes. The rhinopristiform families include Glaucostegidae (Giant Guitarfishes; 6 sp.; all critically endangered), Pristidae (Sawfishes; 5 sp.; all endangered or critically endangered), Rhinidae (Rhynchobatidae, Wedgefishes; 10 sp.; nine of which are critically endangered), Rhinobatidae (Guitarfishes; 41 sp.), Trygonorrhinidae (Banjo Rays; 8 sp.), and Zanobatidae (Panrays; 2 sp.; two species).

ORDER MYLIOBATIFORMES (STINGRAYS)

The Myliobatiformes (*Mylio* = mill, molar) are the stingrays, whose scientific name is based on their grinding teeth. This is a big, diverse group, containing 226 species in 10 families. Myliobatiform rays exhibit a variety of color patterns and body shapes—ovoid, rhomboid, and trapezoid. In stark contrast to the Rhinopristiformes, the myliobatiform rays have no rostral cartilage. All have five gill slits except the family Hexatrygonidae. Their skin is naked except for patches of dermal denticles. They have a slender tail with or without spines. Most possess a dorsal fin-fold, whereas others have a single dorsal fin. This group is viviparous, employing lipid-rich uterine milk. They dominate coastal shallow ecosystems in warmer climates, and members of two families live in fresh water.

Ever since Crocodile Hunter Steve Irwin died from a stingray spine penetrating his chest, there has been renewed interest in this group. Stingrays are not aggressive (except in tourist areas where they are hand-fed). As many as 2000 injuries a year in the US are caused by stingrays. Fatalities, like Mr. Irwin's, are rare and are due to penetration or infection more so than the direct effect of the toxin, which is produced by goblet cells in living epithelial tissue that surrounds the spine, which is a modified dermal denticle. Stingray injuries are described as extremely painful and may be accompanied by nausea, vomiting, and cramps. Treatment includes bathing the wound in water as hot as can be tolerated without burning tissue (< 46°C, or 115°F). If the barb is removed improperly, the epithelium may remain in the wound and continue to produce or release the venom for a short time, which can lead to permanent, serious tissue damage.

Myliobatiform families include Hexatrygonidae (Longsnout Stingray; 1 sp.), Urolophidae (Round Rays; 28 sp.), Urotrygonidae (Smalleyed Round Ray; 16 sp.), Plesiobatidae (Deepwater Stingray; 1 sp.), Dasyatidae (Stingrays; 97 sp.; the largest myliobatiform family), Potamotrygonidae (River Rays; 32 sp.; unique among the myliobatiforms in that the group has secondarily invaded freshwater, all in the Amazon basin; see Chapter 8), Gymnuridae (Butterfly Rays; 12 sp.), Myliobatidae (Eagle Rays; 23 sp.), Rhinopteridae (Cownose Rays; 11 sp.), and Mobulidae (Manta and Devil Rays; 12 sp.).

Is it a Skate or a Ray: A Quick Review

Here is a quick guide to distinguish the orders of batoids. First, rule out the torpediniforms and rhinopristiforms, based on their large caudal and dorsal fins. That leaves the orders Rajiformes (the skates) and Myliobatiformes (the stingrays). If the individual in question has thorns (also called *bucklers*) along its dorsal midline, has a bilobed pelvic fin, rigid rostrum, and a relatively stout tail without spines, and it lays eggs, then it is a skate. If instead, the individual lacks bucklers, has a fleshy snout, single-lobed pelvic fin, and a whip-like tail, often with one or more spines, then it is a ray.

Superorder Squalomorphii (Dogfish Sharks)

The superorder Squalomorphii (*squal* = dogfish) is a diverse group of 174 species in five or six orders and eleven families. They are typically found in colder water (e.g., at mid- to high latitudes and/or in deep water). Squalomorphs date back to at least the Carboniferous (340 mya). The lineage diversified after the Permian Extinction about 200 mya.

Squalomorph sharks are characterized by two synapomorphies (common derived characteristics): loss of an anal fin in all but one group and orbitostylic jaw suspension (box 3.1), an intermediate form that allows a moderate level of jaw mobility (that is, *cranial kinesis*; see Chapter 4). Many have evolved enlarged spiracles. All squalomorph sharks are yolk-sac viviparous, with some employing limited histotrophy through mucoid uterine secretions.

The order Hexanchiformes consists of two primitive families, Chlamydoselachidae (frilled sharks) and Hexanchidae (cow sharks). Some classifications consider the family Chlamydoselachidae as a monotypic family in the order Chlamydoselachiformes. Superficially, this makes sense, since species in the two groups simply do not look like they would be closely related. However, a closer look reveals morphological similarities between both families, including six gill arches and loss of the first dorsal fin. Moreover, these are the only squalomorph sharks with an anal fin. Finally, an analysis of mitochondrial DNA has led to their being united in the order Hexanchiformes. Despite being incorporated into the same order, there are major morphological differences between chlamydoselachids and hexanchids (see below).

Both families are predominantly deepwater, although some hexanchids are found in shallow waters, and others are diurnal vertical migrators.

Families in the order Hexanchiformes:

- Frilled Sharks (Family Chlamydoselachidae; *chlamy* = cloak or garment; fig. 3.6)
    - » two species, the Frilled Shark (*Chlamydoselachus anguineus*) and the Southern African Frilled Shark (*C. africana*), the latter described in 2009 and found from southern Angola to southern Namibia. Both have a head like a reptile, a terminal mouth, big green eyes, an eel-like body, frilly gills that extend beyond the gill slits (hence the common name), distinctive three-pronged pitchfork-type teeth swept backward similar to a python's and unlike those of any modern shark, short rounded pectoral fins, a small dorsal set far back, and an anal fin that is larger than the dorsal. If that is not enough, the first gill slit is continuous across the throat. They likely feed on soft-bodied prey. The lineage to which this group belongs dates back as long as 150 mya. Frilled Sharks have a prominent notochord, but no distinct vertebral centra. They are found in deep water (100–1200 m, or 330–3940 ft) and have small litters (average of six). One study estimated their gestation period at an extraordinary 3.5 years.

Representative Species

Frilled Shark (*Chlamydoselachus anguineus*; fig. 3.6). This is the best-known frilled shark. In both species, dentition is alike in upper and lower jaw and the teeth are distinct from those of all extant sharks, and more similar to the Carboniferous cladodont shark, *Xenacanthus*. Each tooth is pitchfork-like with three fanglike cusps and two smaller cusps between them. These form interlocking rows and point inward. There are about 300 teeth in about 25 rows. If a prey item is unfortunate enough to be grasped by a frilled shark, there is no way it can escape.

The absence of a first dorsal fin may facilitate rolling when feeding, similar to an alligator death roll. This species is commonly seen in Japanese trawls, but may be far more widespread, including the Atlantic.

A

B

Figure 3.6. (A) Frilled Shark. (Paulo de Oliveira / Biosphoto) (B) Unique pitchfork-shaped teeth found in frilled sharks.

- Cow Sharks (F. Hexanchidae; *Hex* = six; *anch* = gill; fig. 3.7)
  - » five species in three genera, *Hexanchus*, *Notorynchus*, and *Heptranchias*. Hexanchids are primitive and are the sister-group to the Chlamydoselachidae. Some have large litters (up to 108 have been seen). Most are deepwa-

A

Figure 3.7. Sharks of the family Hexanchidae. (A) Atlantic Sixgill Shark. (© Katie Grudecki; Bimini, Bahamas) (B) Bluntnose Sixgill Shark with eye partially withdrawn into its socket.

B

ter. They have big green eyes capable of harvesting dim blue-green light at 200–1000 m (650–3280 ft) deep. They also have odd, cockscomb-shaped teeth similar to ancient sixgill sharks from > 200 mya (see fig. 2.1B). These teeth are capable of sawing large chunks of flesh from large prey. This group, along with most other squalomorphs, has retractable eyes (fig. 3.7). Finally, their vertebral column can be called *notochordal* (the vertebral centra are not fused or calcified), and the soft column is so flexible that hexanchiform sharks can bite their own tails.

Representative Species

Bluntnose Sixgill Shark (*Hexanchus griseus*; fig. 3.7). The most well-known hexanchid is *H. griseus* (*hexanchus* = sixgill; *griseus* = gray). These are widely distributed, big (to ≥ 5 m, or 16 ft), stout-bodied predators but with surprisingly weakly calcified jaws and notochord, which may be explained by problems depositing calcium salts at the depth at which they spend much of their time. Hexanchid teeth (see fig. 2.1A) are like mini-saw blades, and the additional flexibility of weakly calcified jaws allows the jaws to bend as they encounter prey, which brings more of the serrated teeth in contact with the flesh, which is more easily sliced and removed: Death by a thousand cuts!

Because of the lack of calcified skeletal structures, this species is lighter and is close to neutrally buoyant. Most are deepwater (300–1000 m, or 1000–3300 ft, a section known as the oceanic *twilight zone*), but may be shallow in some locations. They have a single, spineless dorsal fin set far back along the body, which reduces friction and allows them to spin more easily when sawing chunks of flesh from prey.

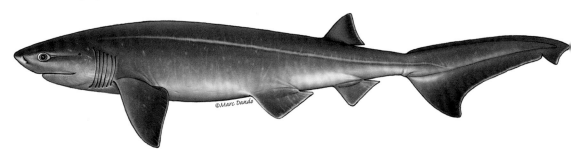

©Marc Dando

Atlantic Sixgill Shark (*Hexanchus vitulus*; fig. 3.7). In 2013, we captured a 1.5 m (4.9 ft) Atlantic Sixgill Shark (*Hexanchus vitulus*) on a longline in approximately 250 m (820 ft) of water in Bimini, Bahamas. Because the shark was dead when retrieved, we dissected the specimen for educational purposes, and discovered that the rectal gland, typically finger-shaped, looked like a kidney bean, and was much smaller than expected. One explanation for the different morphology is that the genus *Hexanchus*, one of the most primitive shark genera, with fossils from the early Jurassic (190 mya), largely has occupied the deep sea throughout its evolution, and thus it is possible that the evolutionary age and degree of isolation of the genus has resulted in different anatomical and physiological systems.

ORDER ECHINORHINIFORMES (BRAMBLE OR PRICKLY SHARKS)

The entire order Echinorhiniformes, which is also a very primitive group, consists of only two species in one family.

The single family in the order Echinorhiniformes:

- Bramble or Prickly sharks (Family Echinorhinidae; *echino* = spiny; *rhinus* = nose or snout, due to large denticles on snout; fig. 3.8)
    » two species of primitive, large, sluggish, deepwater, bottom-dwelling, poorly known sharks. Echinorhinids are distinguished by large, thorn-like denticles and two small, spineless dorsal fins far posterior on the body. They also lack an anal fin. They have distinct teeth with one main cusp and 3–4 smaller cusplets in both the upper and lower jaws (fig. 3.8). Bramble sharks may reach > 3 m (10 ft) and, although stout, they are relatively soft-bodied with flabby musculature.

Representative Species

Prickly Shark (*Echinorhinus cookei*; fig. 3.8). *E. cookei* is found in the Pacific Ocean.

ORDER SQUATINIFORMES (ANGEL SHARKS)

The order Squatiniformes consist of one family of 22 species of ray-like sharks. They are coastal in temperate seas and deeper in the tropics. Although they resemble batoids, with pectoral fins expanded (but not connected to the head) and they in fact have some ecological similarities to batoids, they are true sharks. The single family in the order Squatiniformes:

- Angel Sharks (Family Squatinidae; fig. 3.9)
    » 22 mostly shallow-water species (one-third are deep sea species), all in the genus *Squatina*, with prominent spiracles without valves. Like most squalomorphs, they are cold-water sharks. They are the only elasmo-

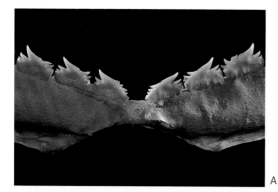

Figure 3.8. (A) Teeth of Prickly Shark. (Courtesy of D. Ross Robertson) (B) Prickly Shark.

A

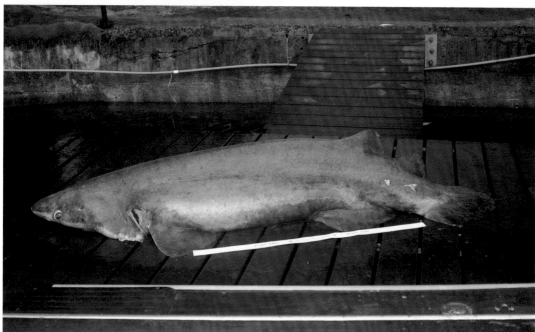

B

Figure 3.9. Common Angel Shark. (LuisMiguelEstevez/Shutterstock.com)

branchs with an inverse heterocercal tail (one in which the lower lobe is larger). Also called a *hypocercal* caudal fin, this type of tail fin was found in some Devonian placoderms. A hypocercal tail elevates the head, which benefits them since angel sharks are lie-and-wait predators with huge gapes on their terminal mouths, and they ambush prey from the bottom. They have been commercially fished since the 1980s and may be served as *monkfish* in some locations.

Representative Species

Angel Shark (*Squatina squatina*; fig. 3.9). Once very common, this species is now listed as critically endangered by the IUCN in the NE Atlantic, Mediterranean, and Black Seas, and overfishing continues.[8] It has been extirpated from the North Sea and parts of the Northern Mediterranean Sea. The only relatively healthy, though presently unregulated, population occurs in the Canary Islands.

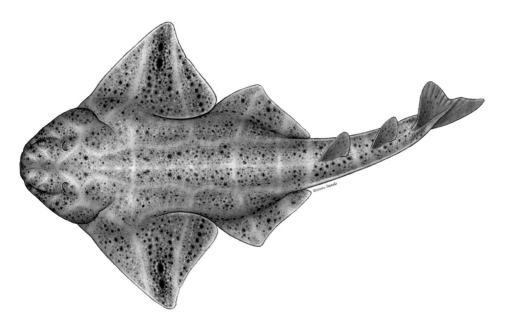

ORDER PRISTIOPHORIFORMES (SAWSHARKS; *PRIST* = COMB; *PHOR* = POSSESSING, OR BEARING)

The order Pristiophoriformes are the sawsharks (not to be confused with the sawfish, which are batoids). There are nine species, all deepwater, with two long electrosensory/chemosensory barbels midway on the tapered rostrum, and two spineless dorsal fins. Like most squalomorphs, they lack anal fins. They grow to 1.7 m (5.6 ft).

It is interesting to note that the sawed rostrum evolved three times in evolutionary history: the pristids (sawfish), the Pristiophoriformes (this group), and a group of sharks that went extinct about 150 mya, the sclerorhynchids.

Figure 3.10.
Eastern Sawshark.
(marinethemes.com/
Kelvin Aitken)

The single family in the order Pristiophoriformes:

- Sawsharks (Family Pristiophoridae; fig. 3.10)
  » eight species in two genera, *Pliotrema* (1 sp.) and *Pristiophorus* (7 sp.). The teeth of the saw typically alternate between large and small and are conical, not bladelike as in the batoid *Pristis*. Like sawfish, pristiophorids are not capable of regenerating rostral teeth that may be lost. Pristiophorids typically feed on fish, squid, and crustaceans, using their cusped, non-flattened teeth. They cruise the bottom, using the barbels and ampullae of Lorenzini on the saw to detect prey buried in the sediment. They then hit victims with side-to-side swipes of the saw, crippling them.

Representative Species

Sixgill Sawshark (*Pliotrema warreni*; see fig. 10.8). This species has six gills and is the only non-hexanchiform shark with this characteristic. It is listed as *Near Threatened* by the IUCN, due to its low reproductive potential and vulnerability to benthic trawls. It is found off Madagascar and nearby to depths of 500 m (1640 ft).

The order Squaliformes consists of 135 species in six families. This diverse order of sharks includes many of the smallest sharks, the lantern sharks (genus *Etmopterus*) and among the largest sharks, the sleeper sharks (genus *Somniosus*). All have five gill slits, two dorsal fins (most with spines, which is not characteristic of the other squalomorphs), no anal fin, a more protrusible jaw than other squalomorphs, and large spiracles. They are typically cold water or deepwater sharks. Only the Spiny Dogfish (*Squalus acanthias*) and the Pacific Spiny Dogfish (*S. suckleyi*) are shallow, cold-temperate species in this order.

Families in the order Squaliformes:

• Dogfish Sharks (Family Squalidae; fig. 3.11)
  » 37 species in two genera, *Squalus* and *Cirrhigaleus*. These true dogfish are found worldwide in tropical, temperate, and boreal seas, in intertidal (the only shallow water squaliforms) to 800 m (2624 ft) or greater. Cylindrical in cross section, they possess two dorsal fins with strong ungrooved spines. Teeth in both jaws are small, similar in size, and are arranged in a single overlapping functional row like a knife blade (fig. 3.11). Their life-history characteristics make them vulnerable to overfishing (see Chapter 11). We include the Spiny Dogfish (*Squalus acanthias*) as the representative species, but it is in fact the oddball since it is one of the few coastal members of the family. The Hawaiian Shortspine Shortspur (*Squalus hawaiiensis*; fig. 3.11) is a more typical squalid shark.

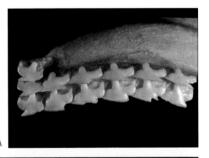

Figure 3.11. (A) Dentition of sharks in the family Squalidae. (B) Hawaiian Shortspine Shortspur.

A

B

Representative Species

Spiny Dogfish, Spurdog, Piked Dogfish (*Squalus acanthias*). This species is named for the presence of spines preceding both dorsal fins. It is intertidal to 900 m (2950 ft) and is the only *Squalus* in the Atlantic which is also coastal. The sister-species in the Pacific, the Pacific Spiny Dogfish (*S. suckleyi*), also has a coastal and deep-sea presence. The Spiny Dogfish age of maturity is about 20 years, and the gestation period of 22–24 months is among the longest known in animals. The IUCN lists it as Vulnerable, with the Northeast Atlantic population considered Critically Endangered.

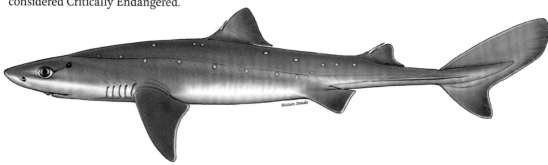

- Gulper Sharks (F. Centrophoridae; *centro* = spine, or point; *phor* = possessing, or bearing)
  - » 16 species of deepwater sharks in two genera, *Centrophorus* (12 sp.) and *Deania* (4 sp.). The family is named for the short, stout, pointed, grooved dorsal spines. Gulper sharks are deepwater sharks found throughout the world's oceans (except in polar regions), with cylindrical bodies, huge green to yellowish eyes, and long, narrow rear lobes of the pectoral fins. They reach 1–2 m (3.3–6.6 ft). Gulper sharks have erect, cusped teeth on the upper jaw and bladelike teeth, like the squalids, on the lower jaw. Their bodies are *very* slimy.

Representative Species

Gulper Shark (*Centrophorus granulosus*). This is a slender shark with green eyes and a long snout. It is among the largest gulpers and is frequently caught on deepwater longlines. The liver of Gulper Sharks has a large volume of oil. Gulper sharks are listed as Vulnerable by the IUCN.

Figure 3.12. Green Lantern Shark (adults and juveniles). Note the photophores (bioluminescent organs) on the underside of the head.

- Lantern Sharks (F. Etmopteridae; fig. 3.12)
  » 51 species in five genera, including among the smallest sharks, named for their bioluminescent photophores, which essentially match the quality of downwelling light and make the silhouette disappear when viewed from below. They range in size from 10 to about 100 cm (4 to 39 in). They are found worldwide, either in the bathypelagic (1000–4000 m, or 3280–13,120 ft) or mesopelagic (200–1000 m, or 650–3280 ft) zones. Etmopterids are also distinguished by two dorsal fins with grooved spines, the second fin and spine usually larger than the first.

Representative Species

Dwarf Lantern Shark (*Etmopterus perryi*). This species is noteworthy because it may be the world's smallest shark, reaching only about 0.2 m (8 in).

Viper Dogfish (*Trigonognathus kabeyai*). This shark is known only from a few localities, including off Japan, Taiwan, and the northwest Hawaiian Islands. It has long, dagger-like teeth. Its morphology is similar to that of the osteichthyan Viper Fish (*Chauliodus*), an example of convergent evolution.

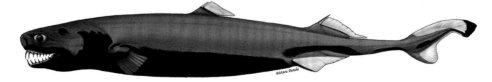

Figure 3.13. Greenland Shark from the North Atlantic. (Courtesy of Charles Cotton and the MAR-ECO project, www.mar-eco.no)

- Sleeper Sharks (F. Somniosidae; fig. 3.13)
  - » (Yawn). 17 species in seven genera of deepwater benthic and oceanic species that are dark brown to pitch black, ranging from small (40 cm, or 1.3 ft) to gigantic (> 6 m, or 20 ft). All have a broad head, flat snout, large spiracles, two small dorsal fins, very small, deep green eyes, and low angular or rounded pectoral fins.

Representative Species

Greenland Shark (*Somniosus microcephalus*; *microcephalus* = small head; fig. 3.13). This is the largest high-latitude shark, reaching 7 m (23 ft) or longer. Although it appears slow moving, its diet consists of active prey (moose and polar bears have been found in its digestive tract). Two scientific discoveries on this species have occurred recently. First, it was observed on camera and ultimately caught at about 1750 m (5740 ft) in the Gulf of Mexico. Given that water temperature at that depth in the Gulf, about 4°C (39°F), is very similar to those in its

known high-latitude home, this discovery was not too surprising. Second, a 2016 study used radiocarbon dating to estimate a mean age of 392 (± 120 years) for one specimen.

- Rough Sharks or Prickly Dogfish (F. Oxynotidae; fig. 3.14)
    » five species, all in the genus *Oxynotus*, found on continental and insular shelves and slopes in the Eastern Atlantic (including the Mediterranean), Western Atlantic, and Western Pacific. They are easily distinguished by their very high and compressed body, which is triangular in cross section, with high, sail-like dorsal fins, each with a large spine that may be concealed by the fin. A lateral ridge is present on the abdomen between the pectoral and pelvic fins. Their skin is very rough (hence, the name), with a luminous organ. They remain lightly studied. Molecular data suggest *Oxynotus* are nested within the family Somniosidae and thus this family is paraphyletic.

Representative Species

Caribbean Rough Shark (*Oxynotus caribbaeus*; fig. 3.14).

Figure 3.14. Caribbean Rough Shark (NOAA).

Figure 3.15. Depiction of American Pocket Shark squirting luminescent fluid. (Courtesy of Mark A. Grace)

- Kitefin Sharks (F. Dalatiidae; figs. 3.15 and 3.16)
  - » nine species in seven genera of dwarf to medium-sized (< 2 m, or 6.6 ft) deepwater sharks, with a narrow head and conical, short snout. Their dorsal fins (or first fin only) are without spines. They are found worldwide. This family includes the pocket sharks, a rare group known only from two species, the Pocket Shark (*Mollisquama parini*) from the South Pacific, and the American Pocket Shark (*Mollisquama mississippiensis*), recently described from a midwater trawl in the Gulf of Mexico.[9] While these are noteworthy because of their rarity and, well, the cuteness of their bulbous head, what stands out about them is their eponymous pockets, bilateral pouches located dorsal to the insertions of the pectoral fins. Unique to this genus, these muscular glands are thought to secrete a luminous fluid (fig. 3.15) that could function to attract prey or fool predators.

Representative Species

Cookiecutter, or Cigar, Shark (*Isistius brasiliensis*; fig. 3.16). This iconic species is known for its unusual lifestyle. Cookiecutter Sharks are neutrally buoyant vertical migrators with luminous organs that attract larger fish and mammals.[10] They possess photophores (bioluminescent organs) on the underside except for a darkly pigmented collar. The light emitted from these photophores matches

Figure 3.16. (A) Mouth of Cookiecutter Shark showing fleshly lips and teeth. (B) Cookiecutter Shark scars on an Opah at a fish auction in Hawaii.

the downwelling light, so the silhouette of the shark disappears from beneath, except for the darkly pigmented collar, which stands out and may trick a potential predatory fish beneath into interpreting that it is a small fish. When the fish investigates, the Cookiecutter Shark accelerates toward the erstwhile predator stealthily and sometimes at great speed using its relatively large and powerful caudal fin. Despite their small size (up to 0.6 m, or 2 ft), Cookiecutter Sharks have powerful jaws and large teeth (fig. 3.16A), which are more heavily calcified than in other squaliforms. In addition to fish and marine mammals (fig. 3.16B), they have been known to attack nuclear submarines, and there are two cases of attacks on people over deep water at night off Hawaii.

Kitefin or Seal Shark (*Dalatias licha*; fig. 3.17). This deepwater shark (200–600 m, or 660–1970 ft) is included here because the morphology of its mouth so closely resembles that of the Cookiecutter Shark that it very likely shares the same mode of feeding, although the Cookiecutter Shark is the only shark recognized for this method of feeding.

Pygmy Shark (*Euprotomicrus bispinatus*). At about 25 cm (10 in), the Pygmy Shark is the second smallest known shark.

A

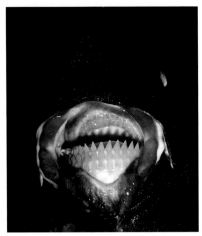

B

Figure 3.17. (A) Kitefin, or Seal, Shark. (B) Mouth and teeth of this species, which show striking similarities to those of the Cookiecutter Shark, and which likely are used to feed in a manner similar to that of the Cookiecutter Shark.

Superorder Galeomorphii (Galea or Galeomorph Sharks)

The galeomorph sharks are a diverse, monophyletic group ranging from small demersal sharks with tiny, cusped teeth to large, pelagic predators with bladelike teeth. These include the beasts that come to mind when people think of sharks, but the group also includes those whose form has diverged. There are about 360 species in 23 families (but subject to ongoing revision) and four orders. They are mostly coastal and pelagic in warmer climates, but the largest family is found in cold water. All have anal fins as well as two well-developed dorsal fins that lack spines (except in the horn sharks). They have highly-protrusible (hyostylic) jaws, and a variety of rostrum shapes. The four orders fall into two clades, with the Heterodontiformes and Orectolobiformes as one sister-group, and the Lamniformes and Carcharhiniformes as another (see fig. 3.1). A major difference between them is the morphology of the snout, with the former group possessing short snouts and no expanded rostral cartilage and the latter with elongated rostral cartilages and larger snouts.

ORDER HETERODONTIFORMES (BULLHEAD OR HORN SHARKS)

The Heterodontiformes (*hetero* = different; *dont* = teeth) include only a single family, the Heterodontidae, and nine widely distributed Pacific and Indo-Pacific, primarily nocturnal, bottom-dwelling species (fig. 3.18). All have heterodont dentition, with cusped teeth in front for grasping and molariform teeth in the rear for crushing prey like sea urchins, using hypertrophied jaw muscles (fig. 3.18; see fig. 4.19). Occasionally the sea urchin spines will pierce the heart, which would likely be fatal were it not for a tough heart and the pericardio-peritoneal canal, which drains blood from around the heart before it fatally compresses it (see Chapters 1 and 7). They have spines on both dorsal fins, a plesiomorphic character not found in other galeomorph sharks. Some have ocular spines (spines over their eyes). They also have prominent supraorbital crests, hence the name *horn shark*. Heterodontiform sharks also possess an *oronasal groove* (see fig. 5.8), which directs seawater from the nares to the mouth, where it can be sensed a second time. They have tiny spiracles and no nictitating membrane. The group is entirely oviparous. They are typically < 1 m (3.3 ft) long.

A

B

Figure 3.18. (A) Port Jackson Shark. (Dirk van der Heide / Shutterstock.com) (B) Lower jaw of a Horn Shark showing the heterodont dentition characteristic of the group. Note the small cusped teeth in the front of the jaws and the flattened, molariform teeth in the rear. (Courtesy of D. Ross Robertson)

Among galeomorph sharks, they are considered basal and are most closely related to the order Orectolobiformes, with whom they share a common ancestor about 200 mya, during the second major radiation of sharks. One example of their close relationship is the coloration of *Heterodontus* and the Nurse Shark, *Ginglymostoma*. Young of both species are vividly spotted, whereas adults are darker brown. Heterodontids superficially resemble the hybodonts, which lived at end of Carboniferous about 360 mya.

The single family in the order Heterodontiformes:

- Bullhead or Horn Sharks (Family Heterodontidae; fig. 3.18)
  » nine species of nocturnally active, sluggish, bottom-dwellers in the Eastern Pacific and Western Indian Oceans, all with a truncated snout with reduced or absent rostral cartilage, which makes the snout appear attenuated (or squished). This shape facilitates eating invertebrates, which they suck into their mouth.

Representative Species

Port Jackson Shark (*Heterodontus portusjacksoni*; fig. 3.18). This species is perhaps the best-known member of its family. It is found along the coast of southern Australia.

Galapagos Bullhead Shark (*Heterodontus quoyi*). This species is found in shallow waters (to 40 m, or 131 ft) in the Eastern Tropical Pacific.

ORDER ORECTOLOBIFORMES (CARPET SHARKS)

The Orectolobiformes (*orecto* = stretched out, spread out; *lob* = lobe, fin), or carpet sharks, consist of about 45 species in seven families, which represents low diversity at the species level but higher diversity at the family level (numerous families with relatively few species). There is strong morphological and molecular support, however, to unite the nurse, zebra, and whale sharks, currently in three separate families, into a single family. There is a wide array of body shapes in the order. This group includes the Nurse Shark (*Ginglymostoma cirratum*), Whale Shark (*Rhincodon typus*), and wobbegong. All except one species are demersal (bottom-associated). They have a short snout, with only a short central rostral cartilage, two spineless dorsal fins, and a very short mouth well in front of the eyes. Like the Heterodontiformes, most have prominent nasoral grooves. Their spiracles are moderate to large and below the eye, except in the Whale Shark. Many have small gill slits, with the fourth and fifth slits overlapping. All except two species are Indo-Pacific. Reproduction ranges from oviparity to yolk-sac viviparity.

Families in the order Orectolobiformes:

- Collared Carpet Sharks (Family Parascyllidae; fig. 3.19)
    » eight species in two genera, all of which are intertidal to high subtidal in the Western Pacific. These are among the few sharks with an eel-like niche, scuttling and slithering using their pectorals. Parascyliids are oviparous.

Representative Species

Collared Carpet Shark (*Parascyllium collare*; fig. 3.19).

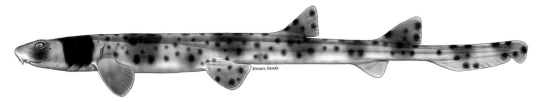

- Blind Sharks (F. Brachaeluridae)
    » two species, both coastal off Australia. Despite its common name, blind sharks have typical orectolobiform vision. These demersal sharks are yolk-sac viviparous.

Representative Species

Blind Shark (*Brachaelurus waddi*). This species is nocturnal and is found in caves and tidepools.

Figure 3.19. Collared Carpet Shark. (marinethemes.com / Kelvin Aitken)

- Wobbegongs (F. Orectolobidae; fig. 3.20)
  » 12 species of demersal, shallow water sharks, most of which occur in the Western Pacific and Indo-Pacific from Japan to Australia. They have a wide head, flattened body, a mouth that is close to terminal, spiracles larger than eyes, and prominent nasal barbels. They are sluggish and eat benthic invertebrates but employ ambush predation to capture fish using very strong jaws. Most are small, but they can exceed 3 m (9.9 ft). They are yolk-sac viviparous.

Representative Species

Ornate Wobbegong (*Orectolobus ornatus*; fig. 3.20). The Ornate Wobbegong is a nocturnal ambush predator.

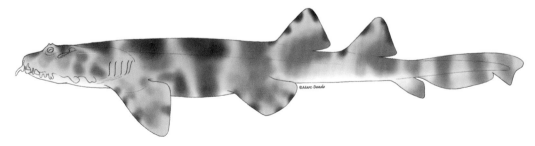

Figure 3.20. Ornate Wobbegong.

- Bamboo or Longtailed Sharks (F. Hemiscylliidae; *hemi* = half; *scyll* = sea monster)
  - » 17 species of small (most less than 0.5 m, or 1.6 ft), sluggish, tropical Indo-Pacific sharks mostly associated with coral reefs, with tails that may be longer than the rest of their body. They are popular aquarium sharks. Most are oviparous.

Representative Species

Epaulette Shark (*Hemiscyllium ocellatum*). Named because of a black spot outlined in white behind the pectoral fins, which resembles military epaulettes, this species is capable of crawling over emergent rocks and coral.

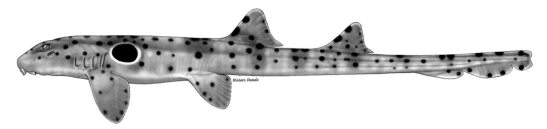

Grey Bamboo Shark (*Chiloscyllium griseum*). This is a common aquarium species and is also eaten by people.

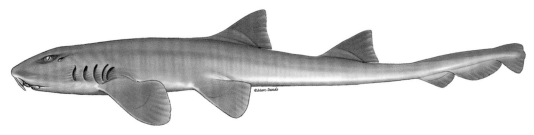

- Nurse Sharks (F. Ginglymostomatidae; *ginglymo* = hinge; *stoma* = mouth; fig. 3.21)
  - » four species, one Atlantic and three Pacific, in shallow, tropical and subtropical waters. They are durophagous (eat hard-bodied prey) and have jaw structure and smaller mouths similar to those of horn sharks but with less hypertrophied muscles, and they lack heterodont teeth. They employ a *suck, crush, spit, repeat* mode of feeding on bottom invertebrates (mostly mollusks and crustaceans).

    They are chocolate brown, with two nearly equal-sized dorsal fins set far back on the body. Their mouth is near-terminal, with long protruding chemosensory barbels. They can reach 3–4.3 m (10–14 ft) and are yolk-sac viviparous, retaining egg cases internally.

Representative Species

Nurse Shark (*Ginglymostoma cirratum*; fig. 3.21). This inshore, bottom-dwelling, largely nocturnal species is robust and reaches about 3 m (10 ft).

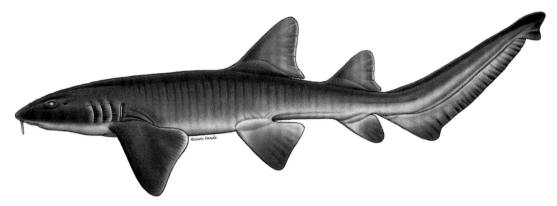

Figure 3.21. Juvenile Nurse Shark. (Courtesy of Chelle Blais)

- Zebra Sharks (F. Stegostomatidae; *stego* = cover; *stoma* = mouth)
  » monotypic, one species in the Indo-Pacific, which occupies a niche very similar to that of the ginglymostomatids. The common name applies to the somewhat zebra-like pattern of juveniles. Adults are spotted, with five prominent longitudinal ridges and with a caudal fin almost as long as the rest of the body. They are common in public aquaria. They are oviparous.

Representative Species

Zebra Shark (*Stegostoma fasciatum*).

©Marc Dando

- Whale Shark (F. Rhincodontidae; fig. 3.22)
  » monotypic, iconic family. The Whale Shark is a huge, planktivorous, filter-feeding shark with a terminal mouth. It is the largest fish (up to 18 m, or 59 ft) and the only pelagic orectolobiform. It is cosmopolitan in tropical and warm-temperate waters. Large aggregations of Whale Sharks have been observed in about 20 so-called hotspots, including off the coasts of Australia, Belize, the Maldives, and the Gulf of Mexico, where they gather to feed (fig. 3.23).

  In some parts of its range, Whale Sharks are killed for meat, liver oil, cartilage, and especially their large fins, which are exported. The 2001 documentary *Shores of Silence: Whale Sharks in India* exposed the unregulated artisanal Whale Shark fishery along Northwest India. The film led the Indian government to list the species on the Indian Wild Life (Protection) Act, with additional conservation measures by non-governmental organizations like the Wildlife Trust of India. The efforts have drastically reduced but not eliminated the Whale Shark catch there.

  Whale Sharks are yolk-sac viviparous, retaining egg cases internally, and one specimen harpooned by fishers contained 300 embryos. Surprisingly, Whale Sharks are maintained successfully in captivity, even after being shipped 8,000 miles.

Representative Species

Whale Shark (*Rhincodon typus*; fig. 3.22).

Figure 3.22. Whale Shark. (Chainarong Phrammanee/Shutterfly.com).

Figure 3.23. Low-altitude aerial photograph of Whale Shark aggregation from 2009. (A) shows over 200 sharks. (B) shows over 68 sharks, along with a tourist boat. (From de la Parra Venegas, R. et al. 2011. PLoS One 6(4): p.e18994)

ORDER LAMNIFORMES (MACKEREL SHARKS)

The order Lamniformes includes only 15 species in seven families (some monotypic); it is diverse only at higher taxonomic levels. This group includes the ultimate predators as well as planktivores. All have two spineless dorsal fins, the second typically much smaller than the first, and usually a more conical snout without any lateral expansion of rostral cartilage (an adaptation for speed). They are all pelagic and mostly inhabit cool waters. They have a variety of tooth morphologies. Lamniform sharks all lack a nictitating membrane. All have anal fins. All species with known reproductive modes are viviparous with oophagy (egg-eating); that is, during development the embryos consume the ova continuously produced by the mother during her pregnancy.

Families in the order Lamniformes:

- Sand Tigers (Family Odontaspidae; *odont* = teeth; *aspid* = shielded)
  - » three species, two of which occur in deeper water (300–400 m, or 985–1310 ft) and are poorly known. Their teeth are like those of the Shortfin Mako (*Isurus oxyrinchus*), but with one or two smaller cusps on the side (hence they are *shielded*). Like the Shortfin Mako, they are predators that grasp their prey.

Representative Species

Sand Tiger (*Carcharias taurus*). The Sand Tiger is the best-known ondontaspid. It is a very slow swimming, big shark (up to 3.5 m, or 11.5 ft) with a long mouth out of which protrude large teeth. The Sand Tiger is found inshore in the Mediterranean, Atlantic, and Indo-Pacific. Its first dorsal fin is set back somewhat on its body.

In aquaria, Sand Tigers appear to hover. They do so by gulping air at the surface and keeping a bubble in their relatively impervious stomach. They are oophagous and embryophagous, meaning that the embryos will consume each other *in utero* until only one remains in each uterus. Sand Tigers are listed by the IUCN as Endangered. Maybe they should just stop eating their siblings!

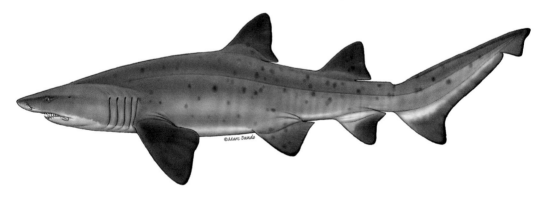

©Marc Dando

- Crocodile Shark (F. Pseudocarcharidae; fig. 3.24)
  - » one species, the Crocodile Shark (*Pseudocarcharias kamoharai*), which grows to 1 m (3.3 ft), and is a mesopelagic, diurnal vertical migrator found in the tropical Atlantic and Indo-Pacific. It has a pointed snout, with large eyes, and dagger-shaped teeth, which may grab squid before swallowing them whole. They often have a white spot on their cheek.

Representative Species

Crocodile Shark (*Pseudocarcharias kamoharai*; fig. 3.24).

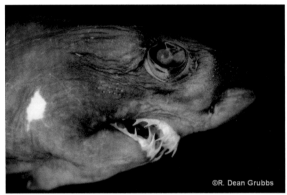

Figure 3.24. Head of Crocodile Shark.

@R. Dean Grubbs

- Goblin Shark (F. Mitsukurinidae)
  » This is a monotypic family whose single member, the Goblin Shark (*Mitsukurina owstoni*; fig. 3.25), has a flabby, elongate body, tube-like nostrils, blade-shaped snout (unique among lamniforms), and extremely protrusible jaws with dagger teeth. Like other sharks with dagger or bladelike teeth, its jaws are relatively weak. The exaggerated rostrum may function to maneuver prey into its underlying mouth. The Goblin Shark lives in deep waters (550 m, or 1800 ft) in both the Atlantic and Pacific, where it feeds on fish and squid.

  This species has an extremely large liver, which makes the species nearly neutrally buoyant. It possibly grows to 3.5 m (11 ft) and is probably yolk-sac viviparous with oophagy. Pictures of the Goblin Shark often depict it with the jaws protruded, which is not the case when they are swimming and is more likely the result of photographing or drawing a dead specimen.

Representative Species

Goblin Shark (*Mitsukurina owstoni*; fig. 3.25).

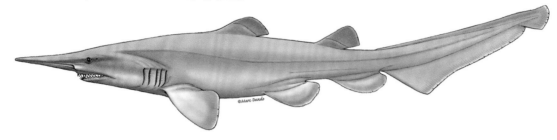

Figure 3.25. Goblin Shark. (Paulo de Oliveira / Biosphoto)

- Megamouth Shark (F. Megachasmidae; fig. 3.26)
  » Another monotypic family, the Megachasmidae was unknown to science until 1976, when a Megamouth Shark (*Megachasma pelagios*), then unnamed, was caught enwrapped in a sea anchor off Hawaii. The fact that a shark of that size and unique appearance had eluded the notice of the

marine biology community was considered astounding. The event is used to this day to support the absurd idea that a *Megalodon* could also be lurking in the deep, undetected by science. (In fact, a more defensible argument is that greatly expanded exploration of the deep-sea would undoubtedly add new sharks, but they would be small and brown). The next specimens were not caught until 1984 and 1987. As of 2016, there were about 67 documented catches. That number exploded to more than 100 the next year as older, previously unknown captures off Taiwan were reported. So much for the *Age of Information*.

Among the largest sharks, measured Megamouths have ranged from 177 to 570 cm long (5.8 to 18.7 ft). The Megamouth Shark is a planktivore, with fleshy lips that were hypothesized to possess luminescent photophores to attract its food source. This would be the only shark outside of the squaliformes to possess photophores and alas—this theory has not been validated. However, the Megamouth is thought to feed differently than other elasmobranch planktivores, by swallowing water and then forcing it forward through the gill rakers, which strain out the plankton.

The species is probably found worldwide in the tropics, but fewer in the Atlantic (only a single one has been caught in the North Atlantic, in Puerto Rico in 2017) and far more in the Philippines and Japan.

Representative Species

Megamouth Shark (*Megaschasma pelagios*; fig. 3.26).

- Threshers (F. Alopiidae; *Alop* = fox; fig. 3.27)
  - » three species, all large (up to 6 m, or 20 ft) with large eyes and small mouths. Their most identifiable feature is their curved, elongate, whip-like caudal fin, whose upper lobe may be one-half the total body length. They reside inshore to pelagic waters (surface to > 350 m, or 1150 ft) in all warm tropical seas, where they consume squid and small pelagic shoaling fish like anchovies and sardines. The mouths of the Common Thresher (*Alopias*

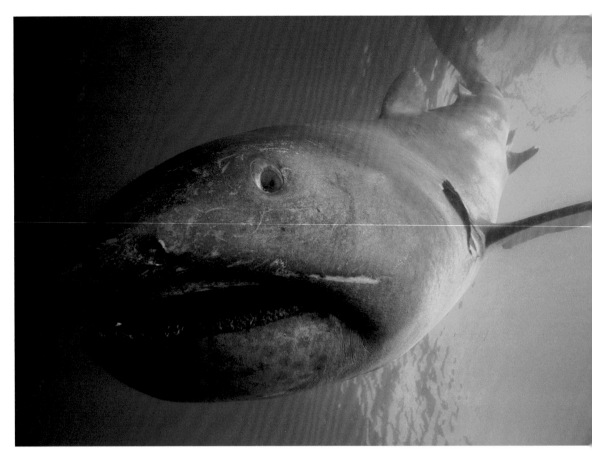

Figure 3.26. Megamouth Shark. (Bruce Rasner / Jeffrey Rotman / Biosphoto)

vulpinus) and the Pelagic Thresher (*A. pelagicus*) are small with weak jaws. The Bigeye Thresher (*A. superciliosus*) eats slightly larger prey and thus has a somewhat larger mouth and larger teeth. They are all yolk-sac viviparous with oophagy.

In feeding, the elongated tail is used to first cause schools of their prey to tighten before swatting and stunning members prior to eating them. They have also been observed swatting the water, which may explain reports by commercial fishers of catching thresher sharks hooked by the tail.

The presence of the elongate caudal fin comes with a hydrodynamic cost. You will see in Chapter 4 that the extended upper lobe of the caudal fin causes the net force exerted by the fin to be upward at about a 45° angle. Thus, to compensate for the elevation of the rear of the body, Thresher Sharks typically have greatly expanded pectoral fins that act as planing surfaces and provide lift.

All three species are listed as Vulnerable by the IUCN.

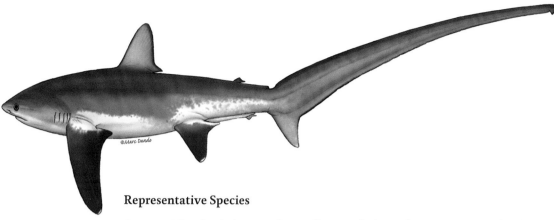

Representative Species

Common Thresher (*Alopias vulpinus*; fig. 3.27A). This is the most common thresher along the US East Coast. They have tiny teeth. *A. vulpinus* gives birth to two to six live pups born at 1.5 m (5 ft) long.

Bigeye Thresher (*Alopias superciliosus*; fig. 3.27B). The Bigeye Thresher is aptly named. Not only is the eye larger, it also is oriented for binocular vision over its head, the only non-demersal shark so adapted. This ability enables them to attack small fish above them.

Pelagic Thresher (*Alopias pelagicus*; fig. 3.27C). The Pelagic Thresher is found only in the Pacific and also has huge pectoral fins, though not as large as those of the Bigeye Thresher.

- Basking Shark (F. Cetorhinidae; *cet* = whale; *rhin* = nose; fig. 3.28)
  - » a monotypic family, containing only the Basking Shark (*Cetorhinus maximus*), the second largest shark (and fish as well), which is located worldwide in cool water, but has been found to migrate to the Amazon River plume at a depth of about 500 m (1640 ft). They grow to as large as 10 m (33 ft). They are planktivorous, with huge gill slits almost encircling their head. They filter plankton using gill rakers, which are deciduous (they fall out and are replaced annually).

    Basking Sharks supported a large fishery in the North Atlantic until the mid-20th century when populations crashed (see Chapter 11). They have recovered globally to the point where they are now listed by the IUCN as Vulnerable. Aerial surveys along the NE US coast beginning in 1980 revealed aggregations as large as 1400 individuals.

    The size of the brain of the Basking Shark is small relative to its body weight,[11] which is attributed to its planktonic feeding, a mode that involves weak swimming and thus lower sensory requirements. They recently have been observed leaping from the water along the coast of Ireland.

A

B

Figure 3.27. (A) Common Thresher Shark off the coast of Virginia. (B) Head of Bigeye Thresher, showing upward-focusing eye. (NOAA Fisheries / Apex Predators Program) (C) Pelagic Thresher, with expansive pectoral fins.

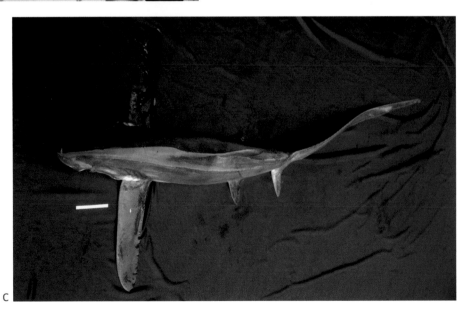

C

Representative Species

Basking Shark (*Cetorhinus maximus*; fig. 3.28).

Figure 3.28. Basking Shark filter-feeding. (Martin Prochazkacz/Shutterfly.com)

- Mackerel Sharks (F. Lamnidae; figs. 3.29–3.31)
  - » five species in three genera, characterized by a large fusiform body (as big as 4–7 m, or 13–23 ft), lunate (superficially homocercal) caudal fin, and pointed snout. They also maintain their body temperature above ambient, which allows them to develop more muscle power and faster processing times for sensory information (see Chapter 8). All are pelagic (typically in 150 to 1000 m, or 490 to 3280 ft) in most temperate and some tropical seas, and they feed on fish, cephalopods, and marine mammals. They are yolk-sac viviparous with oophagy.

Representative Species

Shortfin Mako (*Isurus oxyrinchus*; fig. 3.29). This species is perhaps the most magnificent fish in the sea, with its beautiful coloration (brilliant blue or purple on top, white on the bottom) and its near perfect streamlining. The species is aptly named: *isurus* = equal tail; *oxyrinchus* = sharp snout. It is found globally in temperate and tropical waters.

Distinguishing features include a conical snout, moderately short pectoral fins, origin of first dorsal posterior to the free rear tips of pectoral, lunate tail, dagger teeth, and huge gill slits. They grow to > 3 m (10 ft). They swallow some of their prey whole (bony fishes and cephalopods, predominantly) and are considered opportunistic apex predators. As they age, their teeth become broader and flatter, enabling them to widen their prey options to include organisms too large to swallow whole but from which they can remove a chunk of flesh (e.g., swordfish, tuna, sharks, sea turtles, and marine mammals). Both the Shortfin and the Longfin Mako (*I. paucus*) have been captured or observed impaled with swordfish bills in their head regions. They are yolk-sac viviparous with oophagous embryos. As a result, the developing embryos, as many as 18, have enlarged guts known as egg (or yolk) stomachs (see fig. 6.16A).

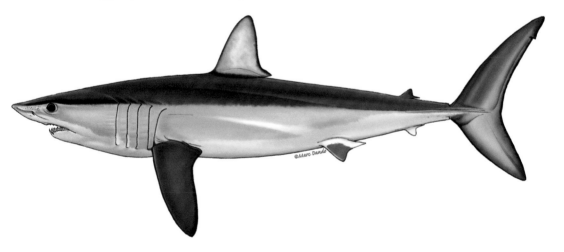

Figure 3.29. Shortfin Mako. (Wildestanimal/Shutterstock.com)

Porbeagle (*Lamna nasus*). The Porbeagle resembles the Salmon Shark (*L. ditropis*; fig. 3.30), including the second caudal keel. However, the Porbeagle has three-cusped teeth (the Salmon Shark has only one additional, small cusplet), and it has a white blotch at the posterior of its first dorsal fin. It is found both in the North and South Atlantic. The Porbeagle typifies overfished sharks (see Chapter 11).

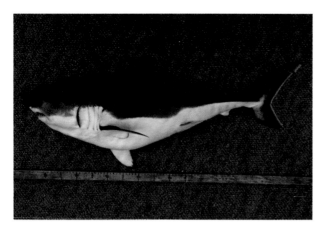

Figure 3.30. Juvenile Salmon Shark, *Lamna ditropis*. (Courtesy of David Itano)

White Shark (*Carcharodon carcharias*; fig. 3.31). What can we say about one of the most iconic species on the planet? Are they overrated? Well, if overrated means that the species receives attention disproportionate to its ecological importance or conservation status, often to the exclusion of more interesting and endangered sharks, then yes. On the other hand, if caring about this species leads to awareness of the plight of other sharks and inhabitants of the planet in general, and thus valuing our natural environment, then no.

Characteristics include its large size (up to 6.1 m, or 20 ft, and 1900 kg, or 4200 lb), a jaw full of triangular serrated teeth for cutting, long gill slits, very black eyes, vivid color changes on its sides, and black tips under the pectoral fins. It is globally distributed in both coastal and oceanic waters of between 12°C and 24°C (54°F and 75°F). Contrary to the public's perception, the population is not declining and in fact has been increasing in many regions the last 20 or 30 years.

The White Shark is perpetually in the news, but one recent report merits attention. The first are reports, first near SE Farallon Island, south of San Francisco, and more recently off South Africa, of White Sharks being killed by Orcas, which apparently removed the liver and possibly other internal organs. Even if this activity is novel, that Orcas may have enlarged the range of prey to include White Sharks would not be an unusual feat for so intelligent an animal.

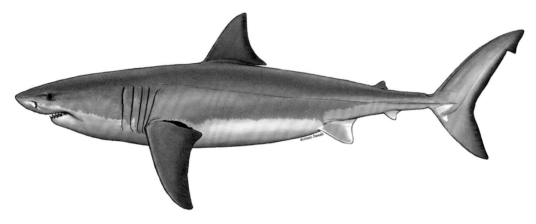

Figure 3.31. White Shark. (Courtesy of Dr. Craig P. O'Connell)

ORDER CARCHARHINIFORMES (GROUND OR REQUIEM SHARKS)

The order Carcharhiniformes contains 293 species currently in eight families, but this is subject to ongoing revision. The order includes the Tiger (*Galeocerdo cuvier*), Bull (*Carcharhinus leucas*), and other sharks with which you are likely familiar. All have two spineless dorsal fins, and an anal fin. They lack nasoral grooves, possess nictitating membranes, and have intestines with a spiral or scroll valve. All three major modes of embryonic development are present, including oviparity, yolk-sac viviparity, and placental viviparity.

Families in the order Carcharhinidae:

- Catsharks (Family Scyliorhinidae; figs. 3.32 and 3.33)
  » 158 species, making this the most speciose family in the order. However, there is strong molecular evidence that this family includes multiple clades and will soon be divided into at least two families. Catsharks are

mostly small (< 1 m, or 3.3 ft), and are found predominantly in cold and deep waters worldwide. They are oviparous. They have elongated, catlike eyes (hence the name) and two small dorsal fins positioned far back on the body. Despite their diversity, you are unlikely to encounter many of these unless you study deep-sea sharks, so we will mention only a few species.

Representative Species

Small-spotted Catshark (*Scyliorhinus canicula*). This species is a shallow water inhabitant and is one of the most abundant sharks in the North Sea, Mediterranean, and adjacent areas in the Atlantic. It has a very slender body and is oviparous.

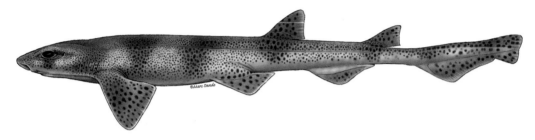

Chain Catshark (often referred to as Chain Dogfish [*Scyliorhinus retifer*; fig. 3.32]). The Chain Catshark is found from 36 to 750 m (118–2461 ft) in the Western North Atlantic, Gulf of Mexico, and Caribbean Sea. It has a reticulated pattern and is commonly displayed in aquaria. Despite the common name, the species is not a dogfish.

Striped Catshark or Pajama Shark (*Poroderma africanum*). The Pajama Shark is found off South Africa and is included here because it is the only shark with horizontal stripes. It is commonly displayed in aquaria.

Blackmouth Catshark (*Galeus melastomus*; fig. 3.33). The Blackmouth Catshark is a small catshark common in the NE Atlantic and Mediterranean Seas. The genus *Galeus* are called sawtailed catsharks because they possess enlarged scales on the anterior portion of the upper lobe of the caudal fin.

Figure 3.32. Chain dogfish at Norfolk Canyon off the coast of Virginia. Note the abandoned fishing net in the background (NOAA).

Ghost or Demon Catshark (*Apristurus*). The genus *Apristurus* includes about 37 species (with more still likely to be discovered) of deepwater catsharks and is also the most common deepwater genus of sharks. There are small, dark black to brown sharks with tiny eyes, which slowly cruise or, depending on the species, sit on the seafloor.

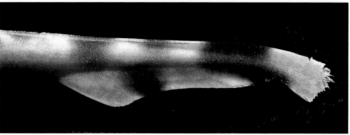

Figure 3.33. Blackmouth Catshark, with inset showing enlarged denticles on the upper lobe of the caudal fin.

- Finback Catsharks (F. Proscyllidae)
  » six species of small, sluggish sharks, sometimes quite common in tropical areas. Some proscyllids are oviparous whereas others are yolk-sac viviparous.

Representative Species

Ctenacis sp.

- False Catsharks (F. Pseudotriakidae; fig. 3.34)
  - » five or six species of these small- to moderate-sized deepwater sharks, found from 300–1000 m (984–3280 ft).

Representative Species

False Catshark or Sofa Shark (*Pseudotriakis microdon*; fig. 3.34). Superficially these flabby sharks look like a gulper shark (black with big green eyes) and they possess large spiracles. However, the gulpers are squalomorph sharks with a tighter jaw suspension and no anal fin, whereas the pseudotriakids are galeomorph sharks, which have a looser jaw suspension and an anal fin. Thus, the resemblance is an example of convergent evolution and is not phylogenetically based. The species was in the news in 2015 when a specimen caught off the coast of Scotland captured the attention of the media. This is the only shark outside the lamniforms known to reproduce using oophagy.

Figure 3.34. False Catshark, or Sofa Shark.

- Barbeled Houndshark (F. Leptochariidae)
  - » a monotypic family, the single species being the Barbeled Houndshark (*Leptocharias smithii*).

Representative Species

Barbeled Houndshark (*Leptocharias smithii*). Found in the Eastern Atlantic, this demersal species is found to depths of 75 m (246 ft). It is yolk-sac placental viviparous.

- Houndsharks (F. Triakidae; fig. 3.35)
  - » 47 species in nine genera of mostly coastal sharks, including the Leopard Shark (*Triakis semifasciata*; fig. 3.35A) and the Dusky Smoothhound (*Mustelus canis*; fig. 3.35B). They all have pavement dentition, a long, pointed snout, and arched mouth. They are pelagic from the surface to > 450 m (1475 ft), where they feed primarily on fish and squid. Triakids are found in the Atlantic, Pacific, and Mediterranean. Interestingly, about half the species are yolk-sac viviparous and half are placental viviparous. Triakids range from 50–200 cm (1.5–6.5 ft).

Representative Species

Spotted Houndshark (*Triakis maculata*). The Spotted Houndshark is a small, harmless coastal shark found in the Pacific off the coast of South America.

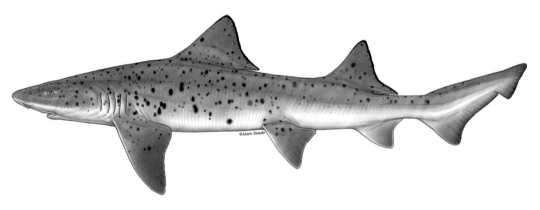

School Shark or Soupfin Shark (*Galeorhinus galeus*). Also known as the Vitamin Shark, this species is easily identified by the large lobe on the upper caudal fin. They have a slender body, long snout, and small second dorsal. Their vitamin A-rich liver led to their being the biggest shark fishery in California in the 1930s and 1940s.

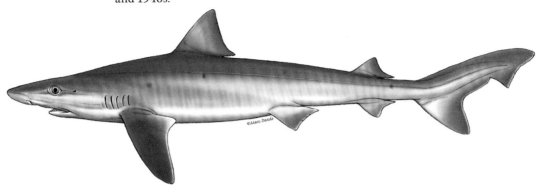

Dusky Smoothhound or Smooth Dogfish (*Mustelus canis*; fig. 3.35). The Dusky Smoothhound (Smooth Dogfish) is a common placental viviparous species in the Northwest Atlantic. There are two subspecies, *M. canis canis* and *M. canis insularis*. The former is most common along the shallow continental shelf from New York to South Carolina, and it pups in shallow estuaries. Oddly, it was recently reported pupping at 400 m in the Gulf of Mexico, which means that the same subspecies utilized two very different depth regimes for giving birth. *Mustelus canis insularis* is the island form of the species and is found in the tropics in deeper waters.

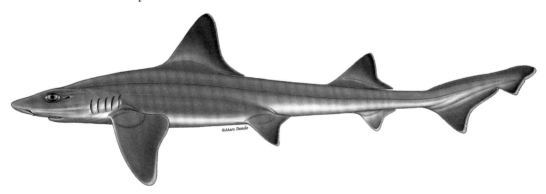

- Weasel Sharks (F. Hemigalidae)
  » eight species, with a robust body and elongate, rounded snout. They live in inshore waters from the surface to 130 m (425 ft), where they feed on fish. They have unusual teeth. The group is placental viviparous and reaches 2.4 m (8 ft). They are all found in the Indo-Pacific.

A

Figure 3.35. (A) Leopard Shark, commonly found off the California coast, at the Birch Aquarium in La Jolla, California. (Courtesy of Matthew Field, www.photography.mattfield.com) (B) Dusky Smoothhound.

B

Representative Species

Weasel Shark *(Hemigaleus microstoma)*. This rarely encountered species is found in southern India, southern China, and parts of SE Asia.

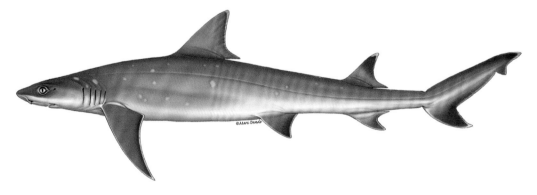

- Requiem Sharks (F. Carcharhinidae)
  - » about 58 species, 34 of which are in the largest genus, *Carcharhinus.* They are the second largest galeomorph family and are the dominant sharks in tropical and subtropical waters, and some even live in freshwater. Carcharhinids have a long, arched mouth with bladelike teeth, which are often broader in the upper jaw, a nictitating membrane, and no spiracle (with one exception). They are found in both inshore and pelagic waters from the surface to 800 m (2625 ft), where they feed on fish, cephalopods, turtles, and mammals. They are all placental viviparous except the Tiger Shark, which uniquely employs embryotrophic viviparity, and which may soon be moved into its own separate family. Carcharhinids occupy all oceans, and many are potentially dangerous. Over 24 species are in the range of 1.7–5 m (5–16 ft).

    Carcharhinids are superficially divided into two broad groups based on the presence or absence of a slightly elevated ridge on the dorsum between the first and second dorsal fins (fig. 3.36). This feature, the interdorsal ridge, is very useful for identifying many carcharhinids, and is the first feature that should be looked for if unsure of the species.

Representative Species

RIDGEBACK SHARKS

Reef Shark or Caribbean Reef Shark (*Carcharhinus perezi*; fig. 3.37). The Caribbean Reef Shark is the most common shark of Caribbean coral reefs and it reaches

Figure 3.36. Conspicuous inter-dorsal ridge characteristic of one group of carcharhinid sharks, in this case a Sandbar Shark. A small ectoparasitic copepod is also shown.

a maximum size of 3 m (3.3 ft). This species eats bony fishes. It occupies a similar ecological niche to the Indo-Pacific Gray Reef Shark (*C. amblyrhynchos*).

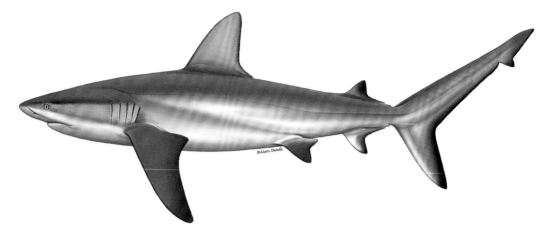

Sandbar Shark or Brown Shark (*Carcharhinus plumbeus*; fig. 3.37). The Sandbar Shark is a coastal water species distributed worldwide in temperate to subtropical waters shallower than 100 m (330 ft). It reaches a maximum length of 2.4 m (8 ft). In addition to having a prominent interdorsal ridge, it can be identified by its oversized first dorsal fin far forward on the body. The Sandbar Shark is the dominant large coastal shark along the US East Coast as well as in Hawaii, but it occupies a deeper water habitat in the latter. The biggest Sandbar Shark nursery (the area where they are born and/or spend their early years) in the world is Chesapeake Bay. In part because of their large first dorsal fin's value to the shark fin soup industry, Sandbar Sharks drove the US East Coast directed shark fishery until they became overfished.

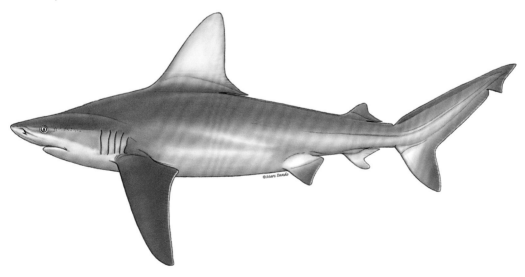

A

Figure 3.37. Ridgeback sharks.
(A) Caribbean Reef Shark. (Courtesy
of Emily Marcus) (B) Sandbar Shark.
(Brandon B / Shutterstock.com)

B

Dusky Shark (*Carcharhinus obscurus*; fig. 3.38). The Dusky and Galapagos Sharks (*Carcharhinus galapagensis*) are so closely related that some geneticists consider them as a single species. The Dusky Shark remains one of the most overfished sharks on the US East Coast, and its full recovery is not expected until after 2100.

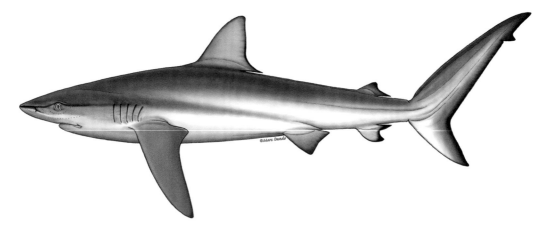

Figure 3.38. Dusky Shark. (© Andrew Raak)

Silky Shark (*Carcharhinus falciformis*). Named because of their small dermal denticles that are smooth to the touch, the Silky Shark can be identified by the convex rear margin of its first dorsal combined with an elongated free tip on the second dorsal fin and long, sickle-shaped pectoral fins. It is one of the most common tropical pelagic sharks in the world. In some areas, Silky Sharks are overfished because of purse seine fisheries for tunas.

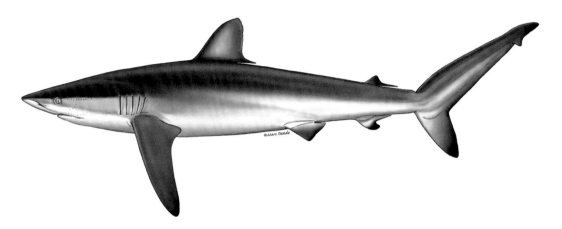

Oceanic Whitetip Shark (*Carcharhinus longimanus*; fig. 3.39). The specific epithet *longimanus* means long hands, which refers to its elongate, paddle-shaped pectorals, and the common name describes the prominent edges of the fins, which appear to have been dipped in white paint. Interestingly, newborns have black tipped pectoral fins. The Oceanic Whitetip Shark is a stocky, large (around 2 m, or 6.6 ft), pelagic shark found in temperate and tropical oceans. It has a reputation of being an aggressive and dangerous shark. The Oceanic Whitetip Shark is inquisitive and may bump people it encounters.

Figure 3.39. Oceanic Whitetip Shark, another ridgeback species. (Photo by Lance Jordan)

NON-RIDGEBACK SHARKS

Blacktip Reef Shark (*Carcharhinus melanopterus*; fig. 3.40A). The Blacktip Reef Shark, whose fins look as if they have been dipped in ink, is often confused with the Blacktip Shark (*C. limbatus*), especially in the media. The Blacktip Reef Shark is found only on Indo-Pacific coral reefs and the eastern Mediterranean, whereas the Blacktip Shark is distributed globally in coastal tropical and temperate waters.

Blacktip Shark (*Carcharhinus limbatus*; fig. 3.40B). The Blacktip Shark is widely distributed and is one of the most common non-ridgeback sharks. It is found both inshore and offshore, in the upper 100 m (330 ft). Along the US East Coast, the species is frequently observed in dense schools of thousands along its migratory route every spring and fall (see fig. 1.19). These are streamlined, swift-moving, fish and squid-eaters, and are often blamed for shark bites along the US Atlantic Coast in the summer.

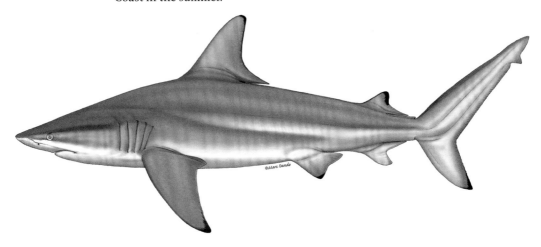

Bull Shark (*Carcharhinus leucas*; fig. 3.41). This is a stout shark with a robust, blunt, rounded snout. It has a large first dorsal fin situated far forward on the body, but without an interdorsal ridge. Its eyes are relatively small, but its teeth are broad and heavily serrated for shearing. The Bull Shark is predominantly found in shallow tropical and temperate waters less than 30 m (100 ft) deep, but it can be found shallower and as deep as 150 m (492 ft).

The Bull Shark is considered dangerous, especially in the developing world where the daily lives of inhabitants find them in Bull Shark habitat, particularly brackish and fresh waters. When Bull Sharks are encountered by a boat in shallow, clear tropical waters, they may go into an agonistic (threat) display, lowering the fins and hunching their back, and when that occurs, they may charge and strike the boat.

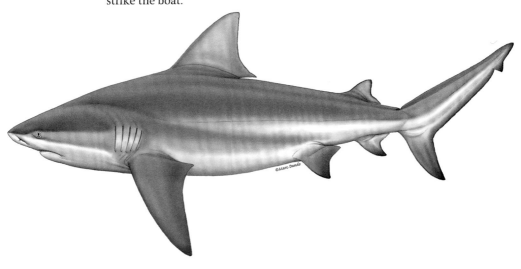

A

B

Figure 3.40. Two non-ridgeback sharks that are often confused with each other. (A) Blacktip Reef Shark. (Courtesy of Michael Scholl) (B) Blacktip Shark. (Courtesy of Laura Claiborne Stone)

Figure 3.41. Adult Bull Shark, a non-ridgeback species. (Andrea Izzotti/Shutterstock.com)

CARCHARHINID SHARKS IN GENERA OTHER THAN CARCHARHINUS

Lemon Shark (*Negaprion brevisostris*; fig. 3.42A). Lemon Sharks are a large (to 3.4 m, or 11 ft), migratory, shallow water, temperate and tropical shark found along the Atlantic and Eastern Tropical Pacific Coasts. Juveniles inhabit mangrove-lined lagoons and adults are reef-associated. They are a muted lemon color with two dorsal fins of approximately the same size. They are philopatric; that is, they return to their birthplace to give birth. Lemon Sharks are a species whose ecology, behavior, life history, and physiology have been extensively studied for over 30 years by Dr. Samuel Gruber and his graduate students and associates at the Bimini Biological Field Station in the Bahamas.

A

Figure 3.42. (A) Lemon Shark. (Fiona Ayerst/Shutterstock.com) (B) The ecologically similar Whitetip Reef Shark. (Courtesy of Lesley Rochat)

Whitetip Reef Shark (*Triaenodon obesus*; fig. 3.42B). The Whitetip Reef Shark, not to be confused with the Oceanic Whitetip Shark, is a common, small (1.6 m, or 5.2 ft), slender, Indo-Pacific shark. Named for its white-fringed fins, the Whitetip Reef Shark sleeps during the day and actively forages on the reef in the evening, an activity facilitated by the flexibility of this species.

B

Blue Shark (*Prionace glauca*; fig. 3.43). Blue Sharks are large (to 3.3 m, or 10.8 ft), pelagic, cold-temperate sharks found in all of the world's oceans. They are one of the most abundant large sharks. They are relatively slender, show countershading (blue on top), and have relatively long pectoral fins. Like other carcharhinids, they are viviparous. They have one of the largest reproduction potentials of the group and are capable of having litters as big as 100.

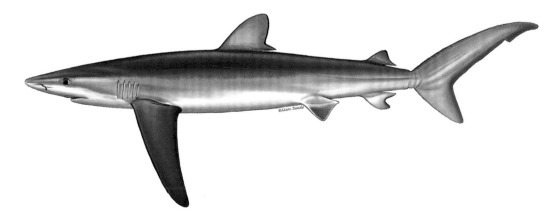

Figure 3.43. Blue Shark. (Wildestanimal/Shutterfly.com)

Atlantic Sharpnose Shark (*Rhizoprionodon terraenovae*; fig. 3.44). The Atlantic Sharpnose Shark is a small (1 m, or 3.3 ft), hugely abundant Atlantic coastal shark, characterized by having a small, black second dorsal fin, a small ridge in front of the anal fin (a preanal ridge), and elongated ampullary pores (like oval slits) on their cheeks behind their eyes. Beginning in May and throughout the summer, in temperate estuaries and nearshore environments, the neonates (newborns) are nearly ubiquitous.

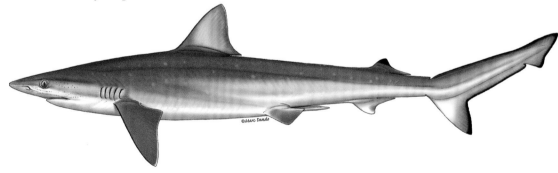

Figure 3.44. Atlantic Sharpnose Shark. Note the black second dorsal and white spots, distinguishing features of this species.

Tiger Shark (*Galeocerdo cuvier*; fig. 3.45). Found worldwide in tropical and temperate coastal waters, this iconic, large shark (5 m, or 16.5 ft), is the only member of its genus, and will likely be moved into its own family because of its unique reproduction, different jaw structure, unique teeth, predorsal ridge (fig. 3.45A), homodont dentition, and keeled tail. Genetic analysis supports this. It is easily identified by its markings, which are most vivid when it is a juvenile, and by its long caudal fin. Like the cow sharks, Tiger Sharks surprisingly have relatively weak jaws.

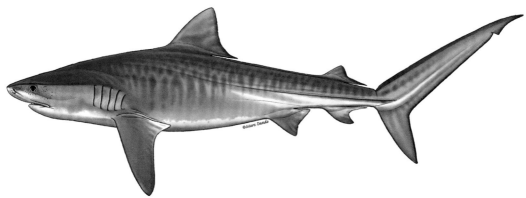

- Hammerheads (F. Sphyrnidae; *sphyrna* = hammer)
  - » Nine species, but it is likely that more will be added to it. The hammer-shaped head, technically known as a *cephalofoil*, ranges in size from the relatively small Bonnethead Shark (*Sphyrna tiburo*) to the bizarrely wide head of the Winghead Shark (*Eusphyra blochii*), an Indonesian species whose head is half as wide as its body length (fig. 3.46). Hammerheads are found in all tropical and warm temperate seas in both inshore and pelag-

A

B

Figure 3.45. (A) Tiger Shark showing its distinguishing features: predorsal ridge, long lateral keel, and wide head. (B) Another view of a Tiger Shark. (HQuality/Shutterstock.com)

ic environments, with animals having been tracked from the surface to nearly 1000 m (3280 ft). They feed on fish, including batoids. Members of the family range in size from the 1.2 m (3.9 ft) Bonnethead to the 6.2 m (20.3 ft) Great Hammerhead (*S. mokarran*). All hammerheads are placental viviparous.

Representative Species

Bonnethead (*Sphyrna tiburo*; fig. 3.47). This small, beautifully speckled shark is extremely common in the Gulf of Mexico and along the US Atlantic coastline. *S. tiburo* is also the only shark with a sexually dimorphic head shape. Males exhibit an anterior bulge on the cephalofoil whereas the head is more rounded in females. Bonnetheads eat crustaceans, especially blue crabs, along with fish and other invertebrates.

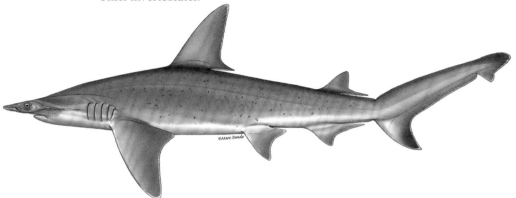

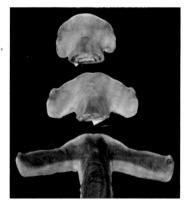

Figure 3.46. Cephalofoils of selected members of the family Sphyrnidae. *From top to bottom*: Bonnethead, juvenile Scalloped, and Winghead Hammerheads.

Figure 3.47. Bonnethead.

Great Hammerhead (*Sphyrna mokarran*; fig. 3.48A). There are no projections on the head of this species, but there is a hump instead of scallops or smooth curve. The first dorsal fin is enormous and there are large pelvic fins as well as a huge upper caudal fin.

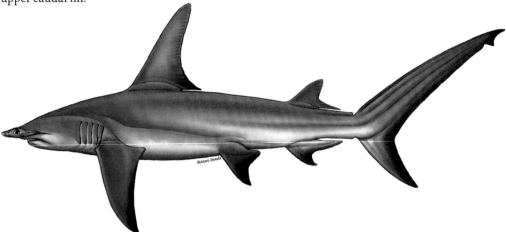

A

Figure 3.48. (A) Great Hammerhead.
(B) Winghead Shark.

B

Slender Hammerhead or Winghead Shark (*Eusphyra blochii*; fig. 3.48B). This exclusively Indo-Pacific, small (1.5 m, or 4.9 ft) species has an extremely wide cephalofoil and is found in murky estuaries.

Concluding Comments

Although not exhaustive, this chapter's overview of the diversity of sharks, which includes iconic species as well as those of ecological and commercial importance, we hope fed your sense of wonder at these magnificent beasts. At the same time, it should have acquainted you with sharks with which you may have been unfamiliar and the characteristics and adaptations that distinguish them from each other. Both of these become important as we discuss the physiology, ecology, behavior, fisheries, and conservation of sharks in future chapters.

NOTES

1. Compagno, L.J. 1984. Sharks of the world: an annotated and illustrated catalogue of shark species known to date. Food & Agriculture Org. (No. QL 638.9. C65).

2. Weigmann, S. 2016. J. Fish Biol. 88: 837–1037.

3. https://sharksrays.org. (Accessed 6/29/19).

4. Recall that a *monophyletic* group simply means that all members of the group are most closely related to each other and are not placed in other groups.

5. Last, P.R. et al. 2016. *Rays of the World: Supplemental Information* 40: 1–10. CSIRO Publishing.

6. Batoids in fact have eight synapomorphies. For details, see: Aschliman, N. 2011. The batoid tree of life: recovering the patterns and timing of the evolution of skates, rays, and allies (Chondrichthyes: Batoidea). PhD dissertation, the Florida State University.

7. Last, P., Naylor, G., Séret, B., White, W., de Carvalho, M., and Stehmann, M. eds. 2016. *Rays of the World*. CSIRO Publishing.

8. https://www.iucnredlist.org/species/39332/117498371. (Accessed 9/9/19).

9. Grace, M.A. et al. 2019. Zootaxa 4619: 109–120.

10. Widder, E.A. 1998. Envir. Biol. Fishes 53: 267–273.

11. Kruska, D.C. 1988. Brain, Behav., Evol. 32: 353–363.

PART II

Adaptational Biology: How Sharks Work

4 / Functional Anatomy of Sharks

Introduction: Shark Skin and Adaptations

In 1996, the swimsuit company Speedo developed a competition, full body swimsuit called the *Fastskin*, and in 2004 they introduced the advanced *Fastskin FSII*. These were both advertised as revolutionary, performance-enhancing body coverings—the *fastest fabric in the world*—that significantly reduced drag, the major frictional force that slows moving objects and the avowed enemy of elite swimmers. Promotional literature from the company emphasized that the design of this line of swimsuits was inspired by sharks.

In advertising campaigns, the company juxtaposed images of swimming sharks and humans that highlighted superficial similarities in form between the two groups. The real design breakthrough, however, according to Speedo, was creating a fabric that channeled water smoothly from front-to-back along the swimmer's body, much in the same way that the ridges and grooves of a shark's dermal denticles were thought to do.

A high-performance swimsuit inspired by sharks would be a cool story, and an exquisite example of *biomimicry* (the act of copying nature's best ideas for human uses), except for the fact that humans swim differently than sharks. In fact, a 2012 paper[1] found no increase in speed in tests using Speedo fabric compared to an average increase of 12.3% using membranes made from shark skin, when each was compared to controls using a *robotic flapping device* (got yours yet?). Humans do not swim like robotic flapping devices either.

Although the Speedo swimsuits may not have allowed wearers to swim like sharks, the company's engineers were correct in their understanding of the drag-reducing role of the dermal denticles of real shark skin. According to the research referenced above, the dermal denticles both decreased drag and increased thrust.

Drag reduction is achieved because the dermal denticles act as *riblets* (small protrusions on the surface). Riblets prevent the formation of little *eddies* (whirlpools), which increase friction by promoting turbulent flow that might slow the shark (fig. 4.1).

Dermal denticles, and the skin of sharks, have additional functions as well. First, they protect sharks. We frequently catch sharks on our experimental fishing gear with bites from other sharks that either do not penetrate the skin or only poke holes but do not tear the skin (fig. 4.2), which is due to the protection offered by the dermal denticles, as well as the thickness and inherent toughness of the underlying skin.

The skin likely also plays a role in reducing adhesion of fouling organisms (like algae, barnacles, and so on) and ectoparasites (see fig. 3.36), whose presence might create eddies as the shark swims, thereby increasing friction and slowing the shark or requiring more energy to swim, not to mention potentially affecting the health of the shark.

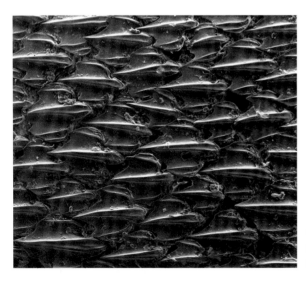

Figure 4.1. Riblets (ridges and grooves) on dermal denticles of Genie's Dogfish shown on a scanning electron photomicrograph. (Courtesy of Charles F. Cotton)

Figure 4.2. Blacktip Shark from experimental longline with a series of puncture marks from the bite of another shark. This shark was successfully released.

The functions that scientists attribute to shark dermal denticles and skin are *adaptations*. Adaptations refer to any features of an organism that enhance its ability to survive and reproduce in its environment *and* which are hereditary (the potential to be handed down to offspring).

Adaptations can be anatomical (structural), physiological (internal function), or behavioral. Though the value of an adaptation to an organism is not always obvious, nature selects for adaptations that promote survival and at the same time, if possible, conserve energy.

Adaptations may entail evolutionary trade-offs; for example, the extended rostrum, or saw, of a sawfish functions in part to detect the presence of prey, and then is remarkably effective at swiftly slashing small prey, either stunning them or impaling them on the sharp rostral teeth. It is also an effective defense mechanism. At the same time, swimming with such an awkward structure protruding from its head must be energetically inefficient for a sawfish, although this has not been studied. Prominent physiological ecologist Albert Bennett put it best: *The organism is a compromise. The result of natural selection is adequacy and not perfection.*

Functional anatomy (or functional *morphology*) is the division of biology that relates an organism's structure to its function. For example, consider the head shape of the hammerhead shark. What functions could a flattened, laterally expanded head possibly serve? Does it make the hammerhead shark a more efficient swimmer (fig. 4.3)? A better predator or migrator?

Figure 4.3. Great Hammerhead in Bimini Bahamas. Note the tight turning radius, enabled in part by using the laterally expanded head as a rudder. (Courtesy of Annie Guttridge)

One of the most important aspects of an adaptation is whether on balance it saves energy. Sharks, like other organisms, require energy for growth, reproduction, and maintenance of the stable internal conditions required for life, like swimming, ventilating the gills, balancing water and salt levels in their bodies, pumping blood throughout their bodies, and numerous other processes at the organ, tissue, and cellular levels. Many adaptations are for acquiring food. And what we are calling *food* may be organisms with their own adaptations to avoid *being food* for something else.

In this chapter, we will examine some anatomical adaptations that make sharks such good predators. In subsequent chapters, we will examine physiological and behavioral adaptations. We start with the Shortfin Mako *(Isurus oxyrinchus)*, which we consider the poster child for adaptive specializations in sharks.

The Ultimate Marine Predator

If you were asked to design a prototype for the perfect marine predator, you might well design a Shortfin Mako. A short list of key adaptations (fig. 4.4) would include:

- A streamlined, almost cylindrical body, with the maximum girth at about 40% of the distance from the snout to the beginning of the tail, positioning which minimizes the forces that act to slow it down.

- Striking countershading, with blue on top and white on the underside, which affords an additional measure of stealth when foraging for prey.

- A caudal fin designed for maximum forward thrust, with a flattened keel at its base, which together facilitate high-speed swimming.

- Metabolic machinery that keeps the body warmer than the water in which it resides, which enables more powerful muscles and faster swimming.

- A high-performance cardiovascular system to make sure the swimming muscles get the energy and oxygen they need and ensure wastes are removed.

- Ram ventilation, big gills, and large gill slits to extract oxygen required to power its metabolic machinery.

- A jaw full of recurved spindle-shaped teeth to grab and hold slippery fish.

- A well-developed sensory system and a large brain for collecting and processing sensory information.

Body Shape

Figure 4.5 shows silhouettes of several kinds of sharks. Notice that several but not all of these sharks possess a streamlined body shape. Given the need to conserve energy, it should be no surprise that this streamlined shape is one of the

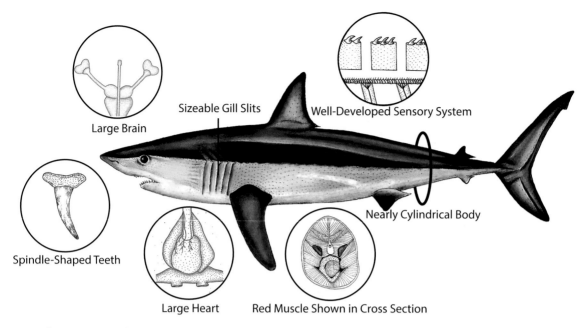

Large Brain

Sizeable Gill Slits

Well-Developed Sensory System

Spindle-Shaped Teeth

Large Heart

Red Muscle Shown in Cross Section

Nearly Cylindrical Body

Figure 4.4. Shortfin Mako showing adaptations for high performance swimming.

most effective for slicing through the water. (But all sharks do not require stream-lining, thus the magnificent horn shark.)

Water is not an easy medium to move through compared to air (it is very diffi-cult to breathe in as well). Water is about 800 times heavier and is 50 times more viscous, or resistant to flowing, than air. But water has one major advantage over air: it supports organisms better. And salt water is more buoyant than fresh water. Larger terrestrial vertebrates must have a heavy skeleton to support their weight on land, whereas most marine vertebrates do not require that extra skeletal mass because of water's buoyant support. (Sharks are heavier than water, but this is due to muscle mass more so than skeletal.)

As a streamlined body moves through the water, a tapered shape minimizes drag that is due to forces that oppose movement. Figure 4.6 depicts two shapes moving through the water. Note that the lines depicting water particles move in straighter lines over most of the body of the more streamlined Crocodile Shark (*Pseudocarcharias kamoharai*), indicating less drag, compared to the Horn Shark (*Heterodontus francisci*), whose body shape disrupts the flow of the water par-ticles more, leading to increased form drag. Which of the shark silhouettes in Figure 4.5 are most streamlined?

In cross-section (a slice through a shark like you would slice a rectangular loaf of bread), most sharks are rounded or slightly laterally compressed. But a few sharks are dorso-ventrally depressed, like the bottom-dwelling Nurse (*Ging-lymostoma cirratum*) and Grey Bamboo (*Chiloscyllium griseum*) Sharks.

In a 1977 study, researchers organized 56 species of sharks into four catego-ries based on body types and tail form.[2] Since then others included batoids and

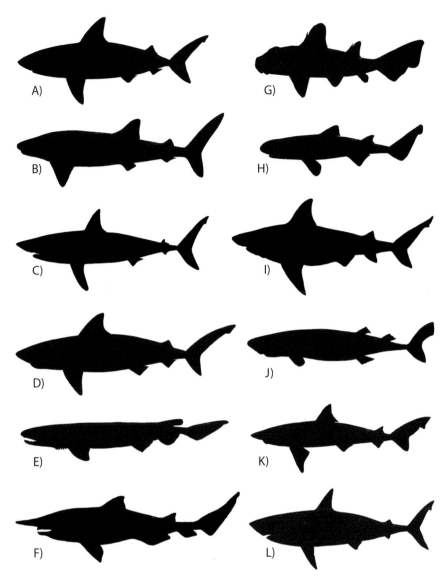

Figure 4.5. Silhouettes of sharks showing diversity of body, head, and caudal fin shapes (body sizes not drawn to scale). Key: (A) Salmon Shark. (B) Whale Shark. (C) Shortfin Mako. (D) Tiger Shark. (E) Frilled Shark. (F) Goblin Shark. (G) Horn Shark. (H) Kitefin Shark. (I) Bull Shark. (J) Cookiecutter Shark. (K) Soupfin Shark. (L) White Shark.

expanded the number of sharks beyond the original 56 and added other morphological characteristics as well as habitat and behavior in their categories[3, 4] (fig. 4.7).

The four main body-type categories are *macropelagic* (type 1), *littoral* (2), *benthic* (3), and *micropelagic / bathic* (*bathic* = deepwater) (4). In addition to body type, these categories include a more comprehensive suite of characteristics. Also, you may conclude, as did the authors making these classifications, that some of the sharks do not seem to belong to their assigned group, and we also agree. Remem-

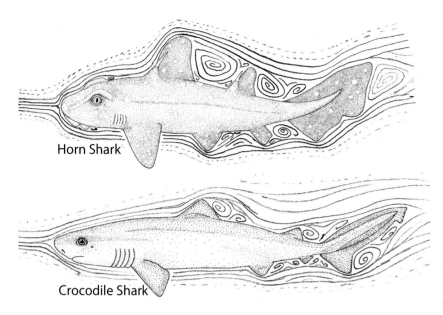

Figure 4.6. Body shape and form drag. Streamlines on a Crocodile and Horn Shark. (Roi Gurka)

ber that organizing > 500 species into only four types will create some unusual bedfellows.

- Type 1: macropelagic sharks, includes members of the families Cetorhinidae (basking sharks) and Lamnidae (mackerel sharks; e.g., the Shortfin Mako) as well as the Whale Shark (*Rhincodon typus*). All are oceanic cruisers with more-or-less conical heads, deep bodies, and caudal fins with high *aspect ratios* (see below), which are associated with sustained increased power and efficiency at high swimming speeds and efficiency alone at lower ones.

- Type 2: littoral sharks, includes members of the families Mitsukurinidae (Goblin Shark, *Mitsukurina owstoni*), Odontaspidae (sand tigers), Alopiidae (threshers), Carcharhinidae (requiem sharks), and Sphyrnidae (hammerheads). Here, *littoral* refers to continental shelf cruisers. Some of these may seem superficially similar to Type 1 sharks, but Type 2 shark bodies are less deep, the head is blunter, and the caudal fin is more asymmetrical with a lower aspect ratio. Swimming speeds of sharks in this category are wide-ranging.

- Type 3: benthic sharks, includes members of the families Heterodontidae (horn sharks), Orectolobiformes (carpet sharks), Hexanchiformes (cow sharks), Pristiophoriformes (saw sharks), and members of the order Carcharhiniformes except those in the family Carcharhinidae. Sharks in this group have a very low tail with a reduced or absent ventral lobe, a large, blunt head, anterior pelvic fins, and more posterior first dorsal fin. Needless to say, sharks thus described are slow-swimming.

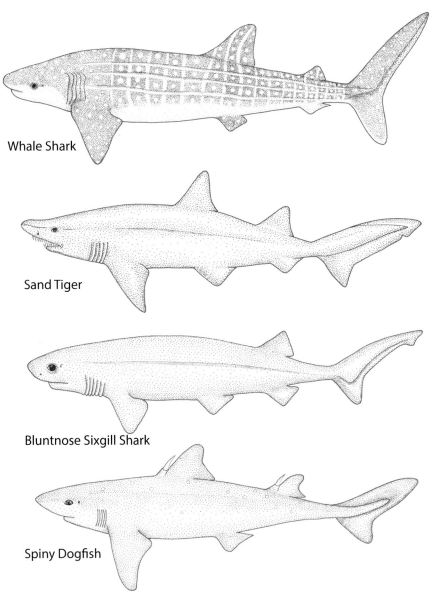

Whale Shark

Sand Tiger

Bluntnose Sixgill Shark

Spiny Dogfish

Figure 4.7. Four basic ecomorphotypes of sharks based on Thomson and Simanek (1977) and Dolce and Wilga (2013). (Thomson, K.S. and Simanek, D.E. 1977. Am. Zool. 17: 343–354 and Dolce, J.L. and Wilga, C.D. 2013. Bull. Mus. Comp. Zool. 161: 79–109)

- Type 4: micropelagic/bathic, includes members of the order Squaliformes. All of these lack an anal fin, have slightly elevated pectoral fin insertions, and possess a large upper caudal fin lobe. These are found in the littoral zone, in the deep sea, and in the shallower water column.

If we expand the list of body types to include dorsoventrally flattened forms, including the remaining batoids and the angel sharks, we could add a fifth body type.

Head

The shark head encases the jaws and serves as scaffolding for most of the senses that account in large part for their prowess as predators. Head shape varies among the > 500 species of sharks (fig. 4.5). The lamnids exhibit a streamlined, conical snout that, in combination with their muscular body and elevated body temperature, allows them to swim more powerfully and faster than other sharks.

The carcharhinids have heads which are wider than high, an adaptation that plays a major role in stable, horizontal swimming, as well as reducing drag when the head moves horizontally (*yaws*) during swimming.

The more laterally expanded head of hammerheads provides both lift and a longer platform for sensory organs. Equally extreme is the Goblin Shark, star of the 2009 sharksploitation movie *Malibu Shark Attack*, which possesses an elongated, flattened snout (see Chapter 3).

Finally, consider the sawsharks and sawfish, beasts with elongated, toothed snouts (sometimes referred to as *bills* but more accurately as *rostra*). Only the former is indeed a shark. Both sawfishes and sawsharks use the toothed rostrum swung side-to-side to disable prey, and the speed and raw power of a swinging sawfish rostrum are not to be taken lightly.

The heads of other sharks may be more pedestrian in that toothiness is not prominent, but these head shapes are equally successful for the particular lifestyles of these sharks. Bullhead (horn) sharks have an attenuated, blunt snout and a jaw with limited gape. These slow-moving, bottom-dwelling sharks eat sedentary or slow-moving bottom invertebrates and do not require streamlining to be happy and well-adjusted.

All of the above sharks have *subterminal* mouths; that is, the mouths are on the lower side of the head (also called underslung jaws). Frilled sharks, angel sharks, the Whale Shark, and the infrequently encountered Megamouth Shark (*Megachasma pelagios*) have *terminal* mouths; that is, at the tip of the snout. The last two of these are filter-feeders in which a large gape allows them to maximize the volume of seawater they intercept as they swim.

Remember that the snout of a Shortfin Mako is conical (see fig. 3.29) and the piglike snout of the Horn Shark (*Heterodontus francisci*) (see fig. 3.18A) is blunted. The external shape is an extension of the internal structure. Figure 4.8 shows tribasic (composed of three parts) snout cartilage of the Shortfin Mako, clearly demonstrating the relationship between the external morphology and the underlying structure.

No section on head shape in sharks is complete without considering a group whose head is among the oddest in larger vertebrates, the hammerheads (family Sphyrnidae). We already mentioned one function: hosting and spreading out an array of senses. Consider one specific sense: vision. In a 2009 study,[5] binocular vision, which results in better depth perception as the brain processes and in-

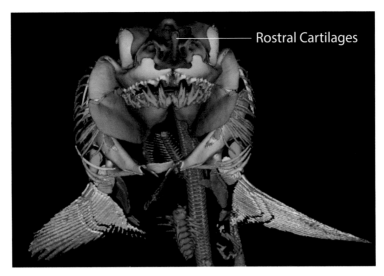

Rostral Cartilages

Figure 4.8. Tribasic snout cartilage (red, *at top*, labeled "rostral cartilages") of the Shortfin Mako shown in this CT scan. (Courtesy of Gavin Naylor/Chondrichthyan Tree of Life Project/sharksrays.org)

tegrates the slightly different view from each eye, was greater on two of three hammerheads (Scalloped [*Sphyrna lewini*] and Winghead [*Eusphyra blochii*]) than that in two carcharhinid sharks. However, the wide separation of the eyes among sphyrnids creates a large anterior blind area close to the head in most species. To compensate, hammerheads swing their heads from side to side as they swim.

The cephalofoil may also increase hydrodynamic lift in sphyrnid sharks, although this conclusion is based on morphology and very surprisingly has not been verified experimentally.

The outline of the hammerhead cephalofoil resembles the wing of an airplane in that it has *camber* (asymmetry between the top and bottom surfaces, creating lift; fig. 4.9A), and its placement brings to mind a canard wing, or forewing (a planing surface anterior to the pectoral fins or a set of smaller wings on an airplane in front of the main wings). Consistent with this interpretation is the relatively small size of the pectoral fins (key planing surfaces in sharks) in sphyrnids (fig. 4.9B). Moreover, the larger the cephalofoil, the smaller the pectoral fins such that the combined surface area of the cephalofoil and pectoral fins remains constant.[6] The cephalofoil may also increase lift in other ways, as described in a 2017 review paper.[7]

The hammerhead cephalofoil may also play roles in both maneuverability and stability. A 2003 study[8] demonstrated that the cephalofoil of Scalloped Hammerheads increased maneuverability and provided hydrodynamic stability compared to the head of Sandbar Sharks (*Carcharhinus plumbeus*). In this study, the hammerheads made sharp turns (those greater than 90°) more frequently

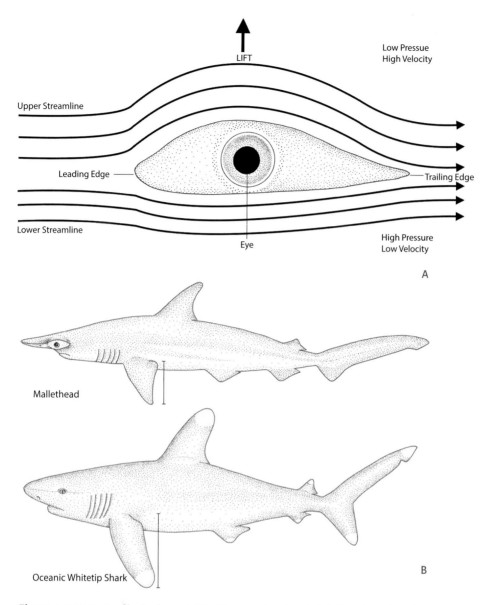

Figure 4.9. (A) Basis of hydrodynamic lift of hammerhead cephalofoil. (Redrawn from fig. 7, Barousse, J. 2009. *Hydrodynamic Functions of the Wing-shaped Heads of Hammerhead Sharks.* Florida Atlantic University) (B) Small pectoral fins of a hammerhead (e.g., Mallethead) compared to those of an Oceanic Whitetip Shark.

than the Sandbar Sharks (see fig. 4.3). Also, their turns were at speeds twice that of the Sandbar Sharks.

The final function of the cephalofoil, prey handling, is based on two published anecdotal observations of this phenomenon in Great Hammerheads (*Sphyrna mokarran*) in which individuals used their heads to pin a Southern Stingray (*Hypanus americana*)[9] and Spotted Eagle Ray (*Aetobatus narinari*)[10] to the bottom before eating them.

Fins

Fins serve a variety of functions, the most noteworthy being stability, maneuverability, and thrust. Refer to Figure 4.10. A shark, or indeed any object in motion in a fluid, may rotate around any of three orthogonal axes (axes at right angles to each other): longitudinal (the long axis of the body), vertical, and horizontal. *Pitch* is the up-and-down movement of the body, like a seesaw or a boat heading through a series of waves hitting it head on. *Yaw* is a side-to-side movement (around the vertical axis). *Roll*, which can make a shark go belly-up, is rotation of the body around the longitudinal axis. Yaw and roll create more problems for a fast-moving body than a slower one.

Sharks possess unpaired median fins and paired lateral fins. Median fins are located along the centerline of the shark, and include one or two dorsal fins, with or without spines, a single heterocercal caudal fin, and a single anal fin (absent in most sharks of the superorder Squalomorpha). The paired fins include the pectorals and pelvics.

The roles of these fins in stabilization are far more complex than it seems at first, and they vary with the size and placement of the fins by species. Moreover, too few studies have been conducted to allow confident generalizations. What has emerged is that the dorsal, and to a lesser extent anal, fins act as keels, allowing quick turns, and primarily controlling roll. The pectoral fins are thought to control pitch and roll, and possibly yaw. The pelvic fins primarily control pitch.

Some of the original work that uncovered the roles of the various fins in sharks was done in what some may consider a gruesome manner by removing one or more fins of a living shark and observing the resulting motion. Other studies have used models of sharks in flumes (water tunnels) or have inferred function based on the underlying structure of the fins or from behavioral observations. These have all contributed to the knowledge base, but until appropriate measure-

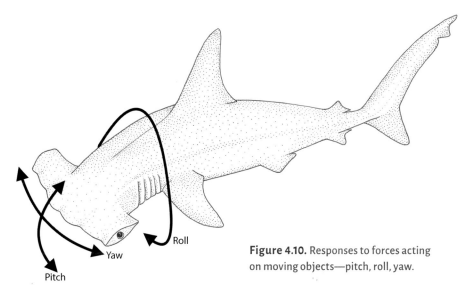

Figure 4.10. Responses to forces acting on moving objects—pitch, roll, yaw.

ments can be made on sharks swimming naturally in their environment, the best information has come from studying sharks swimming in controlled laboratory conditions. The major drawback to this has been the limited number of species studied, which has included principally those most amenable to swimming naturally in captivity, specifically Leopard Sharks (*Triakis semifasciata*), Spiny Dogfish (*Squalus acanthias*), and the bamboo shark *Chiloscyllium plagiosum*.

What is clear from the above studies is that the roles of fins are quite complicated, that studying individual fins separately may not provide the answer to how the fins and body work together to achieve the precise balance between speed, stability, and maneuverability required by sharks and batoids differing in body shape, swimming speed, habitat, and so on.

Basic Structure of Fins

The fins of sharks and rays differ structurally from those of modern bony fishes (figs. 4.11 and 4.12). The latter's fins are composed of internal supporting bony elements (called *pterygiophores*) that support segmented soft rays (called *lepidotrichia*) or harder spines, both of which are derived from scales. The inner supports may be further supported by bones; for example, those in the pectoral and pelvic girdles, or vertebral column.

Cartilaginous skeletal elements also underlie and support the fins of elasmobranchs. The pectoral girdle, which supports the pectoral fins, is composed of the *scapulacoracoid* cartilage (actually, two fused structures, the single *coracoid* bar and paired *suprascapular* cartilages; fig. 4.12). The pelvic girdle consists of the *pubo-ischiadic* (or *ischio-pubic*) bar. The cartilages of both the pectoral and pelvic girdles are not connected to the vertebral column but rather are embedded in and supported by surrounding muscles (in bony fishes, the pectoral girdle typically is attached to the back of the skull). The internal support for the dorsal, anal, and caudal fin in sharks is primarily from the vertebral column.

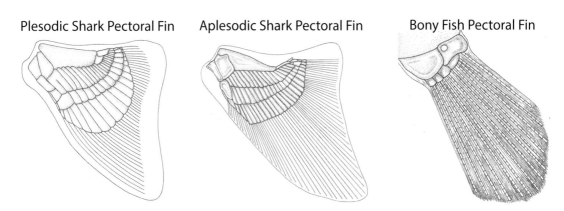

Plesodic Shark Pectoral Fin   Aplesodic Shark Pectoral Fin   Bony Fish Pectoral Fin

Figure 4.11. Fins of bony fishes and sharks, compared. (Redrawn from fig. 5.10, Wilga, C.A. and Lauder, G.V. 2004. In: Carrier, J.C. et al. (eds.). Biology of sharks and their relatives. CRC Press. 139–164 pp)

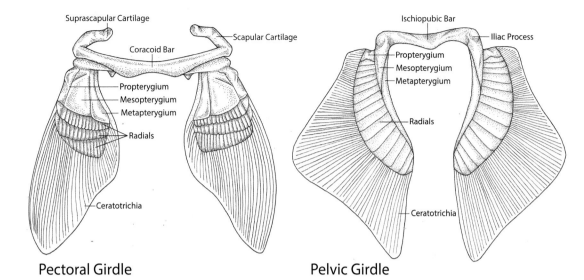

Pectoral Girdle

Pelvic Girdle

Figure 4.12. Pectoral and pelvic girdles of sharks. (Redrawn from Gilbert, S. 1973. *Pictorial Anatomy of the Dogfish.* University of Washington Press)

Moving externally, the fins of sharks have three enlarged cartilaginous *basal* cartilages (termed the *propterygium*, *mesopterygium*, and *metapterygium* and, like bony fishes, collectively called *pterygiophores*) and rows of more distal *radial* cartilages that may vary in length (fig. 4.12; see figs. 1.9 and 1.10).

Most of the area of the fins consists of the unsegmented, soft, flexible fin rays called *ceratotrichia*, which are composed of the elastic protein keratin. In elasmobranch fishes, the ceratotrichia are analogous to the soft rays, or lepidotrichia, of bony fishes. Ceratotrichia are the central ingredient in shark fin soup.

Median Fins

Let us consider the dorsal fin or fins, and the anal fin. We can make quick work of the function of the anal fin, since its precise role is not known. In many fast-swimming sharks (e.g., Type 1), the anal fin (and pelvic and second dorsal as well) is reduced in size, possibly reducing drag. In Type 2 sharks, a group called *continental cruisers* that includes many carcharhinid sharks, the anal fin is moderately sized. These sharks are characterized by being more maneuverable, so it is not a stretch to assume that the anal fin plays a role in maneuverability. Sharks in the order Squaliformes, which are classified as Type 4, lack an anal fin, the functional significance of which is not known.

While the dorsal fin plays roles in stability and maneuverability in all sharks, it is most critical for pelagic sharks; that is, those in the water column like Sandbar, Bull (*Carcharhinus leucas*), and Blacktip Sharks (*Carcharhinus limbatus*). Smaller benthic sharks (e.g., cat and carpet sharks) typically have reduced dorsal fins located more posteriorly on the body, since their bottom-associated lifestyle means they are not continually swimming for long periods.

High speed cruisers like the lamnids have greatly reduced second dorsal fins, while the second dorsal fins of continental cruisers are more moderately sized. This is consistent with the idea that speed and drag reduction are paramount to the former group of sharks, with speed and maneuverability both being important to the latter. An enlarged second dorsal fin could possibly enhance thrust and augment lift.

Consistent with the fact that yaw and roll pose more critical problems to fast-moving objects, slow-swimming sharks like those in Type 3 do not require prominent anterior dorsal fins and thus theirs are more posteriorly placed.

Some sharks (e.g., the frilled and cow sharks) have lost one of their dorsal fins and have their lone dorsal fin far back along the body. The absence of an anterior dorsal fin allows them to roll with less drag when they are feeding and as they roll, they shear pieces of flesh from their prey with their sawlike teeth.

Many bony fishes (e.g., tunas but no sharks) are capable of folding their dorsal fin (and pectoral and anal fins as well) into slots or grooves, and in doing so become more streamlined and decrease drag at higher swimming speeds. These fins can be extended when needed for maneuverability or stability.

Caudal Fin

The caudal fins of bony and cartilaginous fishes alike provide thrust, but there are fundamental differences in both the contribution and direction of this thrust. Recall that the caudal fins of bony fishes are mostly symmetrical (*homocercal*). In some bony fishes, like a barracuda or tuna, the tail provides virtually all of the thrust, whereas in others the caudal fin provides less thrust than other fins (e.g., boxfish, which use pectoral fins; Bowfins, which use the dorsal fin; triggerfish, which use dorsal and anal fins). No matter the degree, the direction of the thrust is horizontal and forward.

Sharks differ in two ways. First, most to all thrust comes from the caudal fin (batoids are another story completely). Second, the asymmetry of the caudal fin influences the direction of this thrust. For the isolated caudal fin, this thrust is generally diagonal; that is, forward and up at about a 45° angle.

If sharks had flexible pectoral fins, then in the same way that when one end of a seesaw goes up, the other end goes down, every stroke of the caudal fin would lift the tail and push the head down, causing sharks to turn summersaults as they moved forward—another case of the tail wagging the dogfish.

But sharks have stiffened pectoral fins, and in many cases horizontally expanded heads, both of which not only resist any downward forces but have been thought to add lift so that sharks swim in a controlled, balanced, straightforward way. Recently, these seemingly safe conclusions have been challenged (see below).

There is considerable variability in the shape of caudal fins. At one extreme are the threshers (see fig. 3.27), whose greatly expanded upper lobe is about half of the length of the body, and is used to herd and stun the small schooling fish that serve as prey. Not surprisingly, the pectoral fins are almost equally exaggerated (see fig. 3.27C) such that the diagonally upward thrust is counterbalanced.

Sharks that rest on the bottom have reduced lower caudal lobes. Sharks like Whites (*Carcharodon carcharias*) and Shortfin Makos have tails that are superficially symmetrical (but still have the vertebrae extending to the tip of the upper lobe), an adaptation that allows them to generate the greater thrust required of these apex predators.

Finally, let us consider the caudal fin of an angel shark (see fig. 3.9). It lives on and over the bottom, and it feeds by surprising its prey from below. A conventional heterocercal fin in which the upper lobe exceeds the length of the lower lobe would be a disadvantage and would tend to send the head in the wrong direction.

This group solved that problem over evolutionary time by essentially flip-flopping the lobes such that the lower lobe is the longer of the two (although the vertebrae still extend to the end of the upper lobe). This is known as a *hypocercal* fin. This adaptation pushes the head up when angel sharks attack from below while resting on the bottom. Evolutionary manipulations like these and others in the animals in this book motivate scientific inquiry and enrich our lives as we marvel at the ingenuity of nature.

Looking more closely into the structure and function of the caudal fin of sharks, let us revisit the concept of *aspect ratios* (fig. 4.13A), which are calculated as follows:

$$A = h^2/s$$

Where,

A = aspect ratio, h = straight line vertical span, and s = surface area of the fin.

Higher aspect ratios provide more thrust relative to drag and are found in species with higher aerobic metabolisms (i.e., high-performance sharks that include lamnids and some carcharhinids). At the same time, caudal fins with higher aspect ratios are not capable of the same level of maneuverability as those with low aspect ratios.

Among all fishes, caudal fins of tunas have the highest known aspect ratios, up to about 7.5 in the Bigeye Tuna (*Thunnus obesus*). Among sharks, aspect ratios vary from 0.76 in the Nurse Shark[11] to > 14 in the Tiger Shark[12] (*Galeocerdo cuvier*; fig. 4.13B).

There is also variation in the aspect ratio of pectoral fins of sharks, with high aspect pectoral fins providing more lift and those with low aspect ratios being superior for slower swimming.

Pectoral and Pelvic Fins

The roles of the paired pectoral and pelvic fins also differ between sharks and bony fishes. Let us consider only the pectorals, since the precise roles of pelvic fins have not been studied in sharks. Among bony fishes, the pectorals are more

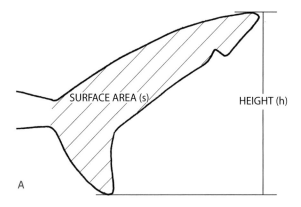

SURFACE AREA (s)

HEIGHT (h)

A

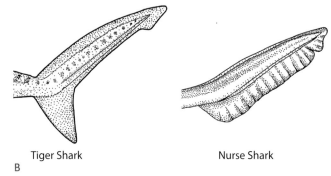

Tiger Shark

Nurse Shark

B

Figure 4.13. (A) Diagram showing how an aspect ratio is determined for a generalized caudal fin. (B) Tiger and Nurse Shark caudal fins, whose aspect ratios are about >14 and 0.76, respectively.

variable in position, are very flexible in most cases, and enable high degrees of maneuverability and, in some cases, propulsion.

In sharks, the pectoral fins are ventrolateral (lower on the flanks) and are supported by sturdy internal structures: the scapulacoracoid cartilage, three enlarged basal cartilages, and more distal radial cartilages (fig. 4.12; see fig. 1.9). Thus, the pectoral fins are relatively stiff. Only the ceratotrichia, which extend outward from the radial cartilages, are slightly more flexible.

Sharks exhibit two different types of pectoral fins, *plesodic* (*ples* = near) and *aplesodic* (*a* = without) (fig. 4.11). In the former, the radial cartilages extend outward more than 50% into the fin, which provides more support and stiffens the fin. In the latter, radials extend less than 50% into the fin, which makes the fin more flexible. Plesodic fins are the more derived type and are found in sharks of the order Lamniformes as well as the carcharhiniform families Hemigaleidae (weasel sharks), Carcharhinidae, and Sphyrnidae, as well as all batoids except those in the families Pristidae (sawfish) and other members of the order Rhinopristiformes (guitarfishes).

Are there functional consequences of having relatively stiff pectoral fins? Yes, although the adaptive advantages are less clear than formerly assumed. What is certain is that the pectoral fins of sharks are incapable of the fine degree of maneuverability of those of most bony fishes.

The traditional view of the functions of the pectoral fins of sharks are that they aid in stability (primarily controlling pitch and roll), that they generate hydrodynamic lift, and that they resist the downward forward forces generated by the heterocercal tail.

However, even these seemingly safe conclusions require some modification. Studies have shown the following:

- While the shark is steadily moving, the pectoral fins play a role in maneuverability by changing the angle of the fin.

- As they increase maneuverability, the pectoral fins may destabilize the body, thus transferring stabilizing of the body to other fins, changing the angle of the entire body.

- Surprisingly, the pectoral fins may not generate lift while the shark is moving. If this turns out to be generally the case, it may imply that most lift is coming from the underside of the body, which is flattened in cross-section anterior to the dorsal fin and more round in cross-section posteriorly in many sharks.

Currently, too few studies, involving only a limited number of species, have been conducted to accept many of these generalities.

Skin and Scales

The dermal denticles of sharks have a number of functions that we have previously listed. Here, we briefly examine how dermal denticles affect the hydrodynamic flow over the body of a shark in motion, a phenomenon that is far more complex than simply channeling water over the body in one direction.

In swimming, the name of the game is achieving the correct balance between thrust and drag while maintaining sufficient lift to stay buoyant. Scales play a central role in that game, along with body shape, metabolic performance, and so on.

Drag refers to the forces that oppose movement, and there are two main types of drag that affect sharks, *pressure* and *frictional*.

Pressure (also called *form*, or *inertial*) drag refers to the resistance caused by a body's shape and profile, and is due to the pressure exerted by moving the water out of the way while swimming, as well as the vortices (whirlpools) formed at the wake from surface projections (fins). Streamlining reduces pressure drag, which increases with swimming speed.

Frictional or *viscous* drag refers to the resistance created by the texture of the skin at the boundary layer formed by water as it flows over the body. The *boundary layer*, typically only one to a few centimeters in thickness in swimming fish, refers to the critical region adjacent to the body in which water flow is affected by friction. Flow that is *turbulent* increases frictional drag, whereas *laminar* (smooth) flow generates less drag.

The main hydrodynamic role of dermal denticles is reducing frictional drag, and the *sine qua non*[13] of frictional drag reduction in sharks are the riblets, or longitudinal ridges creating grooves, of the dermal denticles (see fig. 4.1). Here is how this works.

Frictional drag of a moving object is due to the creation of surface shear. The shear is generated by the interaction of the fish surface with the flow of water. Shear that causes friction also generates vortices that twist, rotate, and interact with each other near the surface of the fish before being ejected or shed into the boundary layer, leading to more chaotic flow (motion-slowing turbulence).

A shark ideally designed so that these vortices would be minimized in size or period, unfortunately, does not exist. One way in which the riblets reduce frictional drag is as follows: The vortices created by the small projections of the dermal denticles are not ejected into the boundary layer but rather continue to interact with the tips of the denticles, as if the tips were sticky, and thus prevent or block turbulent crossflow that might impose additional drag.

It gets more complicated from this point, in part because there are different flow patterns regionally on a shark's body, and in different species as well, due to differences in shape, swimming speeds and styles, as well as the size and distribution of dermal denticles.

It appears that riblets perform two other functions. First, in the posterior part of the body, the dermal denticles actually create turbulence that, counterintuitively, reduces overall form drag. This is accomplished because the total surface area of the body is reduced as the vortices lift off from the riblets and basically create a more streamlined outline for the water to flow more smoothly over the body without separating the boundary later. Secondly, dermal denticles may also enhance thrust.[14]

Variation is the rule when it comes to the size, shape, and pattern of dermal denticles on different species of sharks (fig. 4.14), although dermal denticles all have the same underlying morphology (see fig. 1.7B).

As you might expect, the kinds and distribution of dermal denticles on a shark are associated with its lifestyle.

Studies conducted in 1984[15] and 1986[16] examined dimensions of dermal den-

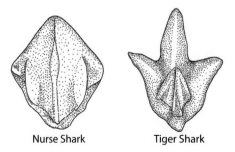

Nurse Shark   Tiger Shark

Figure 4.14. Diversity of scale types: Nurse and Tiger Shark.

ticles in 15 species of sharks separated into three groups based on swimming modes: fast-swimming neritic, intermediate swimmers, and slow-swimming epibenthic.

The dermal denticles of the Nurse Shark (fig. 4.15), a slow-moving species that frequently rests on the bottom, were smaller and lighter than those of continuously swimming species, and the denticles were less-densely packed and did not form the lengthwise channels created by the scale pattern on the other species. Fast-swimming sharks (Shortfin Mako, Scalloped Hammerhead, and Blacktip Sharks) possessed smaller, thinner, and lighter, but more densely packed, dermal denticles. Characteristics of dermal denticles of the intermediate-speed group (Bull, Sandbar, and Blue Sharks, *Prionace glauca*) placed between the other two.

Finally, scales along the flank of the Shortfin Mako were found to be highly flexible, an adaptation the authors of the study[17] that discovered this thought would enhance burst swimming.

How Sharks Swim

Most fish, including sharks, swim by undulating their body in combination with oscillating their caudal fin, a form aptly named *body-caudal fin locomotion*. Among sharks, there is a high diversity of swimming styles and speeds. We already looked broadly at locomotion in each of the four body types above. Here, let us focus on another perspective on modes of swimming.

There are three major modes of swimming in sharks (named for the bony fishes that best exemplify the mode), and additional modes in batoids (fig. 4.15).

1 / *Anguilliform (eel-like)*. A form of axial (along the body) undulation found in slim, elongate sharks like those in the order Orectolobiformes and families Chlamydoselachidae and Scyliorhinidae, in which a wave moves down along the entire body. An entire wave is typically present in the body when swimming fast, with significant lateral head movement (yaw) as speed increases.

2 / *Carangiform / Subcarangiform (jack-like)*. Here, the axial undulations occur in the posterior half of the body, with less than a full wave present. The wave amplitude (the lateral body motion) diminishes as it moves posterior, and there is some head yaw. Subcarangiform swimming is a subdivision and involves more than one half but less than a full body wave, as well as slightly more head yaw than in carangiform. Most sharks are in these categories, including the orders Squaliformes, Carcharhiniformes, and Lamniformes exclusive of the family Lamnidae. Most are pelagic.

3 / *Thunniform (tuna-like)*. Thunniform swimming represents the swimming mode on the opposite extreme. Like a windup fish toy, only the caudal fin and peduncle oscillate, and there is little head yaw. Among sharks, thunniform swimming is restricted to members of the family Lamnidae.

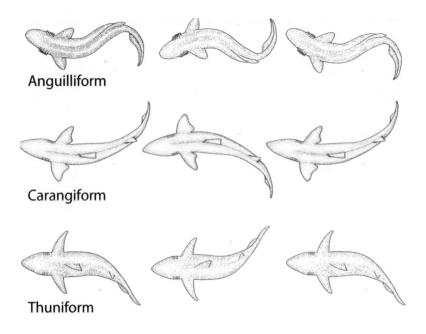

Figure 4.15. Major modes of swimming in sharks and batoids. (Redrawn from fig. 7, Sfakio-takis, M. et al. 1999. J. Ocean. Eng. 24: 237–252)

Undulatory and Oscillatory Swimming of Batoids

For reasons that are self-evident, most batoids are incapable of the swimming modes of any of the preceding categories, except members of the orders Rhino-pristiformes and Torpediniformes, both more basal batoid groups which are axial swimmers, using both their body and tail. Likewise, angel sharks swim more like a torpedinid ray than a shark. Swimming in most batoids is either *undulatory* or *oscillatory*.[18]

In undulatory (also called *rajiform*) swimming, a wave passes down their pectoral fins (wings). Most batoids (all skates, most stingrays) are undulatory swimmers and most undulators are benthic. In oscillatory (also called *mobuliform*) swimming, batoids flap their wings like birds. This small group includes some members of the order Myliobatiformes (cownose, eagle, and manta rays). All are pelagic.

Jaw Suspension

Jaw suspension refers to how the upper and lower jaws connect to the skull and other supporting structures. To get an idea of the importance and role of jaw suspension, place an apple in a big pot of room temperature water and try to pick it up using only your mouth (and not manipulating the apple with your hands)—

an activity called *bobbing for apples*. Not so easy a task for you, is it? Bobbing for apples is not easy for humans, or indeed any terrestrial vertebrates. Why? Because your upper jaw is firmly affixed to your skullcase and does not go anywhere your skull does not also go. This works for you (except when bobbing for apples), given your evolution as consumers of food on land. But would this type of jaw-skull connection work for an aquatic predator like a shark? Of course not. The limited mobility of your human jaw (i.e., its inability to protrude), combined with a rather small gape (opening), limits the scope of your diet. Can you imagine a shark whose jaw is like yours (i.e., solidly attached to its skull), continuously repositioning itself so that its very small mouth is in the right place to catch and bite a prey item in the water? If that were the case, we would very likely be discussing sharks as a minor group or even in the past tense, as evolutionary experiments gone bad, or as dead ends.

How then is the jaw suspension of sharks different from yours (and a chimaera's)? The name of the game in jaw suspension is *cranial kinesis*, which refers to the mobility of the upper jaw, and this is the key to success as an aquatic predator. Refer to Figure 4.16 as you read on.

The jaws of humans and other tetrapods (amphibians, reptiles, mammals, and birds) are *autostylic* (*auto* = self; *styly* = types or styles) or *craniostylic*, meaning that the upper jaw is fused to the cranium. Similarly, the Holocephali (chimaeras) have *holostylic* jaw suspension, where the upper jaw is fused to the cranium and the lower jaw is connected to the upper. Animals with autostylic and holostylic jaw suspension have the least cranial kinesis (jaw mobility) and possess a small gape.

To understand the modes of jaw suspension of sharks, we first need to introduce some anatomical terminology (see fig. 2.10). The upper and lower jaws of sharks are known, respectively, as the *palatoquadrate* and *Meckel's cartilage*. Together, these structures are called the *mandibular arch* (in recognition of their evolution from the first gill arches) but we can simply call them *jaws*. Immediately posterior to the jaws is the *hyoid arch*, which represents the second gill arch that moved forward in evolution to support the jaws to some degree. Even though the first and second arches include the word *gill*, they do not function as true gills, which participate primarily in respiration in sharks.

Here is a brief overview of jaw suspension in sharks and rays, along with the functional significance of each type.

Amphistylic Jaw Suspension

Amphistylic (*amphi* = both) jaw suspension is found in more primitive sharks, including the frilled, cownose, and horn sharks, as well as fossil species.

In amphistylic suspension, the upper jaw (palatoquadrate) is attached by ligaments (strong, fibrous material that connects skeletal elements to each other) at two points to the skullcase (hence the name *amphi*, which means both) and is not tightly fixed to it. Posteriorly, the jaw is provided with very limited support by

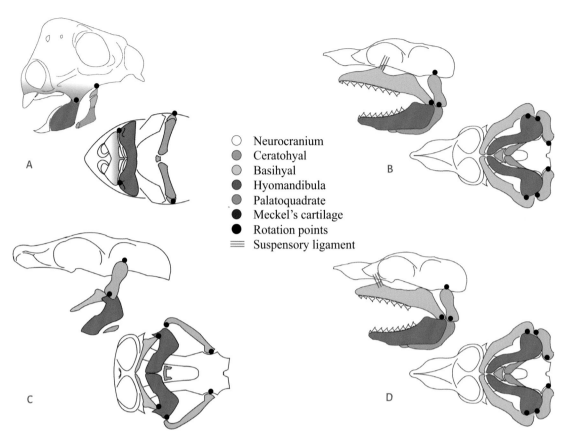

Figure 4.16. Types of jaw suspension in chondrichthyans. (A) Holostylic (*Hydrolagus*, a chimaera). (B) Orbitostylic (*Squalus*, a dogfish). (C) Euhyostylic (*Urobatis*, a batoid). (D) Hyostylic with ethmoid attachments (*Carcharhinus*, a galeomorph shark). Amphistylic suspension is not pictured. (Courtesy of Matthew A. Kolmann)

Legend:
- ○ Neurocranium
- Ceratohyal
- Basihyal
- Hyomandibula
- Palatoquadrate
- Meckel's cartilage
- Rotation points
- ═ Suspensory ligament

the hyomandibular cartilage, part of the hyoid arch. Some of amphistylic sharks (e.g., the sixgills) also possess an orbital process that connects to the skull, which means they are also classified as *orbitostylic*.

The following three types of jaw suspension—*orbitostylic, hyostylic*, and *euhyostylic*—are considered more evolutionarily derived. In these styles, the upper jaw loosens from the chondrocranium and protrudes to varying degrees, which allows the shark to bite from a slightly greater distance and with a significantly enlarged gape.

Orbitostylic Jaw Suspension

Orbit refers to the socket in which the eye resides. Orbitostylic jaw suspension is found in the dogfishes. Here, the orbital process of the palatoquadrate articulates with the chondrocranium, and the hyomandibular arch, while still buttressing the jaw, is shorter.

While the jaw of dogfishes is considered more protrusible than in sharks with amphistylic suspension, particularly the horn sharks, orbitostylic jaw suspension might be expected to result in less jaw protrusibility than in the batoids and the more advanced galeomorph sharks, like Shortfin Makos and Lemon Sharks (*Negaprion brevirostris*), which have greatly loosened suspensoria. However, measurements of upper jaw protrusibility show just the opposite, with the jaw mobility rivaling that of the batoids (see below). There is one drawback to orbitostylic jaw suspension—when the jaw is fully protruded, the maximum gape is substantially reduced, by nearly two-thirds compared to the batoids and galeomorphs.

Hyostylic Jaw Suspension

In modern sharks, including all of the carcharhinid sharks, there is a ligamentous connection (the *ethmopalatine* ligament) between the palatoquadrate and chondrocranium anterior to the orbit, and there is buttressing by the hyomandibular cartilage. This loosening of the jaw allows considerable jaw protrusion, as well as a large gape as the jaw expands both vertically and laterally.

Hyostyly (*hyo* refers to the connection to the hyoid arch) facilitates biting and grasping, shearing, manipulation, and ingestion, and has contributed to the success of modern sharks.

Euhyostylic Jaw Suspension

In the batoids, jaw protrusibility reaches its most extreme case. In euhyostylic (*eu* = true) jaw suspension, there are no ligamentous connections between the palatoquadrate and the skullcase, and the skeletal support for the jaws is provided exclusively by the hyomandibular cartilage at the corners of the jaws.

Euhyostyly equals freedom of jaw movement. A 2014 study[19] showed that batoids exhibit a high level of trophic diversity (different feeding mechanisms). Apparently, not restricting jaw mobility allowed the evolution of muscles to accommodate feeding styles ranging from *suctorial* on smaller, soft prey like worms, to *durophagy*, feeding on hard prey like clams. When we catch batoids on our experimental longlines, we are careful not to pull larger specimens onboard by the leader, lest we damage the rear connections. Imagine being lifted by your tongue, to get an idea of this situation.

Bite Force

Shark bites are among the strongest in the animal kingdom, right? Actually, no and yes. The force of shark bites pretty closely matches that needed to grasp and hold onto their prey, or to take a chunk of flesh from prey, and these forces may or may not be as high as expected. Our expectation of extremely high bite forces likely is based on continuous media reinforcement of that perception on the one hand, or our belief that sharks simply look like they should have high bite forces.

A 1967 scientific study[20] on bite force in sharks used a device called a *gnathodynamometer*, which was encased in approximately 1.4 kg (3 lb) of Bonito flesh, to measure bite force in the adult Tiger Shark and five other species in pens in Bimini, Bahamas. The highest force the researchers found from a single tooth was 60 kg (132 lb) over an area of 2 square mm, which is equivalent to 3000 kg (or more ominously, 3 tonnes) per square cm (42,700 lb per square in). Although these numbers by themselves reflect significant bite force, when they were extrapolated to estimates of the *total jaw area* in the popular press, the numbers grew to nearly 200 tonnes for the total bite force of a 1.8–2.1 m (6–7 ft) Dusky Shark (*Carcharhinus obscurus*). This number is misleading because extrapolating from the force generated by the tip of a single tooth to an entire jaw assumes that *every* square mm of the jaw has the same force as that of the single, measured tooth and that all of these teeth come in contact with a prey item. Recent studies have provided estimates of anterior bite force for a number of species (fig. 4.17).

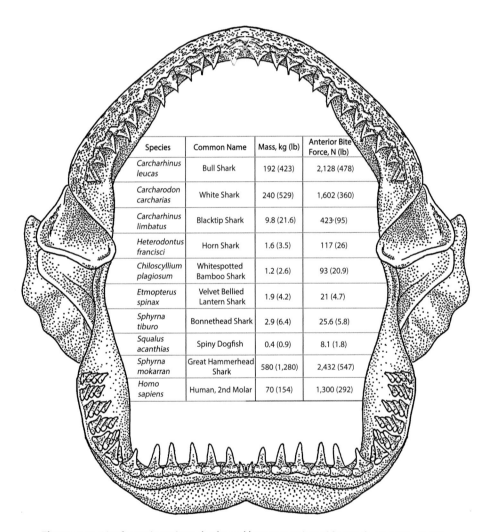

Species	Common Name	Mass, kg (lb)	Anterior Bite Force, N (lb)
Carcharhinus leucas	Bull Shark	192 (423)	2,128 (478)
Carcharodon carcharias	White Shark	240 (529)	1,602 (360)
Carcharhinus limbatus	Blacktip Shark	9.8 (21.6)	423·(95)
Heterodontus francisci	Horn Shark	1.6 (3.5)	117 (26)
Chiloscyllium plagiosum	Whitespotted Bamboo Shark	1.2 (2.6)	93 (20.9)
Etmopterus spinax	Velvet Bellied Lantern Shark	1.9 (4.2)	21 (4.7)
Sphyrna tiburo	Bonnethead Shark	2.9 (6.4)	25.6 (5.8)
Squalus acanthias	Spiny Dogfish	0.4 (0.9)	8.1 (1.8)
Sphyrna mokarran	Great Hammerhead Shark	580 (1,280)	2,432 (547)
Homo sapiens	Human, 2nd Molar	70 (154)	1,300 (292)

Figure 4.17. Bite forces in various sharks and humans. (Adapted from Habegger, M.L., Motta, P.J., Huber, D.R. and Dean, M.N. 2012. Zoology 115: 354–364 and Zhao, Y. and Ye, D. 1994. Journal of West China University of Medical Sciences 25: 414–417)

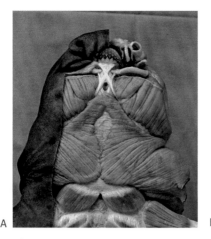

Figure 4.18. (A) Ventral view of a Horn Shark showing hypertrophied jaw musculature. (B) Side view of a Horn Shark. (Physiological Research Lab, Scripps Institution of Oceanography)

How do these forces compare to those of durophagous species, those that eat hard prey and typically possess hypertrophied jaw musculature (fig. 4.18), and teeth designed for grinding? Anterior bite force for a 1.6 kg (3.5 lb) Horn Shark (*Heterodontus francisci*) was only 117.2 N (26.3 lb), but maximum posterior bite force on the posterior molariform teeth was 338 N (76 lb). These seem low compared to measurements for Bull and White Sharks, but when the mass of the Horn Shark is considered, they represent one of the highest among sharks. A 2015 study[21] calculated that the Cownose Ray (*Rhinoptera bonasus*), a member of an entirely durophagous order (Myliobatiformes), would have a bite force of > 500 N (112 lb).

Teeth

Sharks have a variety of methods of ingesting food, and these are associated with variation in tooth morphology. Look at the teeth shown in Figure 4.19. Serrated teeth, found in White and Tiger Sharks, for example, are for shearing, as the shark extends its protrusible jaws into its large prey and swings from side to side, in the process removing a large chunk of the prey, perhaps from a dolphin, seal, turtle, or even another shark.

Some sharks eat smaller fishes that can be swallowed whole, in which case grasping is more important than shearing, and the teeth tend to be long and slender and not serrated (e.g., like those in the Shortfin Mako or Lemon Shark). The scientific name of the genus for the Lemon Shark refers to this: *Negaprion* means *no serrations* in Latin.

Some sharks and rays, including the Nurse Shark, Horn Shark, and Cownose Ray feed on hard-bodied prey, including scallops, clams, snails, crabs, sea urchins, and others. Eating hard-bodied prey (*durophagy*) does not require teeth that tear or snag, but rather teeth that are broader and smaller, and perhaps are

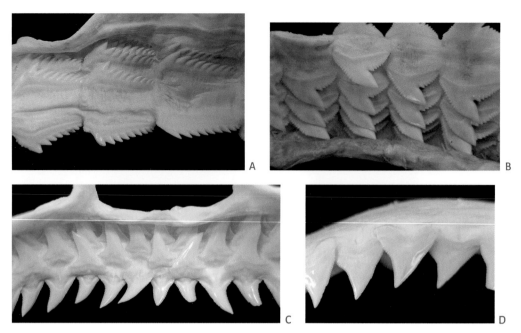

Figure 4.19. Diversity of teeth in sharks and rays. (A) Bluntnose Sixgill Shark. (B) Tiger Shark. (C) Blue Shark. (D) Bull Shark.

even organized into crushing plates. Students in our *Biology of Sharks* course in the Bahamas feed wild Southern Stingrays by hand, and we urge the students to let their fingers linger in the mouths of the smaller rays and feel the power of their crushing jaws, which always loosen their crushing force before it becomes uncomfortable for the student.

In many sharks, the teeth in the upper and lower jaws differ (fig. 4.20). In the Caribbean Reef Shark and most other carcharhinids, the teeth in the lower jaw grasp while those in the upper jaw slice. While not nearly as robust as those of the stingrays, the jaws of these sharks are substantially calcified to limit deformation (bending) during a bite. However, in some species, such as the Tiger Shark and cow sharks, the jaw cartilages are surprisingly weakly calcified and bend easily. This is why if you see the dried jaw of a Tiger Shark, it is usually bent and deformed compared to that of, say, a Bull Shark. Tiger Sharks and cow sharks are relatively homodont (same teeth on the top and bottom jaw) and their teeth possess substantial serrations. The weak calcification permits the jaws to bend across the surface of their prey (e.g., sea turtles in the case of Tiger Sharks), thus allowing most of the functional teeth to make contact. When coupled with shaking of the head or rolling of the body, this permits large chunks to be removed from the prey.

Finally, let us briefly talk about shark bites which, we have said, despite relentless hype to the contrary, are extremely uncommon events.

What exactly does a shark bite look like? Well, it depends. Occasionally, on our research longlines we will pull in a shark or Red Drum with a well-defined arc of flesh removed (fig. 4.21). These clearly are the bites of a shark. We have been bitten a few times while handling small and occasionally large sharks. In contrast, the *wound* on the ankle of the adult film star[22] we referred to in Chapter 1, looks nothing like a shark bite.

Further discussion of the role of teeth in the trophic ecology of sharks can be found in Chapter 9.

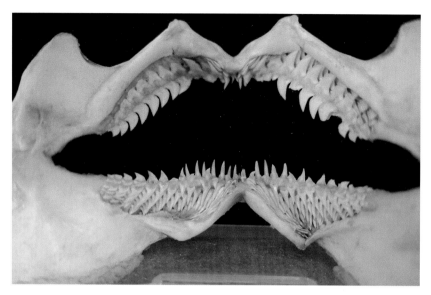

Figure 4.20. Variation in teeth of upper and lower jaw in a shark, in this case a Snaggle-tooth Shark.

Figure 4.21. Typical shark bite, in which arcs of flesh have been removed, in this case from another shark.

Gills

The organs of respiration in sharks are gills (fig. 4.22), which also play a role in ion exchange, pH balance, and excretion of nitrogenous waste products.[23] Here, we will focus on their respiratory function. Like respiratory structures from lungs of mammals to the gills of bony fishes, the gills of sharks have the following general characteristics:

- A short path for diffusion of respiratory gases (oxygen in, and carbon dioxide out)
- A high density of capillaries (i.e., extensive vascularization)
- A large surface area over which the exchange of gasses occurs
- Continuous wetness (a thin film of water in air-breathers)

The subclass Elasmobranchii translates to *plate gills*,[24] referring to the stacked architecture of the gills of this group (fig. 4.22), specifically the platelike *interbranchial septa* and *holobranchs* (gill arches; see below). Here, the gill filaments are connected to these interbranchial septa, unlike the condition in bony fishes, in which there is no septum and the gill filaments are elongated and connected to the gill arch only at one end (fig. 4.22).

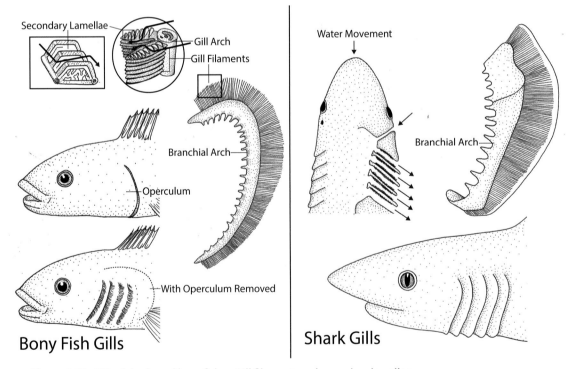

Figure 4.22. Gills of sharks and bony fishes. Gill filaments and secondary lamellae, shown for the bony fish, are also found in the shark gill. (Redrawn from fig. 1, Wegner, N.C. et al. 2012. J. Exp. Biol. 215: 22–28)

The four characteristics of respiratory structures are necessary but not sufficient to extract the required oxygen and eliminate the excess carbon dioxide. The last required ingredients are anatomical specializations within the gills that allow the water in the environment, which is presumably oxygen-rich or at least has a higher oxygen concentration than the blood, to flow over the gills in a direction exactly opposite to the flow of blood in capillaries in the gills.

These specializations, which are found in both chondrichthyans and bony fishes, are the secondary lamellae (also called *respiratory leaflets*), which are the only sites of respiratory gas exchange (fig. 4.22). The mechanism of blood flow through these secondary lamellae opposite water flow on the other side of a thin membrane is the same countercurrent flow principle (fig. 4.23) we see in muscles of endothermic sharks (see Chapter 7). Basically, water enters through the mouth (and spiracles in some) and flows over the secondary lamellae before exiting through the gill slits.

Most elasmobranchs have five bilateral, gill-bearing branchial arches and associated gill slits. This translates into nine *hemibranchs* (literally, *half-gills*, the gill filaments and lamellae on half of each gill arch, or *holobranch*), since the first gill arch (the *hyoid*) has filaments only on its posterior, and arches two through five have filaments on both the anterior and posterior sides. Elasmobranchs with six or seven gills have 11 or 13 bilateral hemibranchs, respectively.

There is variation in gill structure in elasmobranchs. In Collared Carpet Sharks (family Parascyllidae), there is no hemibranch on the posterior of the fifth gill arch.[25] Parascyllids occupy well-oxygenated waters, and have low oxygen requirements, so this hemibranch may not be missed.

The surface area of the gills for gas exchange also varies among elasmobranchs. The strongest correlation with gill surface area is with activity level /

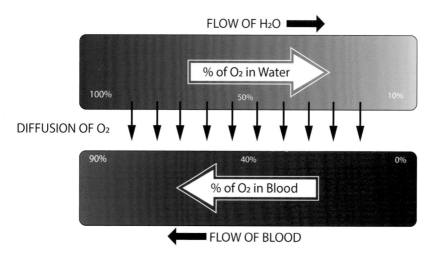

Figure 4.23. Diagram depicting the engineering principle *countercurrent exchange*, as it occurs in the gills of fishes. Values for oxygen percentages represent relative amounts.

metabolic rate. The high-performance lamnid sharks thus have the highest gill surface area to body mass ratios, but the champion is the Bigeye Thresher (*Alopias superciliosus*),[26] perhaps because the latter is a diurnal vertical migrator that spends time in hypoxic waters of the oxygen minimum zone, where a bigger oxygen collector is needed. The gill surface area of Shortfin Makos, expressed as cm^2/10 kg body mass, is about 50,000 cm^2 (7750 in^2), which is close to 2.5 times that of Blue and Dusky Sharks and an entire order of magnitude higher than that of Atlantic Stingrays (*Hypanus sabina*) (2700 cm^2, or 418 in^2),[27] although it is only about half that of a similar-sized tuna.[28]

There are also differences in the dimensions of the water-blood barrier between more and less active elasmobranchs. The former (e.g., the Shortfin Mako) have a diffusion barrier to oxygen as small as one-tenth that of benthic species (e.g., the Small-spotted Catshark [*Scyliorhinus canicula*]).

In addition to increased gill surface area and thinner barriers to diffusion, sharks capable of sustained high speeds, most notably lamnids (mackerel sharks) and some alopiids (threshers), have other modifications. First, their bodies have evolved to allow space in the branchial cavity to accommodate larger gills. For example, the pectoral fins of lamnid sharks are placed more posteriorly on the body, which allows the branchial chamber to be longer and thus more capacious. In alopiids, the gill chamber has expanded more cranially. Fast-swimming sharks also have larger gill slits, all of which are of similar size in each species.

In light of the high swimming speed of makos and their use of ram ventilation, their gills have modifications that stabilize them. They are stiffened by the interbranchial septum, and the filaments have one or two vascular sacs (fig. 4.23). Otherwise the basic support structure is similar to that of other elasmobranchs.

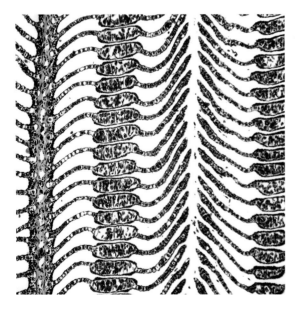

Figure 4.24. Stained section through gill filaments of a 9 kg (19.8 lb) Shortfin Mako showing red blood cells in vascular sacs, which may stabilize the filaments when water movement through the gills is increased during ram ventilation while swimming fast. (Courtesy of Nicholas C. Wegner)

Concluding Comments

This chapter has highlighted only a limited sample of the anatomical adaptations of a small number of sharks, in part because of a lack of research, especially on less common species or those not amenable to captivity. As these deficiencies are overcome, expect extensions and even corrections of the current knowledge of swimming and feeding in sharks and rays.

NOTES

1. Oeffner, J. and Lauder, G.V. 2012. J. Exp. Biol. 215: 785–795.

2. Sambilay Jr., V.C. 1990. Fishbyte 8: 16–20.

3. Compagno, L.J. 1990. Env. Biol. Fishes 28: 33–75.

4. Dolce, J.L. and Wilga, C.D. 2013. Bull. Mus. Comp. Zool. 161: 79–109.

5. McComb, D.M. et al. 2009. J. Exp. Biol. 212: 4010–4018.

6. Thomson, K.S. and Simanek, D.E. 1977. Am. Zool. 17: 343–354.

7. Fish, F.E. and Lauder, G.V. 2017. J. Exp. Biol. 220: 4351–4363.

8. Kajiura, S.M. et al. 2003. Zool. 106: 19–28.

9. Strong, W.R. et al. 1990. Copeia 1990: 836–840.

10. Chapman, D.D. and Gruber, S.H. 2002. Bull. Mar. Sci. 70: 947–952.

11. Nelson, Emily. 2018. Functional morphology and individual variation in movement performance of an apex marine predator. Open Access Theses. 704.https://scholarlyrepository.miami.edu/oa_theses/704. (Accessed 9/9/19).

12. Dean, B. and Bhushan, B. 2010. Philos. T. Roy. Soc. A 368: 4775–4806.

13. This pesky, pedantic Latin phrase rears its ugly head again. Recall that it means *the essential element without which something is impossible.*

14. Oeffner, J. and Lauder, G.V. 2012. J. Exp. Biol. 215: 785–795.

15. Raschi, W.G. and Musick, J.A. 1984. Hydrodynamic aspects of shark scales. Special report in applied marine science and ocean engineering; no. 272. Virginia Institute of Marine Science, College of William and Mary. https://doi.org/10.21220/V5TQ6. (Accessed 9/9/19).

16. Raschi, W. and Elsom, J. 1986. Comments on the structure and development of the drag reduction-type placoid scale. In: Indo-Pacific Fish Biol. Proc. Second Int. Conf. on Indo-Pacific Fishes. Ichthy. So. Japan. 408–424 pp.

17. Motta, P. et al. 2012. J. Morph. 273: 1096–1110.

18. Rosenberger, L.J. 2001. J. Exp. Biol. 204: 379–394.

19. Kolmann, M.A. et al. 2014. J. Morph. 275: 862–881.

20. Snodgrass, S.M. and Gilbert, P.W. 1967. Gilbert, P.W. et al. (eds.). *Sharks, skates and rays.* Johns Hopkins U. Press. 331–337 pp.

21. Kolmann, M.A. et al. 2015. J. Anat. 227: 341–351.

22.https://www.thesun.co.uk/news/3528324/porn-star-molly-cavalli-shark-attack-fake/. (Reluctantly Accessed 9/9/19).

23. An excellent comprehensive reference: Wegner, N.C. 2015. In: Shadwick, R.E. et al. (eds.). *Physiology of Elasmobranch Fishes: Structure and Interaction with Environment.* Fish Physiol. 34: 101–151. Academic Press.

24. This is sometimes interpreted as *strapped gills*, in reference to the tips of the gills not being free but rather are constrained.

25. Goto, T. et al. 2013. Zool. Sci. 30: 461–468.

26. Wootton, T.P. et al. 2015. J. Morph. 276: 589–600.

27. Grim, J.M. et al. 2012. Fish Physiol. Biochem. 38: 1409–1417.

28. Wegner, N.C. et al. 2010. J. Morph. 271: 937–948.

5 / Sensory Biology

Introduction: What is Physiology?

Physiology is the study of biological function. One of the major themes in physiology is *homeostasis* (*homeo* = same; *stasis* = steady or staying), which refers to the roles of an organism's regulatory systems working in concert to maintain a relatively *constant internal environment* or *stable internal function*, even in the face of often-wide changes in the external environment. Physiology is a somewhat neglected topic and one rarely encountered in detail in books for non-specialists. We begin with sensory physiology in this chapter.

Sensory Physiology

You have likely heard that sharks are *swimming noses* capable of detecting a drop of blood from a distance of several miles. This statement is not untrue, but it requires qualifying. Yes, sharks have an extremely well-developed sense of smell, but they are not the only bloodhounds of the ocean, since many bony fishes have similar olfactory sensitivities. For a shark to detect a drop of blood, specific molecules in that blood capable of stimulating the shark's olfactory receptors must physically come in contact with these receptors, and this requires time for the blood to reach the nasal organs of the shark.

Sharks have a suite of senses that allows them to interpret their surroundings and respond appropriately. This suite encompasses those that we humans and other vertebrates possess, and for good measure adds one totally foreign (i.e., the ability to detect vanishingly small electrical fields). These sensors specifically enable a shark to detect prey, predators, conspecifics (members of the same species), other organisms, and obstacles. They also allow a shark to orient itself.

A *sense* can be thought of as a group of specialized cells that work as part of a system to detect some form of physical energy or substance in the environment, and transmit information (signals) about these to the central nervous system, which then may initiate a response.

The five traditional human senses are sight, hearing, touch, taste, and smell. Another way of describing these senses is by the form of energy they detect. Vision is a form of *photoreception* (i.e., it detects energy in the form of light waves). Smell and taste represent types of *chemoreception*, responses to chemical stimuli. And touch and hearing represent *mechanoreception*, a means of detecting pressure or distortion. Sensory physiologists also recognize systems that detect other forms of energy or are specialized forms of the above three types of reception. *Thermoreception* detects temperature. *Proprioception*, also called *equilibrioception*, is the sense of one's own position and equilibrium. *Nociception* refers to the perception of pain. Discussion of these is beyond the scope of this book.

Sharks and rays possess photoreceptors, chemoreceptors, and mechanoreceptors. They also possess *electroreceptors*, sensors capable of detecting electrical fields, and possibly also *magnetoreceptors*, which detect, as you might suspect, magnetic fields.

These sensory receptors all have several factors in common. First, they must be capable of detecting a physical or chemical stimulus. Each kind of receptor responds to its own *adequate stimulus* (i.e., the type and intensity / concentration of stimulus to which the receptor is tuned).

Once the stimulus is detected, cells in the receptor *transduce* the stimulus, meaning that the detection is converted into a nerve impulse or impulses that travel to the brain or the spinal cord. There, the signal is interpreted, processed, and often integrated with information from other senses before an appropriate response is determined, typically nearly instantaneously.

Mechanoreception

In 2015, a Dutch Caribbean Coast Guard helicopter hovered over a shipwreck survivor in the water near Aruba, and was about to rescue him when, according to *The Guardian* newspaper,[1] a shark bit and killed him. Immediately, speculation surfaced that the low-frequency *whomp-whomp-whomp* of the helicopter's rotors played a role in attracting the sharks. Is that possible? What sounds attract sharks? We will answer these and other questions below.

Mechanoreception refers to a response to pressure, distortion, or displacement, including those associated with touch, sound, muscular contraction, and posture. The mechanoreceptors in sharks include the inner ear, lateral line, spiracular organ, touch receptors, proprioceptors, and hair cells. These all share similarities in the morphology of the receptor cells and have been collectively called the *acoustico-lateralis* (or *octavo-lateralis*) system.

Let us start with the ears. Yes, sharks have ears, though you would see them only upon dissection, the only external sign being small, paired *endolymphatic pores* (fig. 5.1A). The endolymphatic pores open into endolymphatic ducts that lead to the cartilaginous otic capsule, which houses the inner ears, each of which is typically described as a *membranous labyrinth*.[2] This labyrinth is structurally divided into upper and lower portions.

The upper part of the membranous labyrinth is composed of three *semicircular canals* arranged at right angles to each other. The lower portion has three sacs, or chambers (or organs), the *sacculus*, *utriculus*, and *lagena*. Each of these chambers has a sensory epithelium called a *macula*. A fourth macula, essentially a spot, called the *macula neglecta* (meaning *neglected spot*) is found at the intersection between the upper labyrinth (the semicircular canals) and the lower labyrinth. All parts are filled with a fluid called *endolymph* (fig. 5.1B).

The three lower chambers of sharks and rays contain *otoliths*, or *otoconia*, calcium carbonate structures the size of sand grains. The otoliths of sharks differ from those of bony fishes and are more specifically called *stataconia* to differentiate them. Stataconia do not lay down concentric rings and thus do not record age as otoliths do in bony fishes.

At the base of each of the upper labyrinth's semicircular canal is a swelling called an *ampulla*. There are bundled sensory hairs, the actual mechanoreceptors (in other words, the *functional unit*) on which the entire system is based, on a ridge called the *crista* inside the ampullae, as well as in the maculae of the three sacs in the lower chamber.

The tops of the hair cell bundles in the semicircular canals are embedded in a transparent structure called the *cupula*, which consists of mucilage, or thick mucopolysaccharide jelly (also called *gel*). Each bundle of hair cells includes a *kinocilium*, which projects the highest, and as many as 60 smaller *stereocilia*. These are connected at their base to sensory nerves, which will convey the information to the central nervous system when these cilia bend in response to a stimulus or stimuli (see below).

Let us discuss the upper labyrinth's semicircular canals first. These are organs in the *vestibular system*; that is, the primary structures concerned with equilibrium or balance (e.g., controlling pitch, roll, and yaw) rather than hearing. The semicircular canals detect angular acceleration; that is, the rate of change of the velocity of a rotating object, the object here being the shark. Thus, these organs sense what the shark itself is doing and control the position of the shark in space, and they do not detect any events in the environment. For pelagic sharks in clear water, vision also plays a role in equilibrium, as does the sense of touch for benthic sharks.

During movement, there is an *inertial lag* of the gel compared to the mechanoreceptive hair cells. Inertial lag, in accordance with Newton's First Law, is a delay

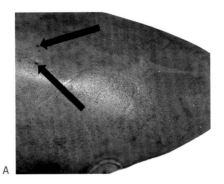

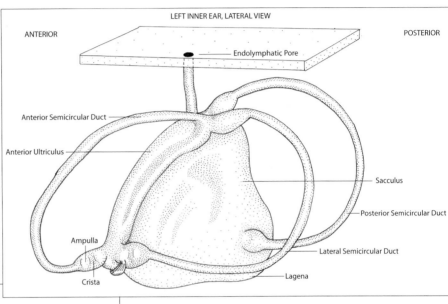

LEFT INNER EAR, LATERAL VIEW

ANTERIOR

POSTERIOR

Endolymphatic Pore

Anterior Semicircular Duct

Anterior Ultriculus

Sacculus

Posterior Semicircular Duct

Ampulla

Lateral Semicircular Duct

Crista

Lagena

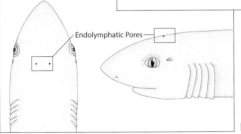

Endolymphatic Pores

Figure 5.1. (A) Endolymphatic pores, external openings of the ears of sharks, from a Blacktip Shark. (B) Internal anatomy of the shark ear. (Redrawn from Gilbert, S. 1973. *Pictorial Anatomy of the Dogfish*. University of Washington Press)

in the flow in response to a force, and it occurs in any fluids or matter of different densities. In sharks, this includes all of the extracellular fluids (e.g., endolymph [including gel], pericardial fluid, and blood), food in the digestive tract, and so on. In this case, in response to movement, the hair cells will move at the same speed as the body (muscles, skeleton, organs, and so on) does, but the endolymph gel *lags behind*. The relatively faster movement of the hair cells embedded in the gel compared to the gel itself causes the hair cells to stick to the gel and to bend, which sends signals to the central nervous system on the rotation of the head such that equilibrium, or balance, is maintained.

The organs of hearing in elasmobranchs are exclusively the inner ear chambers (sacculus, utriculus, and lagena) plus the macula neglecta (bony fishes also have accessory structures, e.g., swim bladders, used for hearing).

Let us turn to the role of the sensory epithelium (the maculae) of the inner ears, which play the primary role in hearing but are also involved in balance. Like the semicircular canals, the maculae also possess a patch of hair cells. These hair cells are embedded in gel surrounding the stataconia that cover the maculae. They act as the inertial elements in the maculae, and function in a manner analogous to the endolymph in the semicircular canals. Mechanical deformation of these produces a signal that informs the central nervous system of straight-line acceleration as well as sound. The macula neglecta, which lacks stataconia but does have gelatinous cupulae, is also important to hearing.

At this point, it is necessary to describe the rudiments of the physics of sound transmission in water. The same physical principles apply to sound in water and air, but there are differences based on their physical properties. Sound travels faster in a solid material than in air. While both water and air are fluids, because water is much denser than air and is negligibly compressible, sound travels about 4.8 times faster in seawater (1530 m sec^{-1}, or 4020 ft sec^{-1}) than air (330 m sec^{-1}, or 1083 ft sec^{-1}) at 35°C (95°F) and at any given frequency the wavelength is about 4.8 times longer.

Sound starts with a vibration or disturbance that propagates within a medium and moves as the energy of one water particle transfers to the adjacent one. With some exceptions, sound in water does not rapidly attenuate because of water's relative incompressibility and thus propagates long distances. Low frequencies travel farther than high ones. Finally, the sources of sounds in water are difficult to localize.

Hearing

The first audiogram in sharks, in which they were presented with sounds of different frequencies, was conducted in 1961 on Bull Sharks (*Carcharhinus leucas*).[3] These sharks could hear frequencies between 100 and 1500 Hz (the range for humans is reported as 20–20,000 Hz) and were most sensitive to sound in the 400–600 Hz range. Audiograms on Lemon (*Negaprion brevirostris*) and Horn Sharks (*Heterodontus francisci*) extended the lower end of the range to 10 Hz.

In the 1980s, advanced neurological methods produced audiograms for a number of sharks and batoids slightly different from those above, with hearing in the range of 40–1500 Hz and highest sensitivity in the 200–400 Hz range.

These studies determined the range of *sensitivities* of the shark and batoid ear, but they did not paint a picture of the acoustic world in which sharks live (i.e., the kinds of sounds sharks respond to in nature). The first study to understand the types of sounds that attracted sharks was conducted in 1963.[4] Noting that sharks were well-known to be attracted to the sounds of struggling fishes, researchers recorded the sounds from fish, most successfully from a Black Grouper, and acoustically characterized them. Then they played sounds like those of the struggling fish on coral reefs where no sharks were visible. The sounds that were successful in attracting sharks were low-frequency (< 60 Hz) pulses, similar to those recorded from the speared, struggling fish. Using these sounds, the researchers attracted Bull Sharks, hammerheads (*Sphyrna* spp.), Lemon Sharks, a Tiger Shark (*Galeocerdo cuvieri*), and additional unidentified carcharhinids. Other sounds were not attractants. Subsequent studies confirmed these results.

From how far away can sharks detect sounds? Let us revisit the physics of sound in water again. Sound has both particle motion and pressure components. The former produces sound that is rapidly attenuated as it moves from its source and is known as *acoustic near field* sound. Depending on the frequency, this is usually under 15 m (49 ft). Sound pressure extends up to hundreds of kilometers from the source and is thus *acoustic far field* sound.

The architecture of the inner ear of elasmobranchs is poised to hear only near field sounds (i.e., it is sensitive to particle motion). Bony fishes with swim bladders are also capable of detecting sound in the far field.

Thus, we are left with a paradox—the contradiction between the verified observations that sharks are capable of responding to sounds in the kilometer range, and the structural limitations of their anatomy that apparently limit their range even more. Some researchers have concluded that sharks orient to distant sound sources using a system that is more sensitive to particle motion than had been thought.

Irregular, low frequency sounds attract sharks, but do any sounds repel them? Yes, some sounds represent aversive stimuli. Lemon Sharks withdrew if sounds that might have been otherwise attractive were initiated when an individual was in sight and approaching. Similarly, Silky (*Carcharhinus falciformis*) and Oceanic Whitetip Sharks (*Carcharhinus longimanus*) withdrew if the intensity of the sound was changed when played back. Sounds simulating vocalizations of Orcas, natural predators of some sharks, likewise initiated withdrawals in Silky, Lemon, and other carcharhinid reef sharks, but not White Sharks. That naivete may have doomed the White Sharks (*Carcharodon carcharias*) killed by Orcas in South Africa recently.

Sharks thus are attracted to sounds of prey and some are repelled by sounds that may signal threats. How do they use sound during their daily activity when these sounds are absent? Put another way, do sharks also use hearing, perhaps in concert with other senses, to continually assess their environment? Sharks live in an environment that can be relatively noisy. Sounds under a frequency of 1 kHz in the marine environment that a shark might encounter and perceive include those associated with swimming fish schools, fish sounds, and waves.

Back to the helicopter and the unfortunate swimmer. While the shark that killed the shipwreck victim could have been attracted to the swimmer using any of its suite of senses, yes, the sounds of the hovering helicopter unfortunately could have played a role.

Other Mechanosenses

You learned above that water conveys sensory information from mechanical disturbances (sounds and movements) at much greater distances than in air. Sharks and batoids, and bony fishes as well, are poised to sense this information via a series of external mechanoreceptors. In elasmobranchs, these include the lateral line, pit organs, spiracular organs, and vesicles of Savi (batoids only).

In all of these mechanoreceptors, the basic unit is the *neuromast*, which is a hair cell covered by a gelatinous cupula, similar to the auditory and vestibular hair cells in the inner ear. These mechanoreceptive neuromasts exist in an array of configurations that include canals with pores connecting to the environment, canals lacking pores, free (or superficial) neuromasts in grooves or pits, and subdermal or spiracular pouches. Together they allow the shark to detect predators, prey, other organisms, and conspecifics; to detect water flow and other physical characteristics of its surroundings; and for proprioception (perception of aspects of its own body). We discuss these other mechanoreceptors in sequence below. Note there is overlapping, integrated function among these sensory systems.

The lateral lines of sharks may either have pores, and thus be connected to the environment, or they may be non-pored (sealed and not open to the water). The pored canals run bilaterally along the flanks from the posterior of the head near the endolymphatic pores to tip of the tail (fig. 5.2). In batoids, the pored lateral line is on both dorsal and ventral surfaces but is concentrated on the former. Along the lateral line, the neuromasts are lined up in a water-filled tube, or canal,

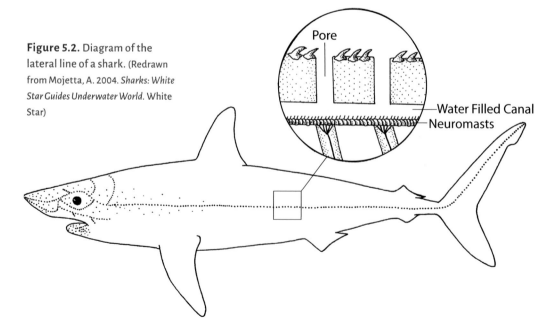

Figure 5.2. Diagram of the lateral line of a shark. (Redrawn from Mojetta, A. 2004. *Sharks: White Star Guides Underwater World.* White Star)

under the skin. The non-pored canals of sharks and other elasmobranchs are located only on the head region of sharks and the ventral surface of batoids.

Differences in function between the pored and non-pored canals are a direct result of the former's connection with the environment. Lateral lines with pores are said to sense *distant touch*, an experience terrestrial organisms lack. *Distant* in this case is a bit of exaggeration, since the lateral line is thought to detect only near-field signals (i.e., those within one of two body lengths of the source). Hair cells of the neuromasts in the lateral line canal bend in response to a variety of stimuli, including low frequency sounds, vibrations, and water movement. This enables the shark to detect prey, predators, other organisms, and conspecifics, as well as to determine the velocity and direction of water flow around its body. Sensing water movement is what enables *rheotaxis* (the ability to orient with or against the current).

The non-pored components of the lateral line cannot react to external water movement. Instead, these neuromasts respond to internal fluid velocity, which requires a tactile external stimulus that deforms the skin, the kind that might occur on the head in sharks during mating, feeding, or when the body contacts the substrate in batoids.

Free, or unspecialized, neuromasts, which are also known as *pit organs*, are found on the tops and sides of head, trunk, and tail, as well as surrounding the spiracles and near the gills and ventral to the mouth (fig. 5.3). In batoids, pit organs are found in grooves on the skin whereas in sharks they are typically in shallow pits between modified scales. Pit organs are thought to detect external water movement. Active pelagic sharks (e.g., carcharhinids) have more pit organs than sluggish demersal species (e.g., heterodontids).

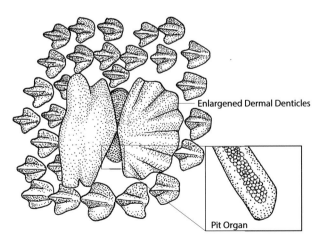

Figure 5.3. Pit organ, free neuromasts (sensory hair cells embedded in gelatinous cupula) in a small depression, or pit, between two enlarged dermal denticles. In some cases, the ridges of the dermal denticles direct water to the pit organ. Inset shows sensory surface of the pit organ. (Adapted from fig. 2, Peach, M.B. and Marshall, N.J. 2009. J. Morphol. 270: 688–701)

Spiracular organs, which are derived from lateral lines, are part of the first visceral pouch, or gill slit (also known as the *spiracular slit*), which in elasmobranchs is reduced to the spiracle. Their basic unit is the neuromast and their function is quite specialized; spiracular organs respond to movement of the cranial-hyomandibular joint. Although their function is enigmatic, they may serve as proprioceptors.

Vesicles of Savi are found on the ventral surface of three families of batoids (Torpedinidae, Narcinidae, and Dasyatidae). They are blind pouches (not connected to the environment) thought to be tactile receptors useful in prey detection or communication with conspecifics by sensing vibrations in the substrate or from direct contact with prey or conspecifics.

Chemoreception

In our annual *Biology of Sharks* course at the Bimini Biological Field Station in the Bahamas, one of the students' favorite activities is attracting and hand-feeding juvenile Lemon Sharks[5] in an isolated tidal lagoon. We spread out into a half-moon shape, and a *chum bag* (a mesh sack containing minced fish) is positioned on a stake in the sediment where the incoming tide slowly carries pieces of it deeper into the lagoon. After a period ranging from a few minutes to perhaps 30 minutes, we begin to see ripples representing the dorsal and caudal fins of the juvenile sharks breaking the water's surface as they wend their way toward us. After conditioning the sharks to take the squid or herring, the students each hand-feed one or two sharks.

This exercise is a microcosm of how a shark senses its environment. Since the lagoon is < 100 m (330 ft) long, the sharks likely hear us, and initially move away. Vision, electroreception, and other forms of mechanoreception cannot come into play at such a distance. What remains is chemoreception, specifically olfaction (smelling), gustation (tasting), as well as a third form known as *common chemical sense*.

Chemoreception, which is the most ancient of the senses, having evolved over 500 million years ago, refers to detection of environmental chemicals. Chemoreception in water is quite different from chemoreception in air, since in the former the chemicals must be dissolved as opposed to in the gaseous state in the latter. Moreover, water is heavier and more resistant to flowing than is air. Thus, diffusion is slower in water, and the medium itself often moves more slowly than air. On the other hand, odors dissipate more quickly in air than in water, since water moves as parcels that are less likely to break apart than those in air.

Chemoreception is most important for feeding and reproduction. Chemical molecules sensed in aquatic environments are typically of small molecular mass, like amino acids and some steroids.

Back to the lagoon. The chemosensory system of these juvenile Lemon Sharks brings in the sharks. They then use their other senses in the final stages of locating the food source.

Let us start with the sense of smell. The organs of olfaction are paired *olfactory sacs* (also known as the *olfactory apparatuses*) on the ventral surface just anterior to the mouth, in cartilaginous nasal capsules. Water is typically channeled into the nares, the channels that lead into the olfactory sac (you might call these nostrils) by flaps on the lateral (outermost) incurrent, then out through the medial (innermost) excurrent openings (fig. 5.4). The small distance between incurrent and excurrent openings might mislead you to think that the water that enters travels only a short way before it exits. In reality, the water takes a circuitous path through a sensory epithelium consisting of lamellae analogous to those of the gills (fig. 5.4C). Called the *olfactory rosette*, this organ immensely increases the surface area for sensing environmental chemicals, and the slow transit time ensures that compounds in the water come into contact with the sensors in the rosette. Although the olfactory rosette contains cilia that may propel water and mucous through it, most sharks must be moving, or there must be a water current, in order for water to enter through the incurrent opening.[6] In sharks with nasoral grooves (e.g., Small-spotted Catshark), most of which frequently rest on the sea floor, the respiratory current (pumping water over the gills) also plays a role in water movement into the rosette. Olfaction in sharks is not linked to respiration and, unlike humans, they do not need to respire to smell, meaning that sharks

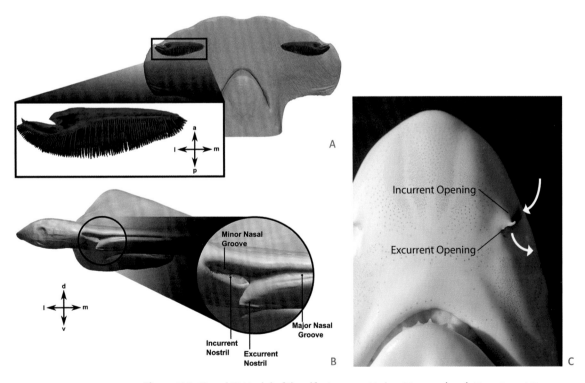

Figure 5.4. (A and B) Model of the olfactory rosette in a Hammerhead. (From Rygg, A.D. et al. 2013. PloS One 8: p.e59783) (C) Incurrent and excurrent openings in olfactory sacs of a shark. White arrows depict direction of water flow.

smell their environment whenever water moves through their olfactory rosette, and thus the appellation *swimming noses.*

Molecules to which the shark is sensitive (e.g., those in mucus sloughed off of the skin of prey) stimulate receptors called *olfactory receptor neurons* (ORNs) on the epithelium of the olfactory lamellae. Cells comprising the ORN also add to the increased surface area through little projections called *microvilli.*

Stimulation of ORNs sends olfactory information via the short olfactory nerve to the *olfactory bulb*, an anterior extension of the brain that is located within the olfactory sac (fig. 5.5). Processing of the olfactory information takes place first in the olfactory bulb. Final processing of olfactory signals occurs in the olfactory area of the forebrain, or *telencephalon*, which functions in processing a broader range of sensory information than solely olfaction, as well as playing a role in complex behaviors.

There is variation in both the olfactory rosette and olfactory bulb among elasmobranchs. As you might expect, this has been interpreted to correlate with the degree of reliance on olfaction by different species or by habitat. One study[7] that looked at 21 species of elasmobranchs concluded that lamellar number and surface area did not correlate with phylogeny but that benthopelagic (living near the bottom) species had more lamellae and higher surface area than exclusively benthic species.

The size and shape of olfactory bulbs was also shown to vary.[8, 9] For example, the largest olfactory bulb known thus far, at 18% of total brain mass, belonged to the White Shark, while the smallest bulbs, at 3% of total brain mass, were found in species of *Carcharhinus*, as well as Blue Sharks (*Prionace glauca*) and Shortfin Makos (*Isurus oxyrinchus*).

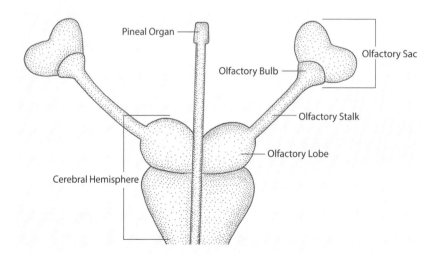

Figure 5.5. Olfactory nerve and bulb, anterior extensions of the brain, of a shark. (Redrawn from Gilbert, S. 1973. *Pictorial Anatomy of the Dogfish.* University of Washington Press)

The above disparities in relative size of the olfactory bulb raise questions about the correlation with reliance on olfaction, especially in the White Shark and Shortfin Mako, confamilials sharing numerous physiological and anatomical characteristics for their high-performance lifestyle. Researchers studying this have speculated that differences in the diets of these two predators may explain the difference. White Sharks primarily consume marine mammals, which produce copious amounts of odorous defecation. Thus, the larger than average olfactory bulb of White Sharks may have evolved as an adaptation to detect marine mammal prey via their smelly emanations. This may also explain why humans gravitate toward fast food.

In general, trends in the size of the olfactory bulb are similar to those in lamellar surface area (i.e., they parallel ecological niche more so than taxonomic relationship). The largest relative olfactory bulb sizes occurred in migratory pelagic sharks, and deep-sea sharks also had large olfactory bulbs, implying the importance of smell for both groups. Reef-associated carcharhinids, hemiscyllids (longtail carpet sharks), and benthic batoids had the smallest olfactory bulbs.

The Role of Chemoreception

In an early 20th century study,[10] crushed crabs wrapped in cheesecloth (to minimize visual cues) fed to captive Dusky Smoothhounds (*Mustelus canis*) led to changes in their behavior, specifically a series of turns until the bait was located, seized, and removed, when the sharks moved to within a few feet of the crab packet.

When cotton plugs were inserted into both nares of the sharks, they ignored the crab packets. When cotton plugs were placed in only a single nostril, the sharks' turns were to the side opposite the plug. The study concluded that the presence of food elicited directed movement by the sharks, that odors in the food were the stimuli, and that the nasal organs detected the stimuli when they were in the odor stream.

This study also investigated responses to natural foods, exposing Tiger Sharks to decayed shark flesh and tuna extracts, and Blacktip Reef (*Carcharhinus melanopterus*) and Grey Reef Sharks (*Carcharhinus amblyrhynchos*) to a wide variety of potential prey, and concluded that all of the items tested served as attractants, but there were variation in responses to some items by some species. Extracts from oily-fleshed fish (e.g., tuna, grouper, and eel) elicited more activity in the sharks than extracts from snappers and other fish with drier flesh, a finding that shark fishers routinely exploit.

A variety of organic biomolecules serve as attractants to elasmobranchs, but amino acids (components of proteins), particularly natural ones, and nucleotides (the building blocks of the genetic material) seem to be most important.

Is the sense of smell in sharks as prodigious as commonly asserted? Studies have shown that elasmobranchs are capable of detecting concentrations of some odorants as low as one billionth of a mole per liter (10^{-9} mol L^{-1}), which represents an extremely high sensitivity. However, many bony fishes have similar sensitivities, so if sharks are swimming noses, then bony fishes are as well.

Before we move on to taste, let us consider hammerheads, which have widely spaced olfactory organs and deep narrow grooves running to these organs along the anterior margin of the cephalofoil (fig. 5.4). This degree of separation gives this group the potential to compare odor on one side to the other more effectively or at a greater distance than in other elasmobranchs. A 2009 study,[11] showed that overall Scalloped Hammerheads (*Sphyrna lewini*) were as or more sensitive to odorants than other sharks and bony fishes, although the authors cautioned that more comprehensive data were needed before conclusions could be drawn.

Gustation and Common Chemical Sense

Although there is debate about whether the sense of taste should be differentiated from olfaction in aquatic organisms, we consider it separately here. Gustation is assumed to have a role secondary to olfaction and is more specialized in its function, specifically to assess food quality; however, since taste is among the most poorly studied senses, its function is not well understood.

Taste receptors in elasmobranchs are located on taste buds found inside the mouth and on the gills. Modified placoid scales, known as *oral denticles*, are also present lining the mouth,[12] limiting the space for taste receptors. The taste buds sport papillae (fig. 5.6), which increase the surface area on which the receptors are located and thus increase sensitivity. There are also microvilli on the taste papillae, small projections that further increase the surface area for taste. Unlike in mammals, papillae are not concentrated on the tongue.

Studies[13, 14] in the Brownbanded Bamboo (*Chiloscyllium punctatum*) and Blue Shark showed that taste papillae were spread throughout the oropharyngeal cavity (gill arches and rakers, and dorsal and ventral surfaces of the mouth), but the highest densities of taste papillae were found on or near the oral valves, folds of tissue in the anterior of both jaws (maxillary and mandibular valves) behind the teeth. A 2011 study[15] concluded that taste papillae were more evenly distributed throughout the oropharyngeal cavity in benthic sharks than in pelagic species.

These different distributions of taste papillae in the oral cavity make good sense. Benthic species (e.g., Nurse Sharks [*Ginglymostoma cirratum*] and Brownbanded Bamboo Sharks) often manipulate their prey in their mouths prior to swallowing, during which time assessing the palatability of the ingested item may occur.

In pelagic sharks, the location of the highest density of taste papillae in areas immediately adjacent to the teeth and the oral valves also appears adaptive. This area is the first point of tactile contact when an elasmobranch bites into its targeted food item. At this point, the predator must assess whether the item represents food on its menu and is worth continuing to consume, a process that may be instantaneous.

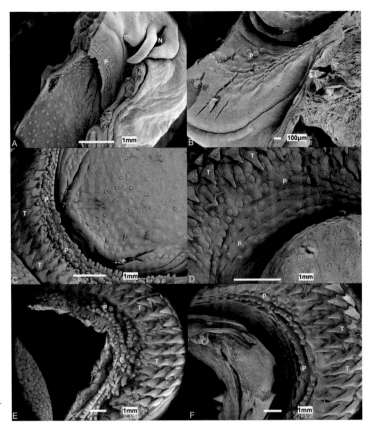

Figure 5.6. Scanning electron images of taste papillae (P) on the taste buds of a Brownbanded Bamboo Shark, *Chiloscyllium punctatum*. N = nostril (nare), T = tooth. (From Atkinson, C.J. et al. 2016. Biology Open: bio-022327)

The concentration of oral papillae near the mouth also explains *bite-and-release* (also called *bite-and-spit*) shark bites. Along the SE US coast, during the summer months, there may be several reported shark bites that leave many of the victims with minor punctures that may not even require a trip to the emergency room. The culprits in these cases are thought to be 2 m (6.6 ft) long or smaller Blacktip Sharks (*Carcharhinus limbatus*), streamlined, swift-moving, bite-first-ask-questions-later fish-eaters. The waters in which these interactions occur are often quite murky, so much so that your hand disappears when immersed only inches. Add swirling schools of menhaden and mullet, so-called *baitfish*, nearshore to the mix of bathers and fast-moving feeding sharks, and you have a foolproof recipe for mistaken identity: Blacktip Shark bites bather.

Fortunately, what happens next most of the time, turns what might have been a fatal interaction into an *ouchie!* and a badge of courage. What likely occurs is what you might experience when you bite into spoiled food. Blech! Spit it out! Exactly what the Blacktip Shark does, and does so almost instantaneously, thanks to the evolution of taste receptors near the mouth and a rapid-fire nervous system that receives and processes the information and sends out a *stop!* signal to the jaws. While this may save victims of bites from Blacktip Sharks, if the perpetrator is a White Shark, and it mistakes a person for, say, a seal, then even if the same receptors indicated that, *oops!*, the shark made a mistake, the outcome will sadly be more than an ouchie.

Some elasmobranchs are capable of sampling water twice for olfactory, gustatory, and common chemical sensory information (fig. 5.7). Members of at least four families of sharks, Heterodontidae (horn sharks), Orectolobidae (carpet sharks), Parascylliidae (collared carpet sharks), Hemiscylliidae (longtail carpet sharks), and numerous batoids (e.g., Rajidae, Myliobatidae, Dasyatidae, Torpedinidae, and Urolophidae) possess paired *oronasal grooves*, channels that funnel water from the nares into the mouth, where taste papillae and common chemical sense receptors are free to sample it.

Lastly, we consider the mode of chemoreception called the *common chemical sense*. The receptors for this type of chemical sensitivity are not specialized cells or groups of cells, but are rather unspecialized, free nerve endings. Humans apparently use this sense when we feel that the air after a storm is crisp. The function of this sense in elasmobranchs is somewhat enigmatic, although most specialists feel that the common chemical sense serves to protect sharks from harmful chemicals. Chemical shark repellents and deterrents, both synthetic and natural (e.g., the skin of the Moses Sole) are thought to act on receptors of this system.

Orienting to Chemical Senses

We have now discussed the placement, morphology, and sensitivity of the three types of chemical senses in elasmobranchs. But how exactly do sharks zero in on the source of the chemicals (fig. 5.8)? Movements in response to a stimulus, in this case odor, are known as taxes (singular = taxis).

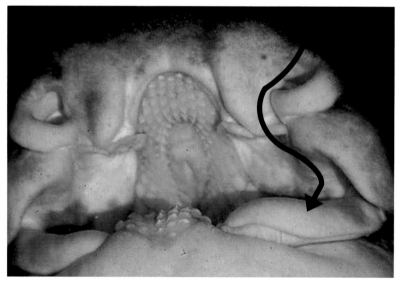

Figure 5.7. Oronasal groove in a Horn Shark. Arrow denotes movement of water through oronasal groove.

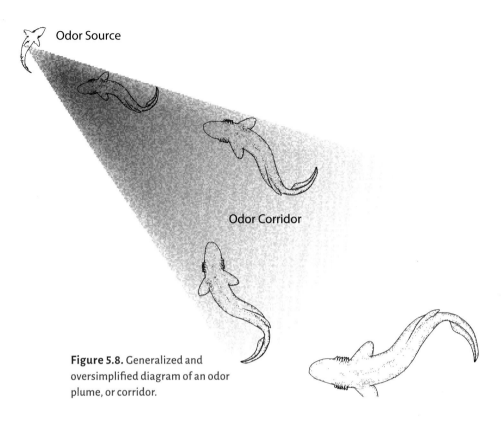

Odor Source

Odor Corridor

Figure 5.8. Generalized and oversimplified diagram of an odor plume, or corridor.

The first taxis, sampling the environment and turning toward the stimulus, is divided into two separate, distinct subcategories, *klinotaxis* (*klin* = to cause to lean) and *tropotaxis* (*tropo* = turn or response), which are differentiated by their temporal nature. Klinotaxis involves *successive* comparison of stimulus intensities between disparate spatial points (e.g., such as would occur by side-to-side head movements), whereas tropotaxis involves *simultaneous* comparison of stimulus intensities on two sides of the body (that is, at the same time).

Early work exploring this phenomenon pointed toward sharks being capable of simultaneously assessing odor intensities and turning toward the higher concentration (i.e., tropotaxis). However, recent captive experiments with the Dusky Smoothhounds revealed for the first time that sharks steered using bilateral odor *arrival time* rather than concentration differences. And even with delayed pulses of higher concentration the sharks continued to turn toward the side stimulated first.

Another taxis employed by some elasmobranchs is *rheotaxis* (*rheos* = stream), in which an organism swims with or against a current. Rheotaxis is similar to tropotaxis, except instead of locating a chemical stimulus with search patterns, the individuals turn in the direction of the current and swim upstream. Lemon Sharks, for example, were shown to respond positively to chemical stimuli using rheotaxis and subsequently followed the current upstream.

Photoreception

The main organ of photoreception in sharks is, of course, the eye. Did you know that the visual system of sharks, as well as that of most other lower vertebrates, also includes another organ of photoreception, called the *pineal organ* (fig. 5.9)? All vertebrate groups except the hagfish possess the structure, which is more widely known as the pineal *gland*, because its primary function is secreting the hormone *melatonin*, which regulates both an organism's daily cycles (its *circadian rhythms*) and its reproductive hormones.

In lower vertebrates (cartilaginous, bony fishes, and amphibians), the pineal organ / gland is located near the top of the brain, where it receives light that enters through specialized areas on the top of the head, which explains a name the structure is no longer called—the *third eye*. That the morphology of the shark cranium is specialized to transmit more light to the pineal gland than surrounding tissues (fig. 5.9), up to seven times more, was discovered in 1975 by the shark biologist Sonny Gruber and associates.

The shark eye has been the focus of studies for 250 or more years. For most of that time, the subject of these studies was its anatomy. Sharks have a typical vertebrate, image-forming eye (fig. 5.10). The eyeball sits in a socket called the *orbit*, which is composed of the cartilaginous *sclera* posteriorly and the clear *cornea* anteriorly. The cornea has the same refractive index (ability to bend light) as seawater and thus serves only to protect the eye and not to focus the incoming light as it does in humans and other terrestrial vertebrates, where as much as 75% of the total refraction occurs. The tough, protective sclera surrounds and protects the eye.

Three additional specializations found in some sharks serve to protect the eye when faced with mechanical damage (e.g., during interactions with prey or predators). In lamnid sharks, the entire eye rotates 180° and the tough, fibrous

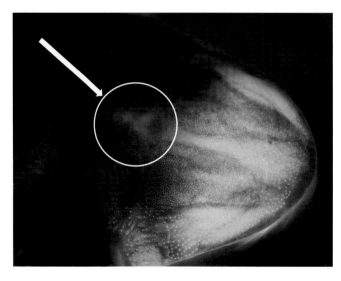

Figure 5.9. Pineal window (circle and arrow) of an Atlantic Sharpnose Shark illuminated by shining a bright light through its open mouth. (Courtesy of Robert Johnson)

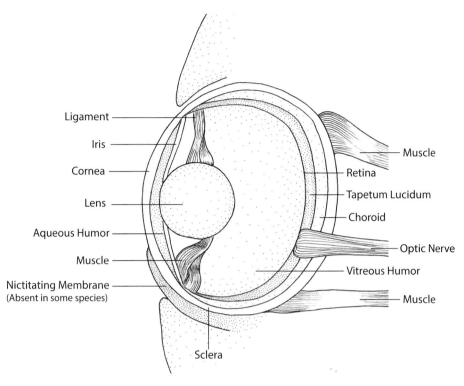

Figure 5.10. Anatomy of the eye of sharks. (Redrawn from fig. 13.3, Hart, N.S. et al. 2006. Com. in Fish. 2: 337–392)

sclera is positioned where the cornea had been. Hexanchid sharks use a different strategy. They use extraocular muscles to pull the entire eye back into the head several centimeters (see fig. 3.7B).

Sharks also possess upper and lower eyelids, though they are fixed in most species and moveable in only a few, including Nurse and Swell Sharks (*Cephaloscyllium ventriosum*). In sharks of the order Carcharhiniformes, a fold of the lower eyelid develops into a tough, opaque, thin, third eyelid called the *nictitating membrane* (see fig. 1.12), whose external surface is covered with placoid scales and which moves upward to protect the eye in a manner analogous to the rotation of the sclera in lamnids.

There are six muscles attached to the outside surface of the back of the orbit (and are thus called *extraocular* muscles), which is typical of vertebrates. These serve to allow coordinated movements of the eyes.

Lining the sclera is a layer called the *uvea tract*, which functions to deliver oxygen and nutrients to the tissues of the eye. The uvea tract includes a darkly pigmented layer called the *choroid* and parts of the *ciliary body*. The ciliary body and other ciliary structures perform a variety of functions, including supporting the lens, controlling the amount of light that enters the eye, and secreting the *aqueous humor* in the anterior portion of the eye between the cornea and the lens.

Part of the choroid layer lying behind the retina is the *tapetum lucidum*, a tissue composed of guanine crystals, which reduces internal glare and scattering of light, and improves vision in low-light conditions. This is the same tissue that is responsible for the *deer in the headlights* reflection from the eyes of cats, racoons, and so on. The tapetum lucidum of a shark is twice as efficient as that of a cat.

The pupil, which regulates the amount of light that will strike the retina, comes in a wide variety of shapes in elasmobranchs (fig. 5.11). These include the typical vertebrate round shape, as well as horizontal, vertical, and oblique slits. Oddly shaped pupils are found in some benthic sharks and batoids, although any specialized functions of these pupils are not always clear.

The pupil in most elasmobranchs is capable of dilation and contraction. This is accomplished by means of a *dynamic* iris, the muscular structure controlling the diameter of the pupil. In some deep-sea species (e.g., the Ghost Catshark [*Apristurus*]), the pupil is fixed in the fully dilated state, which makes sense in light of its dark surroundings. In species with mobile pupils, the size of the pupillary opening correlates with light levels. In high light, the pupil contracts which, in addition to reducing the light entering the eye, also increases the visual depth of field, just like in a camera.

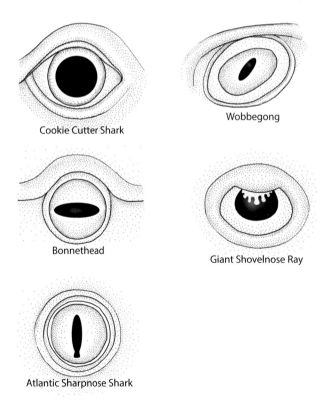

Cookie Cutter Shark

Wobbegong

Bonnethead

Giant Shovelnose Ray

Atlantic Sharpnose Shark

Figure 5.11. Diversity of pupillary shapes in sharks.

The lens (fig. 5.10) in elasmobranchs is a transparent structure responsible for all of the refraction (bending of light so that it can be focused on the retina). It is composed of water and structural proteins. There is also variability of the lens of elasmobranchs. Shapes include spherical, near-spherical, and lenticular (ellipsoidal). The lens in most species is large and powerful. Unlike the lens of terrestrial vertebrates, accommodation, or maintaining focus at various distances, is achieved by moving the lens backward (for far vision) or forward (for near vision) to focus light on the retina rather than changing shape. This movement is accomplished by the six extraorbital (outside of the eye socket) eye muscles.

The retina is the eye's photoreceptor and is the tissue responsible for converting the image that the lens focuses on into nerve impulses that it sends to the brain via a thick, white, ophthalmic nerve passing through the rear of the orbit. Sharks and rays have duplex retina; that is, the retina houses both rods and cones, the photoreceptive cells that are responsible for vision in low-light (*scotopic*) and bright light (*photopic*) environmental conditions. The presence of cones containing multiple visual pigments implies a basis for color vision, which has been experimentally verified in several species of sharks.

The size of eyes varies as well, most notably with depth distribution and habitat.[16] Epipelagic and upper mesopelagic sharks (e.g., Bigeye Thresher [*Alopias superciliosus*]) have the largest eyes, along with benthic and benthopelagic species (e.g., Portuguese Dogfish [*Centroscymnus coelolepis*]) that live in the mesopelagic zone. From an ecological perspective, benthopelagic and pelagic species whose prey are mobile and active also have larger eyes than less active species. The smallest eyes are found in coastal pelagic sharks and especially sluggish benthic sharks and batoids.

The visual field, defined as the space visible from the stationary eyes, also varies with species, specifically with differences in head morphology, and eye position and mobility. The extent of the visual field is important for prey detection, predator avoidance, and mating. In most batoids, and in some benthic sharks (e.g., angel sharks [*Squatina* spp.], wobbegongs [*Orectolobus* spp.], and horn sharks [*Heterodontus* spp.]), the eyes are located dorsally. In most sharks and a few batoids the eyes are laterally placed.

Recent studies[17, 18] measured aspects of the visual fields of several sharks and rays. They concluded that visual field was a function of phylogenetic age, habitat type, and locomotory patterns. The fastest-swimming and most evolutionarily derived batoid in the study, the Cownose Ray (*Rhinoptera bonasus*), had the largest binocular overlap and vertical visual field, which is consistent with the need for enhanced visual information in the rapidly changing landscape it encounters during swimming, while schooling, and while foraging on bottom invertebrates, primarily mollusks.

The three demersal (bottom-associated) batoids studied all bury themselves and had good overhead visual fields. Even though these species consume benthic infauna, they lacked good ventral vertical visual fields and instead relied on sensitive tactile (touch) receptors and electrosenses.

Among sharks, binocular overlap (an index of depth perception) in hammerhead species increased with the expansion of the head, and in the Winghead Shark (*Eusphyra blochii*) was nearly four times larger than in two carcharhinids. Anterior horizontal binocular overlap in the Bonnethead (*Sphyrna tiburo*) was smallest, a product not only of its smaller cephalofoil, but also because of the median placement of the eye in this species compared to the anterior position in the other two hammerheads in this study (see illustrations of these sharks accompanying species descriptions in Chapter 3). All five hammerhead species tested had a 360° vertical visual field around the head.

Electrical and Magnetic Senses

Another favorite activity of students in our annual *Biology of Sharks* course in Bimini, Bahamas, is snorkeling with Caribbean Reef Sharks (*Carcharhinus perezi*; fig. 5.12).

Figure 5.12. Snorkeling with Caribbean Reef Sharks. (Courtesy of Bimini Biological Field Station)

Typically, within minutes of our arrival, sharks arrive, having been attracted by the sounds of the boat engine and the water slapping at the hull, as well as the movements of the staff member in the water positioning the anchor. When a piece of bait hits the water, the sharks use their hearing and other mechanosenses to approach the source of the sound. Some detect the odor in the eddies of the water current and move upstream in the direction of the source. Then in the clear topical waters, the sharks zero in on the bait using their vision. When they are in the last stages of acquiring the food, as they open their mouths, their nictitating membranes move upward to cover and protect their eyes. At that point, they rely on a sense foreign to humans, electroreception, to direct them to the piece of bait. They sense the minute electrical fields that surround all living organisms, and which are amplified if they are bleeding, since blood is an electrolyte and as such conducts electricity.

This suite of senses often works exquisitely for the first shark at the bait, but sometimes in the dance of the feeding sharks, two sharks will converge on the bait from different locations simultaneously. Then, both sharks will trust the > 400 million years of evolution leading up to that moment and, when they are less than a meter (3.3 ft) away from the bait, they will open their mouths almost in unison, deploy their nictitating membrane, and allow their electrosensory systems to guide them to the bait. At this point the systems fool them, because the electrical signal they detect is not that emanating from the bait, but rather the one from each other. As the bait drifts safely away, at least until the next shark senses it, the two sharks will attempt to bite each other, apparently confusing the other shark with the bait. Fortunately, we have never witnessed any damage done, perhaps because of the gustatory receptors located near the teeth that inform each shark of its mistake before the bite is completed.

The electroceptive system that guides a shark to its destination consists of receptors called *ampullae of Lorenzini* (fig. 5.13), that are concentrated on the head of sharks and batoids and are capable of detecting very weak electric fields (1nV/cm–0.1μV/cm, the strength of electrical field within about 0.5 m, or 1.6 ft, of the source in salt water) of other organisms and even inanimate objects. Each ampulla consists of a bulb under the skin, which opens to the environment and connects to other ampulla via a canal. The ampullae are lined with a sensory epithelium, and the canal and the bulb itself are filled with a gel capable of conducting electricity. The tissues underlying the ampullae are insulated from the current by structures called *tight junctions* between adjacent cells.

The pioneering studies that illuminated that sharks possessed an electrosensory system were conducted by Adrianus Kalmijn, based in part on observations in 1935 by the Dutch scientist Sven Dijkgraaf that the Small-spotted Catshark (*Scyliorhinus canicula*) was very sensitive to metallic objects (a rusty steel wire). In the 1960s, Kalmijn was able to demonstrate that electric fields with voltage gradients as low as 0.01 microvolt per centimeter elicited a physiological response (decreased heart rate) and Kalmijn and others showed that the organs responsi-

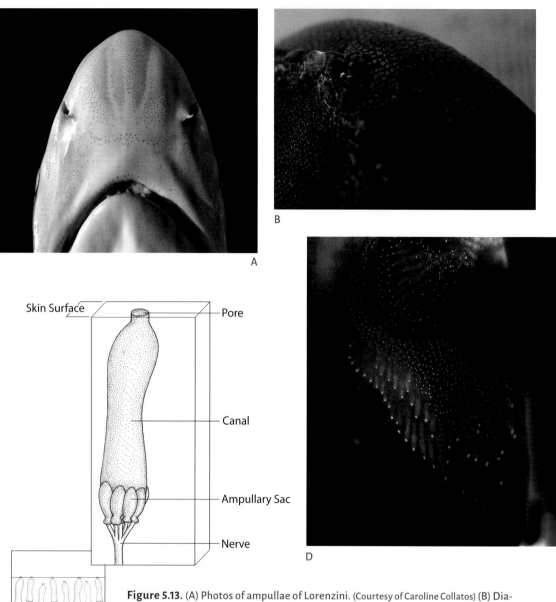

Figure 5.13. (A) Photos of ampullae of Lorenzini. (Courtesy of Caroline Collatos) (B) Diagram of a single ampulla. (Adapted from fig. 13.2, Hart, N.S. et al. 2006. Com. in Fish. 2: 337–392) (C) Gel squeezed from the ampullae on the snout of a shark. (Courtesy of Caroline Collatos) (D) Ampullary canals illuminated on the right half of the head of an Atlantic Sharpnose Shark by shining a light through its mouth. (Courtesy of Robert Johnson)

ble for electroception were the ampullae of Lorenzini, which were also sensitive to temperature and mechanical stimuli.

The salient question that required attention was this: What is the biological significance of this electrical sensitivity? Answering that question first required learning if electrical fields at the level of sensitivity of the ampullae were present in the environment and, if so, how were these fields used by sharks and batoids.

At the time, it had already been established that freshwater bony fishes were capable of detecting weak electrical currents, and that some of these had *electro-genic organs* capable of producing electrical discharges that they used to communicate with each other or sense their environment. No such organs were present in sharks, however.

This brings us to Kalmijn's pivotal study involving two experimental subjects, the Small-spotted Catshark and Thornback Skate (*Raja clavata*), and a prey species to serve as the stimulus, the European Plaice or Flounder (*Pleuronectes platessa*). This landmark paper was published in 1971 in the *Journal of Experimental Biology*[19] (fig. 5.14), and it is summarized below.

Response to Live Prey: A Sense of Plaice

If they had recently fed, the sharks and rays in the study largely ignored the European Plaice, which had buried themselves in the substrate and were hidden from view. If the sharks and rays were stimulated by the addition of a few drops of whiting juice, a *more frenzied* feeding response ensued in both predators. When the shark was within 15 cm (0.5 ft) of the buried fish, the shark would turn to the source, suck up the substrate covering the fish and eject it through the gill slits, then grasp and eat the plaice. The ray behaved similarly, with differences in manipulating the prey.

Using an Agar Chamber to Attenuate All but Electrical Stimuli

A flat chamber was constructed out of agar, a gel that prevented the predator from detecting optical, chemical, or mechanical cues from the prey, but which was transparent to electrical stimuli. A Plaice was placed in the gel chamber, which had incurrent and excurrent openings to allow circulation of seawater, and the chamber was placed on the bottom and covered with sand. Whiting juice was introduced into the pool, which aroused the sharks and skates. When the sharks and skates were within 15 cm (0.5 ft) of the chamber, they behaved similarly to the situations in which the prey was buried beneath the substrate (i.e., they cued in on it and attempted to dig it up).

Allowing Optical, Chemical, or Mechanical Cues

Since the Plaice was always buried, either by the substrate alone, or in the chamber covered by the substrate, vision was eliminated as a cue required to locate the prey.

To test for an olfactory (chemical) response by the elasmobranch predators, Kalmijn replaced the Plaice in the chamber with a mesh bag containing pieces of whiting. The sharks and rays detected the odor and went to its source, which was the end of the excurrent tube, not the chamber itself.

The final test was for mechanical cues that could be detected by the acoustico-lateralis system of the shark and ray. Kalmijn could not rule out that movements of the Plaice within the chamber produced vibrations that the predators detected and used to locate the prey. To eliminate vibrations, Kalmijn modified the chamber slightly by covering it with a thin polyethylene film. Then, he repeated the test after motivating the predators with whiting juice. Even though stimulated by the whiting, neither the sharks nor the rays were able to find the prey in the chamber, thereby eliminating their sensing of mechanical cues.

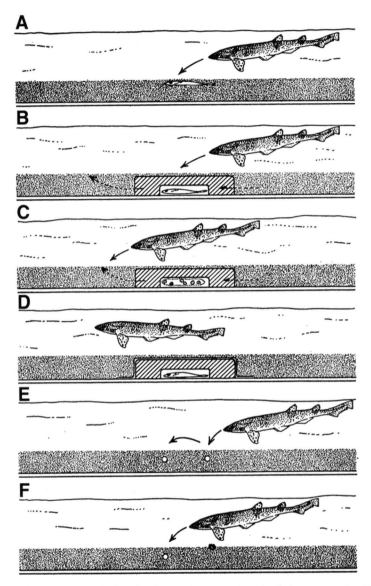

Figure 5.14. A Kalmijn's classic experiment on the role of electroreception. The figure depicts the feeding responses of the shark *Scyliorhinus canicula* to (A) Plaice under sand, (B) Plaice in agar chamber, (C) pieces of Whiting in agar chamber, (D) Plaice in agar chamber covered with plastic film, (E) electrodes producing electric dipole field, and (F) piece of Whiting and electrodes (only one shown). Solid arrows = responses of shark; dashed arrows = flow of seawater through agar chamber. (Reproduced with permission from Kalmijn, A.J. 1971. J. of Exp. Biol. 55(2): 371-383)

Confirmation

While the suite of tests eliminated visual, olfactory, and mechanical sensing of the Plaice in the agar chamber, and strongly suggested that the predators used electroceptors to locate the prey, Kalmijn performed one last set of studies. He buried electrodes that generated a small (1 Hz) current, added some whiting juice, and observed that the sharks and rays approached the current source and removed the overlying substrate, just as they had done to the prey not contained in the agar chamber. Moreover, they would ignore pieces of whiting on the substrate in favor of the electrodes, indicating that electrical fields were more potent attractants than visual or olfactory signals.

In addition to uncovering that the electroreceptors of the two predator species in his classic study were biologically meaningful in locating prey, Kalmijn suggested that elasmobranchs might also use their electrosensory system to sense conspecifics. He also identified inanimate sources of electrical potential gradients in seawater, including ocean currents, tidal currents, and waves.

Kalmijn and others went on to demonstrate that elasmobranchs could detect the weak electrical fields of water currents and possibly the Earth's magnetic fields, which they could use for short-distance navigation.

Hierarchy of Senses

Although the entire suite of senses often integrates with each other to paint a complete sensory picture of the environment, different senses in elasmobranchs come into play at various distances and for specific functions (fig. 5.15).

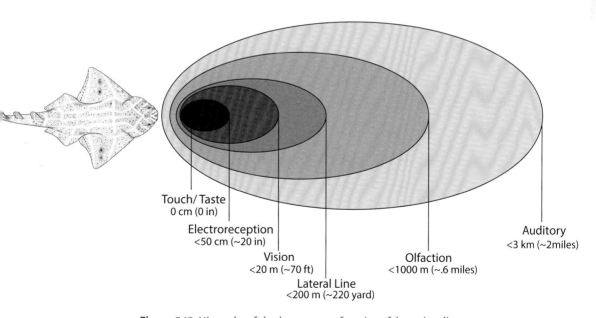

Touch/ Taste
0 cm (0 in)

Electroreception
<50 cm (~20 in)

Vision
<20 m (~70 ft)

Lateral Line
<200 m (~220 yard)

Olfaction
<1000 m (~.6 miles)

Auditory
<3 km (~2miles)

Figure 5.15. Hierarchy of shark senses as a function of detection distance.

Concluding Comments

The sensory world in which sharks live is rich in information, but at the same time can be confusing. As you have seen in this chapter, sharks possess an exquisite set of sensors, and they collect and integrate sensory information to make sense of this sensory information, which in large part has contributed to their success.

NOTES

1. https://www.theguardian.com/environment/2015/dec/22/man-killed-by-shark-in-caribbean-while-being-rescued-by-coast-guard. (Accessed 9/9/19).

2. *Labyrinth* is an anatomical term that refers to structures with communicating cavities or canals.

3. Kritzler, H. and Wood, L. 1961. Science 133: 1480–1482.

4. Nelson, D.R. and Gruber, S.H. 1963. Science 142: 975–977.

5. We recognize that hand-feeding wild animals is a controversial activity, and there are good reasons to oppose the activity. We consider this a valid educational activity, and it is done infrequently such that the likelihood of long-term impact is low.

6. Theisen, B. et al. 1986. Acta Zool. 67: 73–86.

7. Schluessel, V. et al. 2008. J. Morphol. 269: 1365–1386.

8. Northcutt, R.G. 1978. In: Hodgson, E.S. and Mathewson, R.F. (eds.). Sensory biology of sharks, skates, and rays. US Office of Naval Research. 117–193 pp.

9. Demski, L.S. and Northcutt, R.G. 1996. In: Klimley, A.P. and Ainley, D.G. (eds.). *Great White Sharks*. Academic Press. 121–130 pp.

10. Parker, G.H. 1914. Bull. U.S. Bur. Fish. 33: 61–68.

11. Tricas, T.C. et al. 2009. J. Comp. Physiol. A 195: 947–954.

12. Atkinson, C.J. and Collin, S.P. 2012. Biol. Bull. 222: 26–34.

13. Atkinson, C.J. et al. 2016. Biol. Open, bio-022327.

14. Collin, S.P. et al. 2015. In: Fish Physiology 34. Academic Press. 19–99 pp.

15. Atkinson, C. 2011. The gustatory system of elasmobranchs: morphology, distribution and development of oral papillae and oral denticles. PhD dissertation, University of Queensland.

16. Lisney, T.J. and Collin, S.P. 2007. Brain, Behav. Evol. 69: 266–279.

17. McComb, D.M. and Kajiura, S.M. 2008. J. Exp. Biol. 211: 482–490.

18. McComb, D.M. et al. 2009. J. Exp. Biol. 212: 4010–4018.

19. Kalmijn, A.J. 1971. J. Exp. Biol. 55: 371–383.

6 / Reproduction

Introduction: The Unparalleled Diversity of Chondrichthyan Reproduction[1]

Shark sex. It does not get any better than that, especially if the amazing array of reproductive styles and strategies piques your biological interest.

As a whole, the diversity of specializations for reproduction in this group is nothing short of astonishing, especially in the ways in which developing embryos are nourished. Some sharks lay eggs, while others are born live, but the myriad ways in which these are accomplished will amaze you. Sharks vary from most bony fishes in that all fertilization is internal. Instead of producing thousands to millions of eggs like bony fishes, few of which have a chance to hatch and survive until adulthood, sharks produce a small number of larger eggs, and the newborn shark is born at a larger size that gives it a greater likelihood of surviving. Fecundity, the number of offspring per female pregnancy, ranges from one to > 300 in this group. And, unlike some species of bony fishes, there is no parental care after birth.

Sexual Dimorphism

Sexual dimorphism refers to observable differences between males and females of a species. *Secondary sexual characteristics* refer to those beyond external sex organs and may include size, color, markings, or even behavior.

Sexual dimorphism usually comes into play in courtship and reproduction. Charles Darwin recognized this phenomenon and called it *sexual selection*. If natural selection can be summarized as *survival of the fittest*, then sexual selection can be thought of as *survival of the sexiest*. This means that in some species one sex, typically the females, selects the male with whom she will mate based on some sexually dimorphic characteristic.

Among chondrichthyans, the most obvious difference between males and females is the presence of claspers (see fig. 1.5) on the males, which is a *primary* sexual characteristic. Claspers, also called *mixopterygia*, are the intromittent organs, which is the general term describing the structures responsible for sperm transfer for organisms using internal fertilization.

Claspers are the rearward tubular extensions of the inner margin of the pelvic fins. Claspers of immature elasmobranchs are pliable and soft, but they calcify and harden during maturity. A single clasper is inserted into the cloaca (the common urogenital and anal opening) of the female (fig. 6.1), followed by insemination.

Figure 6.1   facing page

A

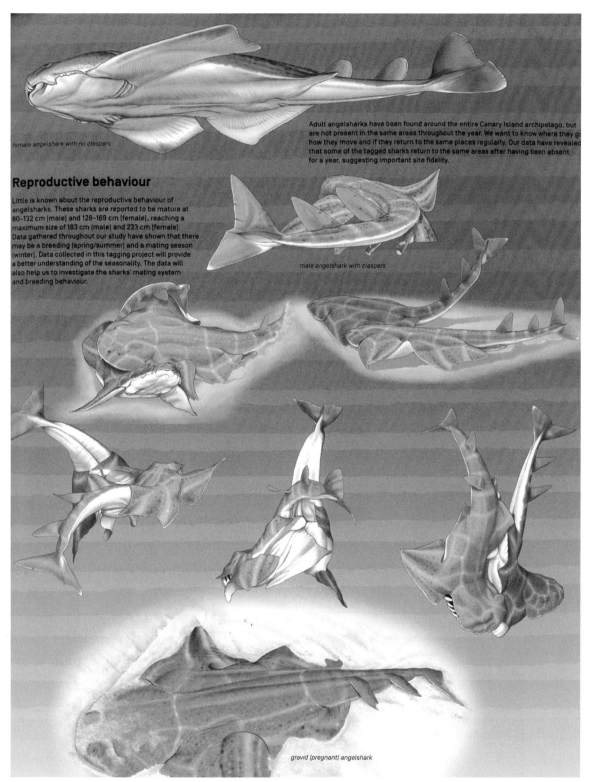

Adult angelsharks have been found around the entire Canary Island archipelago, but are not present in the same areas throughout the year. We want to know where they go, how they move and if they return to the same places regularly. Our data have revealed that some of the tagged sharks return to the same areas after having been absent for a year, suggesting important site fidelity.

female angelshark with no claspers

Reproductive behaviour

Little is known about the reproductive behaviour of angelsharks. These sharks are reported to be mature at 80–132 cm (male) and 128–169 cm (female), reaching a maximum size of 183 cm (male) and 233 cm (female). Data gathered throughout our study have shown that there may be a breeding (spring/summer) and a mating season (winter). Data collected in this tagging project will provide a better understanding of the seasonality. The data will also help us to investigate the sharks' mating system and breeding behaviour.

male angelshark with claspers

gravid (pregnant) angelshark

B

Figure 6.1. (A) Photo of clasper inserted into cloaca of a female Nurse Shark. (Copyright Jeffrey C. Carrier. All rights reserved. Used with permission.) (B) Graphic showing mating in Angel Sharks. (Marc Dando/Used with permission from Save Our Seas Foundation)

Once the clasper is inserted into the female, its tip splays out and, depending on the species, features a series of structures (spines, hooks, and/or spurs), which hold the clasper in place as mating occurs. See Copulation below for how the claspers transfer sperm.

In sharks and batoids, the second most obvious example of sexual dimorphism is a difference in size. In viviparous species, where those that carry their developing embryos inside their body and give birth to live young (about half of all chondrichthyan species), as opposed to laying eggs, the females are larger (fig. 6.2). This makes sense. During pregnancy, irrespective of whether a female gives birth to a single neonate, as in the case of the False Catshark (*Pseudotriakis microdon*), or 40–60 young, as occurs in the Tiger Shark (*Galeocerdo cuvier*), the embryos take up an increasing amount of space as they grow. And in preparation for pregnancy (i.e., between pregnancies), females in many species fatten themselves, since there is considerable maternal input of calories and nutrients into the developing embryos.

Among egg-laying chondrichthyans, there is no size dimorphism, since eggs are typically deposited early in the development of the embryos and thus do not require the addition of space or maternal resources beyond those to the point of deposition.

Another sexual dimorphism, though it is not externally visible, is the evolution in females of thicker skin on the flanks. This is a strictly defensive adaptation, evolved in response to bites from males during courtship in some species (fig. 6.3). During mating season, females of numerous species can be found with mating scars on their pectoral fins, head, and body, ranging from minor abrasions to severe wounding. Some of these bites would be capable of inflicting serious damage were it not for this thicker skin, which is often just deeper than the size of the teeth of the male, thus protecting the internal organs from potentially lethal puncture wounds or exposure.

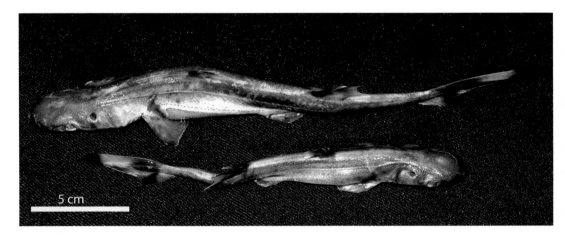

5 cm

Figure 6.2. Sexual dimorphism in Green Lantern Shark. *Top* = female.

Figure 6.3. Mating scars (male courtship bites) on the flanks of a Large Bull Shark. (Courtesy of Joel Blessing)

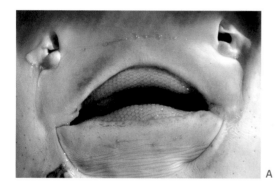

A B

Figure 6.4. Examples of sexual dimorphism in the Clearnose Skate. The teeth of the female (A) are specialized for feeding, whereas those of the male (B) are for both feeding and grasping the female.

Male rajiforms (skates), and some myliobatiforms (stingrays), exhibit a sexual dimorphic characteristic also related to biting females (fig. 6.4). In these species, female teeth are designed exclusively for feeding, where male teeth are enlarged and cusped and are used for grasping onto the female in addition to feeding. These teeth are lost after mating season and are replaced with non-cusped teeth.

In female rajiforms, there are rows of large, pointed, sharp spines, or thorns, on the midline of their dorsal surface and tail. These spines, which are enlarged dermal denticles, are also present on the males, but are much smaller. Females use them to discourage mating attempts by some males, since mating events may entail pursuit by several males at a time.

Not to be outdone, male rajiforms also have a sexually dimorphic characteristic, a series of modified dermal denticles called *malar* and *alar* spines on the cheeks and pectoral fins, respectively. These spines are used to anchor the male to the female after the male bites the posterior margin of the female's pectoral fins. Alar and malar spines are deciduous and drop off following mating season.

Internal Fertilization

All chondrichthyans use internal fertilization, whereas members of only three families of bony fishes do. Internal fertilization results in offspring that are not larva or postlarva, that is, born at a very small, often microscopic size, like bony fishes, but rather are basically miniature to small adults, poised to fend for themselves from birth. Having a higher survival probability is central to the chondrichthyan way of life, since there is no parental care.

The evolutionary success of chondrichthyans does not come without costs. This mode of reproduction involves a high investment of resources by the female. Because of these resources, which include nourishment for the developing embryos and physical space for them to develop, fecundity (the number of offspring) is lower than in bony fishes.

Copulation

Males of all chondrichthyan fishes have two claspers. Only one of the claspers, however, is inserted into the female at any time. Why, then, are there two? The answer lies in how the clasper is inserted into the female, and the competition among males (fig. 6.5). Before a clasper is inserted, it must be inflated with seawa-

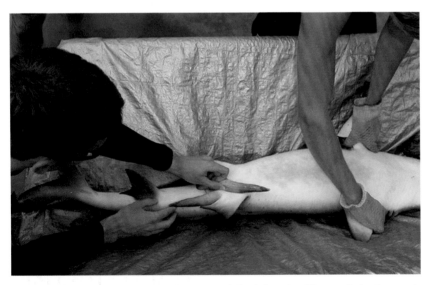

Figure 6.5. The right clasper of a Whitetip Reef Shark flexed as if it were being inserted into a female on its left. (Courtesy of Nick Whitney)

ter. The inflated clasper then rotates 90° or more across the body before insertion. The clasper on the shark's left side is thus only useful if the male is on the left of the female, and vice versa. Thus, if multiple males are attempting to copulate with a single female, they have access from either side.

During copulation, the orientation of the male and female varies with the species. In many kinds of sharks, the male or males will swim alongside the female and attempt to insert one of the two claspers while biting onto the pectoral fin or flank. In the family Scyliorhinidae (catsharks), the male grabs the flank or even the first dorsal fin of the female and envelops her entire body, surrounding her like a donut. Some batoids, like Cownose Rays (*Rhinoptera bonasus*), often copulate belly-to-belly while in others the male is situated on the female's back. Southern Stingrays (*Hypanus americana*) employ both methods of copulation.

Transfer of sperm from the male to the female in chondrichthyans occurs via a groove in the claspers. When the clasper bends and folds across the body, the tip splays open and anchors itself into the cloaca, a method so effective that females must often shake vigorously to disgorge the clasper.

The shape and size of claspers and the mechanism of anchoring vary by species. In squaliform sharks, the tips of the splayed claspers are very sharp hooks used for anchoring. In skates, there are no well-developed hooks but rather the splayed open clasper just enlarges sufficiently to hold the clasper inside the cloaca. The cow sharks, a primitive group, have a simple clasper with two cartilages that are inserted with almost no protrusions for anchoring. These sharks are also unusual in that the clasper sits in a sheath analogous to the foreskin of the human male.

Reproductive Anatomy

What follows is an overview of what we have called the *reproductive plumbing*, the organs of reproduction, their location in the body, and their function. Like other systems in sharks and their relatives, there is some variation. The reproductive tracts of egg-layers, for example, differ from that of live-bearers, as you might have already concluded. But some aspects of the process are conserved irrespective of the species. Ova produced by the ovaries in the process of oogenesis are fertilized by sperm made in the testes by the process of spermatogenesis. The developing embryo receives its initial nutrition from yolk deposited in the egg. After the yolk is absorbed, the variation (see below), begins.

All chondrichthyans have two entire reproductive tracts (i.e., two male or two female). In some species (e.g., sharks in the genus *Carcharhinus*), both tracts are simultaneously functional, and a single copulation event, or sequential fertilization by two males, can fertilize ova in one or both uteri. In other species, only one of the two ovaries is functional and it supplies ova to both reproductive tracts (i.e., to two uteri). In the Cownose Ray (*Rhinoptera bonasus*), only a single uterus is functional.

Anatomy of the Male Reproductive Tract

The male reproductive system of elasmobranchs consists of testes, a series of genital ducts where the sperm mature as they move through the system, a urogenital papilla, siphon sacs, and claspers (fig. 6.6).

The organ of sperm production is the testis, and males have paired testes in the anterior portion of the body cavity attached to the dorsal body wall at the anterior end by a thin tissue layer called the *mesorchium* (fig. 6.6). The testes of some species may be fully or partially encased in the *epigonal organ*, a structure unique to sharks and rays (and whose function is described below).

Testes in chondrichthyan fishes are relatively simple and are similar to those of the primitive bony fishes like sturgeon and gar. Chondrichthyans have three

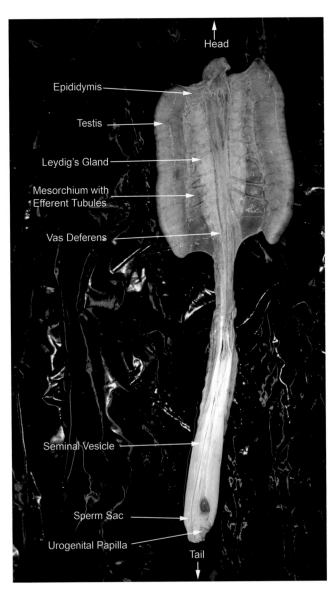

Figure 6.6. Labeled photo of the male reproductive tract from the Atlantic Sixgill Shark.

basic types of testes,[2] which differ in where the sperm cells form and mature. The most common form of testes in elasmobranchs are *diametric* (located at the diameter). In diametric testes, which are found in all squalomorph and galeomorph sharks (excluding lamniforms), developing sperm are located along a strip of the testes, and these sperm go through two phases to mature as they migrate across each testis. The other types are *radial* (in lamniform sharks) and *compound* (in a few batoids).

Irrespective of the type, all testes in elasmobranchs are associated with the epigonal organ (not pictured in fig. 6.6), a structure not found in other vertebrates (including chimaeras), which is responsible for hematopoiesis (production of both red blood cells and those of the immune response system).

Within each testis are microscopic, spherical structures known as *spermatocysts*. Within each spermatocyst are immature little hairlike germ cells called *spermatids*, which are destined to become sperm or, more accurately, *spermatozoa*. The spermatocysts also contain clumps of Sertoli cells, which produce hormones that help in sperm maturation. Once mature, chondrichthyan sperm (fig. 6.7) look like wiggly hairs (due to their helical head), have a flagellated tail, are long (> 30 μm), and lack the expanded head characteristic of mammalian sperm.

Once sperm have neared maturity in the testes, they swim through a series of efferent ducts (also called *ductus efferens*; *effere* = to carry out) in the thin mesorchium into the next genital duct, the paired *epididymi* (fig. 6.6), highly convoluted structures just ventral to the kidneys, which are flattened, ribbon-like, and embedded in the dorsal body wall. Each epididymis is lined with secretory cells whose discharges promote sperm maturation and vitality.

Posteriorly, the epididymis leads to another sperm storage organ, the *ductus* (or *vas*) *deferens*. At this stage in mature male sharks, the anterior portion of the kidney has been modified into an organ known as Leydig's Gland, which produces hormones that it secretes into the ductus deferens to nourish the sperm. The ductus deferens also has specialized secretory cells as well as ciliated cells. Their job is to continue to promote sperm viability and to create currents that move the sperm along.

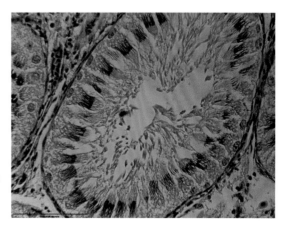

Figure 6.7. Mature sperm (dark lines) from a Finetooth Shark. (Courtesy of Amanda Brown)

Sperm make their way down the ductus deferens to a *seminal vesicle*, where they are stored until mating begins. At that point, a sphincter muscle relaxes and the sperm move into the *sperm sac*, which is also known as the *alkaline gland*. This gland is packed with secretory cells that create an alkaline (high pH, from 8.9 to 9.2) environment inside the sac by releasing carbon dioxide and hydroxide ions. What is the reason that the sperm are bathed in a non-neutral solution? The answer is twofold. First, the next structure in the system is a single urogenital papilla, into which the sperm sac empties, as does the urinary system. Thus, the sperm are exposed to the acid characteristic of urine, which could damage or destroy them were they not protected by the alkaline contents of the sperm sac buffering the urine's acid. Secondly, research has shown that the alkaline gland contains proteins that function at high pH and increase sperm motility, which becomes critical to their success at fertilizing ova.

The next step in the copulation process is to get the sperm from the sperm sac into the clasper, which is not connected to it. This is where the rotation of the claspers comes into play. Each clasper has a groove, and this groove on the clasper contacts the urogenital papilla in the cloaca.

There still is one major hurtle to overcome: how to flush the sperm from the sperm sac to the female, since there is no mechanism for ejaculation within the male reproductive tract. To solve this problem, sharks and batoids (but not chimaeras) have another trick, use of a structure called the *siphon sac* (fig. 6.8). The siphon sac is a dead space on the ventral surface of male sharks running almost the entire length of the abdominal space, which translates into about one third of the body length, between the skin and the underlying muscles. This subcutaneous space has only a single opening, a pore located near the urogenital papilla. Prior to copulation, seawater is sucked into the space.

Figure 6.8. View from the rear of an exposed siphon sac from a male Blacknose Shark on its back. (Photo by Joshua Bruni)

In some of the species whose siphon sac has been examined, there are cells that produce and secrete the chemical 5-hydroxytryptamine (5-HT). This chemical has been shown to stimulate the uteri to contract repeatedly, which facilitates transport of the sperm to the ova. Then during copulation, the male bends, compressing the siphon sacs and forcing the contents through the pore, through the cloaca, and into an opening on the now-erect clasper (now rotated across the body) called the *apopyle*. Simultaneously, sperm pass through the urogenital papilla, and into the apopyle, where they are entrained by the seawater from the siphon sac and flushed into the female reproductive tract. This method of intromission is unique among vertebrates.

Anatomy of the Female Reproductive Tract

The female reproductive system consists of ovaries, genital ducts called *oviducts* (which include shell glands and uteri),[3] a vagina, and cloaca (fig. 6.9).

The organs that produce the ova, or egg cells, in females in the process of oogenesis, as well as hormones that control their development, are the ovaries (fig. 6.9). Females have paired ovaries in the anterior of the body cavity, although in some galeomorph genera (e.g., *Prionace, Scyliorhinus, Pristiophorus, Carcharhinus, Galeus, Mustelus,* and *Sphyrna)* the left ovary atrophies and only the right ovary is functional. In some myliobatiform batoids (e.g., *Urolophus*), the right ovary becomes vestigial and the left ovary is functional. In some species (e.g., *Prionace glauca*), the ovaries are embedded in the epigonal organ. In immature species, the ovaries are reduced to a small strip, whereas in mature females they are large and globular, with bright yellow follicles containing the ova, on the surface. Like the testes, ova in elasmobranchs are similar to those in primitive modern bony fishes.

There are two types of ovaries. Lamniform sharks have a single, *internal* ovary (the right one). *Internal* refers to the fact that this ovary lies within the epigonal organ. In non-lamniform sharks, the ovary (either the single functional one, or both) is *external* (sometimes referred to as *naked*, or *gymnovarium* type), either on the surface of the epigonal organ or attached to the body wall by a mesovarium, a thin mesentery that connects the ovaries to the dorsal wall of the body cavity. The size and number of ova vary between these two types, with the former producing numerous small ova (3–5 mm, or 0.1–0.2 in, diameter) whereas external ovaries produce fewer but larger ova (as big as 50–60 mm, or 2–2.4 in, diameter). In both types of ovaries, the developing ova (known as oocytes) are *vitellogenic* oocytes (full of yolk). The differences in size and number of ova of each ovary type are important to the mode of embryonic nutrition employed by each group (see below).

The ova produced by each ovary next move into the *oviduct* (also called a *Müllerian tube*), through the *ovarium*, a two-layered tissue resembling plastic wrap, similar to the male mesorchium.

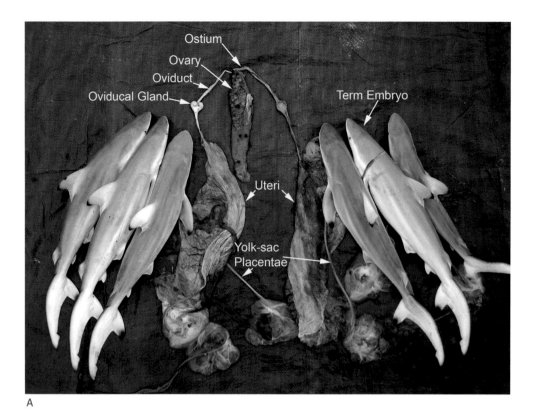

A

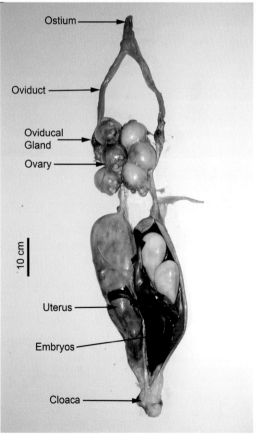

Figure 6.9. Labeled photos of the female reproductive tracts from (A) Blacktip and (B) Gulper Sharks.

B

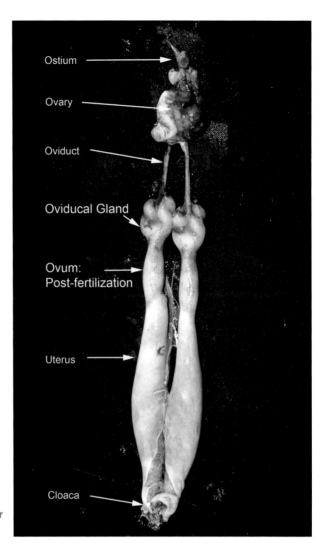

Ostium

Ovary

Oviduct

Oviducal Gland

Ovum:
Post-fertilization

Uterus

Cloaca

Figure 6.10. Labeled photo of the reproductive tract of a female Atlantic Sharpnose Shark showing ova exiting the oviducal glands after being fertilized.

The oviduct varies among species, but in all cases it is essentially a long tube with some differentiation along its length into a funnel called the *ostium* (or *falciform ligament*), a shell gland (also known as a *nidamental* or *oviducal gland*), a connecting segment (the *isthmus*), and the *uterus*. These components function to collect and transport the ova, form a membrane or egg case (also known as a *mermaid's purse*) around them, receive and store sperm, and protect and nourish to varying degrees the developing embryo. Primitive bony fishes like sturgeon have a similar architecture, whereas in advanced bony fishes, the ova are shed straight into the environment.

Single ova are shed from the ovary through the ostium into the oviducal gland. In egg-laying (*oviparous*) species, the oviducal gland is well-developed and it is in this location that the eggshell is deposited. In those species that bear live young, the oviducal gland is small or even vestigial, and typically deposits only a membrane.

The oviducal gland is also the site of fertilization. Sperm swim up the uterus into the oviducal gland where they are stored and maintained viable for periods that have been documented to be seven years or longer. The phenomenon of sperm storage in elasmobranchs is not well-known. It has been observed in numerous species, but apparently not in lamniforms. Biologist John Morrissey provided fertilized ova (viable embryos) to researchers for over seven years from female Chain Catsharks (*Scyliorhinus retifer*) that he had maintained in captivity and had caught in the wild, but which had never been in captivity with males. Multiple parentage is known from many elasmobranchs as well, and this may also involve sperm storage (see below).

The fertilized egg, now called an embryo, moves into the isthmus and then to the uterus, primarily through contractions of smooth muscle. In oviparous (egg-laying) species, the uterus is not well-developed and serves mainly to convey the egg to the cloaca and then to the environment. In contrast, the uterus of viviparous (live-bearing) species is expanded and houses the embryo while it develops. Additional roles of the uterus in the nutrition of developing embryos are discussed below.

Figure 6.10 shows a reproductive tract of the Atlantic Sharpnose Shark (*Rhizoprionodon terraenovae*) after fertilization.

Reproductive Cycles

Sharks and their relatives exhibit one of three main reproductive cycles: *continuous*, *seasonal*, or *punctuated* breeding.[4]

Continuous breeding involves year-round reproduction in which females are pregnant for up to a year or more, with vitellogenesis (development of ova) occurring during pregnancy in preparation for mating and subsequent fertilization. Females typically mate soon after parturition (birth), with no resting phase. Continuous reproduction is characteristic of many live-bearing sharks where there is limited to no matrotrophy (maternal input); for example, deep-sea squaliforms that are continuously pregnant, and some batoids.

In *seasonal* breeders, reproduction is restricted to certain times of the year (typically spring and summer) with shorter pregnancies and annual reproduction frequency. Seasonal breeding occurs in most myliobatiform rays, egg-laying skates, and some live-bearing sharks.

The final category is *punctuated* breeding, in which females are pregnant often for a year or more, then are dormant for an intervening period before becoming pregnant again. In other words, they experience a two-year or even three-year reproductive cycle. Dusky Sharks, for example, have a three-year cycle that includes a 20-month gestation period, followed by a 16-month gap. Punctuated breeding is found in placental live-bearing sharks where maternal input to offspring is high.

Endocrine Control of Reproduction

The production of sex cells and preparation for pregnancy are under the control of the endocrine system, the complex series of glands that secrete hormones that regulate physiological function in organs and organ systems throughout the body.

Endocrine glands like the adrenals, pituitary, thyroid, and ovaries and testes are highly conserved among vertebrates over evolutionary history (i.e., they are present in the entire extant lineage). You, a trout, and a shark all possess the same glands, although trout have some unique glands, and the chemicals produced in different taxa are also conserved, though they are not identical.

The protein or steroid hormones produced by endocrine glands and secreted into the bloodstream are chemical messengers that go to other glands and signal them to start or stop producing other hormones, or they go to organs and signal them to produce an enzyme or other chemical. They bind to receptors in the cell membranes of target tissue, and this binding often triggers biochemical response within the cell (e.g., production of another hormone or maturation of a cell). Regardless of the system, there is a hormone cascade associated with development. The timing of the release of hormones is influenced by environmental cues, the most important being temperature and day length.

Reproduction in all vertebrates, including sharks, is regulated by what is known as the HPG (*hypothalamus-pituitary-gonadal; gonadal* refers to testes and ovaries) axis. These three separate glands function in unison in regulating the reproductive process.

How to Assess Reproductive State in a Shark

In order to understand important aspects of the life histories of sharks, it is vital to be able to tell if a shark is mature or immature. If the shark is a mature female, it is even more useful to know if she is pregnant or postpartum. Fortunately, there are ways of making these determinations, though some require chemical analysis or specialized instrumentation like ultrasound machines. For many commonly encountered sharks, researchers have used some of these techniques to publish sizes of maturity for males and females. These published values, however, should be used only as rough estimates, since populations from different areas mature at different ages and sizes, or the timing of reproduction varies. The methods below can be used to determine the stage of maturity and reproduction in an individual, or for an entire population.

In males, the first method, examining the claspers, is a non-invasive one (i.e., it does not require an incision or removal of muscle tissue or blood). In immature males, the claspers are short and flexible, the latter due to the lack of calcification, although claspers lengthen and become less flexible as a shark approaches maturity. If you examine a large number of male sharks or batoids over a broad range of sizes, you can assign them to two categories based on the above criteria. Then, you can use the vertebral centra (by a very invasive technique we describe in Chapter 9) to determine the age at which the claspers become calcified and the shark is sexually mature.

Another way, also invasive, involves measuring levels of reproductive hormones in the bloodstream. In males, testosterone levels in the blood begin to increase during spermatogenesis and remain elevated until after mating.

In females, unless they are visibly pregnant, there is no external method of determining maturity comparable to that used with claspers. The main nonchemical method of determining maturity of female sharks is to sacrifice them or use animals killed by cold-shock or even as bycatch in fisheries, and examine their reproductive tracts, which can indicate whether a shark is mature or not, whether she has mated before, and whether she is or was pregnant.

Alternatively, levels of circulating reproductive hormones can also be measured, although there are fewer peaks in the levels in some cases. In some species, levels of female hormones begin to become elevated during the vitellogenesis stage of ova production, when the eggs begin to become yolkier. Then these levels may decrease during copulation, since by that time they are fully developed, and no additional hormonal signals are required.

Finally, a method increasingly being employed to assess pregnancy in sharks is high-resolution ultrasound (see fig. 6.15).

Hermaphroditism and Virgin Births

Hermaphroditism (the presence of both male and female reproductive organs in an individual) occurs but is uncommon in the animal kingdom, except in bony fish, where it is the norm in more than 20 families (e.g., wrasses, groupers, and sea basses). Among bony fish, the hermaphroditism is sequential (one sex transforms into the other) and not simultaneous, except in the mangrove rivulus, *Kryptolebias marmoratus*, in which self-fertilization occurs and the offspring become clones of the parent.

Hermaphroditism has been found in some sharks as well, although only in the case of the deepwater catshark *Apristurus*, an oviparous (egg-laying) group, have hermaphrodites been shown to be fertile. Hermaphroditism may have developed in this group of deep-sea sharks to compensate for the small population size and thus to increase reproductive success, but this is purely speculative. How frequently does hermaphroditism occur in sharks? Not very frequently, either within a species or among species. To date, about 12 cases of hermaphroditic elasmobranchs have been reported.

In the lantern shark (*Etmopterus unicolor*) researchers[5] examined 70 specimens and found about 30% were hermaphrodites, 30% gonochoristic males, and

47% females. The authors speculated that the cause was likely *environmental contamination* (e.g., radioactivity).

Another reproductive phenomenon is *parthenogenesis*, or virgin birth. In many groups (e.g., in some plants, insects, and bony fish), parthenogenesis occurs as a natural form of asexual reproduction. Parthenogenesis was unknown in sharks, however, until it was reported for a captive Bonnethead (*Sphyrna tiburo*)[6] and has now been described in multiple species of captive galeomorph sharks and one myliobatid ray.

There is also one report of parthenogenesis occurring in the wild and leading to viable offspring in the Smalltooth Sawfish (*Pristis pectinata*).

Modes of Embryonic Development

You may have learned that animals either lay eggs or give birth to live young, which is true as far as it goes. Chickens are *oviparous*, as are most bony fishes. And mammals are *viviparous*. But these terms are named only for the point at which either the egg or live young leave the mother. If you look backward from the point to the time the ovum is fertilized (except in organisms like bony fishes where fertilization is external) until either the eggs are deposited or young are born, there is much diversity in the developmental pathway among vertebrates, especially in how the developing embryo is nourished. And no vertebrate group shows greater variety in the modes of embryonic nutrition than the sharks and their relatives. The term *lecithotrophic* (*lecitho* = yolk; *trophic* = feeding) denotes that embryos receive their nutrition exclusively from yolk, a fat-, protein-, and often vitamin-rich substance produced by the mother during oogenesis. In contrast, in *matrotrophic* (mother feeding) species, most of the nutrition throughout gestation comes from the mother. Among the chondrichthyans, modes of embryonic nutrition exist on a continuum between these extremes but fall into four major categories and here we divide these into 10 subcategories.

The modes of embryonic nutrition in elasmobranchs are:

- **Oviparity** (extended oviparity, lecithotrophic oviparity)
  - » Single (immediate)
  - » Retained (delayed)

- **Yolk-sac Viviparity** (previously called *ovoviviparity*)
  - » Lecithotrophic viviparity
  - » Limited (mucoid) histotrophy
  - » Embryotrophy

- **Oophagy**
  - » Lamniform oophagy
  - » Embryophagy
  - » Carcharhiniform oophagy

- **Lipid Histotrophy**
  - » Lipid histotrophy using trophonemata
  - » Placental viviparity

A Few Words About Oviparity

Oviparity refers to egg-laying. Here, the source of nutrition throughout the entire developmental period is the yolk deposited by the mother during vitellogenesis. Thus, once the yolk is deposited, there is no additional maternal contribution to embryonic nutrition.

An important consideration in embryonic development, one that has implications for both the mother and the newborn, is whether the embryo gains or loses weight from either the point the eggshell is deposited in oviparous species or from when the egg is fertilized to hatching in viviparous species, to when the neonate is born. Researchers determine the weights of embryos and neonates by drying them in an oven at 60°C (140°F) until a constant weight, called the *dry weight*, is reached. At this point all of the water has been evaporated and all that remains is dry organic matter.

In oviparous species, since all of the nutrition that the developing embryo can receive is limited to what is contained in the yolk in the sealed egg, the end-weight cannot be greater than the original. It cannot be even the same, since the transfer of energy from yolk to body mass is not perfect.

Why is this? It is thermodynamics 101, the same reason why you do not ultimately gain a half-pound when you eat a half-pound of food, since the conversion of the energy contained in that half-pound of food to your body weight involves loss of over 50% of the energy as heat, as well as use of some of that energy to fuel your body's routine functions and the elimination of waste products.

Similarly, the developing embryos are using energy for respiration and metabolism, and they also defecate and urinate. So, thermodynamically speaking, there must be a decrease in the dry weight of the neonate prior to birth, and this decrease is typically in the 20% to 25% range.

Here are some facts about both categories of oviparous sharks and their relatives:

- Young are smaller than in viviparous species.
- 40% of sharks and rays are oviparous, including the largest family of sharks, the Scyliorhinidae (catsharks), and all skates (i.e., half of batoids).
- All chimaeras are oviparous.
- Ova are large. After being fertilized, they are enclosed in an egg case while in the oviducal gland, are deposited, then hatch after periods of 12 weeks to 15 months, during which time all development is external of the mother.
- Eggs are laid on the substrate or attached to bottom structures.

- The embryo is nourished solely by the yolk sac.

- Oviparous sharks are primarily small and benthic.

- Body functions of the embryo (e.g., urination and defecation) occur in the egg.

- Nearly halfway through development, a seam opens in the egg case allowing the flushing of waste and inflow of oxygenated water into the protective shell.

Single Oviparity (Immediate)

In the case of single oviparity, *single* does not refer to the number of eggs. It means that, in contrast to retained (or delayed) oviparity, described below, the eggs are not retained within the mother during the entire developmental period but rather are deposited soon after being fertilized.

Single oviparity is characteristic in two major groups, the Rajiformes (skates), which have about 290 species, and the Scyliorhinidae (catsharks), the most speciose family of sharks (> 150 species), as well as Chimaeriformes, all nine species of Heterodontiformes, and also some Orectolobiformes.

Let us look at development in the Clearnose Skate (*Rostroraja eglanteria*; fig. 6.11), which takes about 12 weeks from fertilization. The egg case is sealed off from the environment when it is deposited, but the accumulation of waste products and the decrease in dissolved oxygen would eventually make for a toxic, lethal brew if it were not removed.

At between five and six weeks, a seam along the horns of the egg case dissolves and allows for the influx of seawater and the efflux of the gradually toxifying contents. Replacement of the fluid in the egg case is facilitated by the presence of a very long tail, whose motion sets up a current that flushes out the tainted

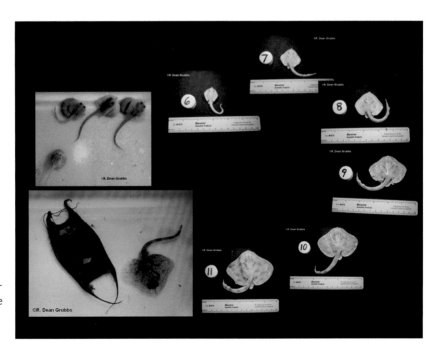

Figure 6.11. Development in the Clearnose Skate, an example of single oviparity.

Figure 6.12. Corkscrew egg of a Horn Shark, and near-term embryo.

water. This means that at 5–6 weeks of development, the seawater environment is identical inside and outside the egg case.

Horn sharks (heterodontids), which typically inhabit rocky coastal substrates, lay individual corkscrew-shaped eggs (fig. 6.12), which are adapted to become lodged between rocks so they are protected and do not get washed onto the beach. Horn shark mothers have been observed carrying the eggs in their mouths and positioning them among the rocks.

Retained (or Delayed) Oviparity

Here, the shell is put around the eggs in the oviducal gland, but the eggs are retained in the mother, so egg-laying is delayed. Similar to the situation in single oviparity, since there is no additional maternal nutrient input, there is a 20–25% weight loss experienced by the time of hatching or birth.

In several scyliorhinid and orectolobid sharks, the fertilized eggs are retained in the oviducts of the mother for a portion of development and are then deposited on the seafloor prior to hatching. In some cases, the eggs are not laid but instead are incubated within the mother until hatching occurs inside her uteri and the offspring are born live. A classic example is from the Nurse Shark (*Ginglymostoma cirratum*), described in a beautifully illustrated paper.[7] As the embryos reach about the halfway point of development, they hatch from their eggs within the uteri and the now-empty egg case remnants are shunted out of the mother. The remaining development occurs in the mother. The external yolk sac has already been absorbed before that point, but it has not been used up. Instead there is an internal yolk stomach that contains the yolk, which is still available as a source of nutrition until birth.

Whale Sharks (*Rhincodon typus*) were once considered as exhibiting single oviparity based on the discovery of egg cases containing embryos; however, closer examination of the embryos revealed an egg case that likely would not hold if deposited on the substrate, suggesting that they may have been aborted. Subsequently, Taiwanese researchers cut open a pregnant Whale Shark that was being sold for meat and found about 300 embryos in egg cases at various stages of hatching, similar to the reproductive mode described in the Nurse Shark.

After reading the next section, you may wonder why we categorized Nurse Sharks and Whale Sharks as oviparous and not viviparous, given that young are born alive. This illustrates that the modes of embryonic nourishment we describe are not always discrete but exist on a continuum. Nurse Sharks and Whale Sharks are often described as exhibiting lecithotrophic live birth. However, other live-bearing species do not have a thick egg case around the embryo during early development, and hence the reason for the inclusion of Whale and Nurse Sharks among species exhibiting retained oviparity here.

A Few Words About Viviparity

Viviparity refers to live birth, where the developing embryos are protected within the mother while they develop and grow, and embryos are typically born at a size that increases their potential to survive. In your readings, you will sometimes find shark viviparity divided into *aplacental* and *placental* modes. We avoid the term aplacental because it is uninformative since it describes what it is not (i.e., not placental) and it masks diversity developmental modes, including some that are equal in matrotrophic inputs to that in placental live-bearers. We broadly divide viviparity in elasmobranchs into three groups based on the major source of nutrition, either a yolk sac, the mother's unfertilized eggs, or lipid-rich histotroph secreted from the mother's uterus (uterine milk), though in all cases, the developing embryo initially gains its nourishment from the yolk sac.

Yolk-sac Viviparity (Ovoviviparity)

The most widely utilized type of live birth in the elasmobranchs is yolk-sac viviparity, where the developing pup receives the bulk of its nourishment from its yolk sac and is thus heavily lecithotrophic. In yolk-sac viviparous species, the oviducal gland is reduced and unspecialized, since no true shell is deposited, and sperm storage is likely limited. Instead of a shell, a membrane is deposited around the developing eggs. In some cases, a single membrane is deposited around all eggs fertilized in the oviducal gland, forming a structure referred to as a *candle* (see fig. 6.13C). The candles move into the two uteri and serve to maintain intra-uterine conditions required for the embryos to develop. The embryos break free of this membrane about halfway through development and are then free in the uterus. Overall, about 40% of extant species of elasmobranchs practice yolk-sac viviparity. This includes all 160+ species of sharks of the superorder Squalomorpha (i.e., angel, frilled, saw, dogfish, gulper, sleeper, lantern, and cow sharks), plus a few of the Carcharhiniformes (e.g., some smoothhounds) and Orectolobiformes (wobbegongs). No members of the orders Heterodontiformes or Lamniformes utilize yolk-sac viviparity. Among batoids, the Torpediniformes (torpedo rays) and Rhinopristiformes (guitarfish and sawfish) are yolk-sac live-bearers.

Lecithotrophic Viviparity and Limited Histotrophy

Most yolk-sac viviparous species are strictly lecithotrophic, meaning embryos receive all nourishment from the yolk sac, similar to what occurs in oviparity, but there is no eggshell. Over the course of their development, the embryos exhibit a 20–25% weight drop, due to the absence of additional maternal nutrition. In species in which there is no additional maternal input, females do not require a resting phase (see below) and will physiologically prepare for their next pregnancy while already pregnant. Thus, it is not surprising that most squalomorph mature females are almost always pregnant. Interestingly, the squalomorph sharks employing yolk-sac viviparity have among the longest gestation periods of any vertebrates, two years or more, and some, like the Little Gulper Shark (*Centrophorus uyato*) produce only a single pup following this long gestation.

Moreover, since most of the sharks of the superorder Squalomorpha are deep-sea, where seasonal cues for the timing of reproduction are absent, it is not unusual to find sharks in all stages of development at any time (fig. 6.13A and B.).

Lecithotrophic viviparity is best-known in the genus *Squalus* (fig. 6.13). In *Squalus*, all of the embryos, typically three or four, are placed in a single candle. They continue to absorb their yolk while in the candle, and they break out when they are one-third to halfway developed and continue to absorb their yolk.

In the Gulper Shark (*Centrophorus granulosus*), the mean dry weight loss between egg and neonate is 19.5%, but in the Great Lantern Shark (*Etmopterus princeps*), the decrease is only 8%. This suggests that the Gulper Shark embryos get all nourishment from the yolk sac (i.e., are lecithotrophic), and the Great Lantern Shark was receiving some maternal contribution, likely through mucoid secretions from the mother's uterine wall (termed *limited histotrophy*). The maternal contribution may be *limited* compared to those employing uterine milk, but in representatives of some families (e.g., Triakidae) there is weight gain up to 350%, which implies this additional maternal contribution of nutrients can be significant. In these triakids (and some other species) there are frilly extensions of the uterine wall that secrete a mucous that has nutritive ability and may also aid in respiration by providing additional oxygen.

Facing page

Figure 6.13. Lecithotrophic viviparity in two members of the genus *Squalus*: Genie's Dogfish (A) and Hawaiian Shortspine Spurdog (B) with mature embryos and ova. Note the variation in fecundity between two closely related species. (C) Two candles from Genie's Dogfish, showing seven embryos.

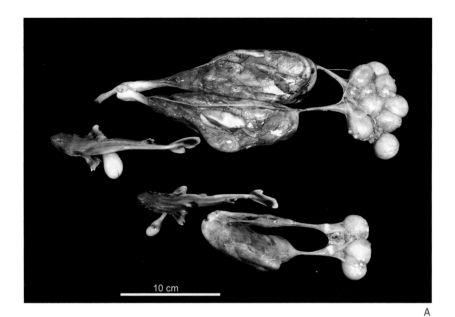

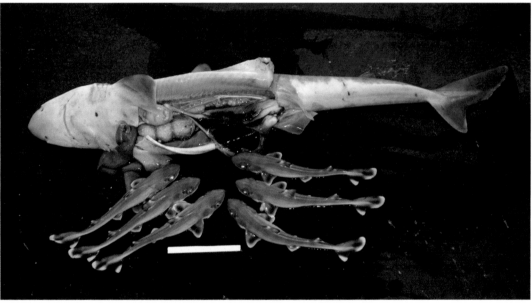

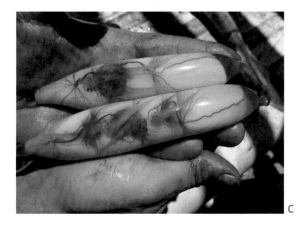

Figure 6.14. Sheath around the rostrum of neonate Smalltooth Sawfish.

Some batoids, such as the electric rays, guitarfishes, and sawfishes, are yolk-sac live-bearers; however, it is unknown if these are strictly lecithotrophic or limited histotrophs. Speaking of sawfish, ever wonder how female sawfish avoid getting wounded by the toothed rostra of embryos *in utero*, or when they give birth to young? Here is how they do it: the toothed rostrum of each embryo and newborn is covered by a gelatinous sheath that slowly dissolves over the course of days after birth (fig. 6.14).

Embryotrophy

Embryotrophy, described recently by Castro and others,[8] occurs only in Tiger Sharks and is a mode of yolk-sac viviparity in which embryos are sequestered in sacs containing a clear fluid. The fluid, termed *embryotroph*, is not the protein- and fat-rich *uterine milk* we discuss next, but it has nutritive value, leading to dry weight gains of as much as 1092% in this species. In contrast to lecithotrophy, in which nutrition is provided by the yolk, embryotrophy is a form of *matrotrophy* (*matro* = mother); that is, the nutrition comes from the mother.

Embryotrophy represents a novel form of nutrition among sharks, and a deviation from that of other carcharhinid sharks, which all use placental viviparity. This is one of the reasons researchers have proposed removing Tiger Sharks from the family Carcharhinidae and placing them in their own monotypic family.

Embryotrophy has contributed to the reproductive success of this species. Tiger Sharks are known for having among the largest litters, typically 45–60 pups (fig. 6.15), of all sharks. Fecundity of most carcharhinids is limited because maternal input of nutrition is required throughout pregnancy, and females cannot store energy reserves large enough, or eat enough food, to supply these needs to a large number of embryos. In Tiger Sharks, the nutrition is deposited in the early embryo.

Lamniform Oophagy

Oophagy means egg-eating, and it occurs only in lamniform sharks (fig. 6.16) and the small carcharhiniform family Pseudotriakidae (false catsharks). Here, like all other elasmobranchs, embryos begin development using the stored energy of their yolk sac. In lamniform oophagy, at about the same time that most of the yolk has been absorbed by the embryos, the mother begins ovulating unfertilized ova. The embryos swim within the uteri and consume the new ova and develop large egg stomachs. This matrotrophic means of embryonic nutrition results in dry weight gains of > 1000%. All lamniform sharks utilize oophagy for embryonic nourishment and the high matrotrophic contribution through unfertilized eggs typically results in few offspring. For example, oophagy in the deep-sea Crocodile Shark (*Pseudocarcharias kamoharai*) (fig. 6.16B) typically results in only four young (two per uterus) and the Pelagic Thresher produces only two offspring. It has been hypothesized that some if not all lamniform sharks may also use lipid histotrophy through uterine secretions relatively early in development. Indeed, a recent paper[9] suggested that the uterine wall of White Sharks contains villi similar to the trophonemata we will discuss later, and that these villi may produce a lipid-rich histotroph (uterine milk) that nourishes embryos after the external yolk-sac has been absorbed but before unfertilized ova have been ovulated into the uteri. This is another example of how the categories we use to classify modes of embryonic nourishment are not always discrete.

Embryophagy (Adelphophagy)

The lamniform Sand Tiger (*Carcharias taurus*) takes oophagy to the next level, in that the fastest-growing embryo in each uterus eats its siblings, a form of nutrition called *embryophagy* (embryo-eating), *adelphophagy* (sibling-eating), or *intra-*

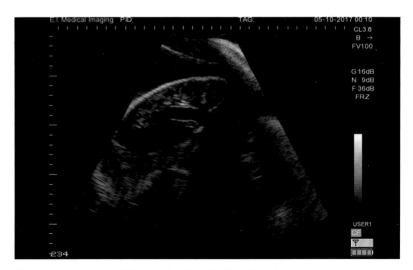

Figure 6.15. Ultrasound image of a Tiger Shark embryo *in utero*. (Courtesy of Matthew Smukall)

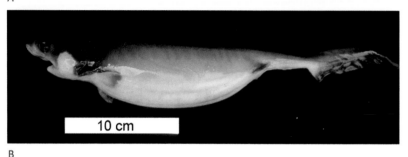

Figure 6.16. (A) Embryo of Salmon Shark exhibiting its yolk stomach filled with ova it has consumed in utero. (© Kenneth J. Goldman, PhD) (B) Crocodile Shark Embryo.

uterine cannibalism. In embryos of most shark species, teeth are among the last major anatomical structures to develop. In most cases they remain internal and do not break through the gums until slightly before birth. Sand Tiger embryos, however, exhibit precocious tooth development, an adaptation for intrauterine predation. When the largest embryo with precocious tooth development reaches about 12 cm (4.7 in), it starts eating as many as a dozen other embryos in its uterus and stores the energy in its yolk stomach. The result is that females give birth to two large young (one in each uterus), each > 1 m (3.3 ft) and thus formidable predators with a high survival probability at birth. Talk about sibling rivalry!

Carcharhiniform Oophagy

The carcharhiniform family Pseudotriakidae (false catsharks) use a form of oophagy that differs from that of the lamniforms. During ovulation and fertilization, a batch of unfertilized eggs are deposited in the same egg envelope with a fertilized egg. In the uterus, a single embryo lies along with unfertilized ova, which are ingested by the embryos after the ova begin to break apart. There is evidence that pseudotriakids also have limited histotrophy. Dense villi extend from the uterine wall and secrete a clear fluid that is believed to contribute to limited mucoid histotrophy.

Histotrophy

Histotrophy is a very strongly matrotrophic form of nourishment, meaning the embryos rely primarily on maternal energy sources. As in all sharks and rays, initial nourishment to the developing embryo is supplied from the yolk sac. After yolk is depleted, the mother secretes copious protein and lipid-rich histotroph or uterine milk. There are two primary types of histotrophy based on the mode of maternal transfer of this histotroph—through uterine villi called *trophonemata* or through a yolk-sac placenta.

Lipid Histotrophy Using Trophonemata

Trophonemata (*tropho* = feeding; *nema* = thread), which occur only in a single order, the Myliobatiformes (about 220 species), refers to the presence of long villi, or filaments, in the uterus called *trophonemata* (fig. 6.17). These are referred to as

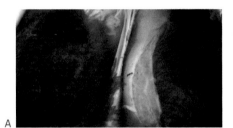

Figure 6.17. (A) Trophonemata in the myliobatiform Smooth Butterfly Ray in late term of pregnancy. (B) Mother and term embryos in the Yellow Stingray, also a trophonematic species.

A

10 cm

B

placental analogues because they mimic the function of placenta by secreting uterine milk, although there is no direct connection between the uterine wall and the embryo. Myliobatiform ray embryos begin with a yolk sac, but the mother develops trophonemata, from the epithelium of the uterus. These trophonemata secrete a lipid-rich (fat- and oil-filled) nutritive histotroph (uterine milk). The trophonemata are bright red before producing milk, then turn pink when milk is produced.

Because the weight gain is the largest of any vertebrate (up to 5000%), trophonematic embryonic nutrition may be the most advanced type of development strategy in the group. The embryos swim in this rich nutritional medium that bathes every orifice and exchange surface, thus they cannot *avoid* nourishment.

Myliobatiformes produce relatively small litters, as few as one (e.g., Cownose Rays), but they are very large and their survival probability is high. Also, Cownose Rays have frilly extensions on the yolk stalk, but this vilified structure and its function have not been fully studied. The single pup of Cownose Rays is born after a 12-month gestation period. There is a huge increase in weight for the developing embryo.

Myliobatiform rays that possess barbs are born with a fleshy tip on their barb, thus avoiding a sticky situation at birth. The tip disintegrates quickly, with a fully calcified spine ready to go, if needed.

Placental Viviparity (Yolk-sac Placental Viviparity or Placentotrophy)

As the name implies, *placental viviparity* refers to a form of live birth in which the developing embryo receives nutrition from a structure (called a placenta in mammals) that connects the embryo to the mother through the uterine wall. However, the placenta in sharks is analogous (similar in function) rather than homologous (similar in derivation or origin) to the mammalian placenta. It thus might more appropriately be given the name *yolk-sac* placenta, although this might create confusion with the lecithotrophic form of embryonic nutrition which, you will recall, is also known as *yolk-sac* viviparity. We will stick with the name placental viviparity for this category, although you may see this referred to as *yolk-sac placental viviparity* or *placentotrophy* in the literature. No matter the name, placental viviparity represents the opposite extreme to oviparity.

In sharks with placental viviparity, after the ova are fertilized, they are encased in a membrane in the oviducal gland, and then delivered into the uterus, where the membrane-encased sharks are stacked (fig. 6.18). Once in place, in many species the uteri compartmentalize into crypts, thus separating the developing embryos. During early development, as is the case in all of the forms of embryonic nutrition we have described, the embryos rely on the yolk sac for nutrition. After the yolk is exhausted, the empty yolk sac interdigitates with the heavily vascularized uterine wall, forming the placental connection. The mother then feeds the embryos directly through uterine secretions, rather than through the blood stream as in placental mammals. The placental connection separates from

Figure 6.18. Examples of placental viviparity. (A) Bignose Shark female and pups with intact placental connections (umbilical cords). (B) Scalloped Hammerhead embryos with placental connections.

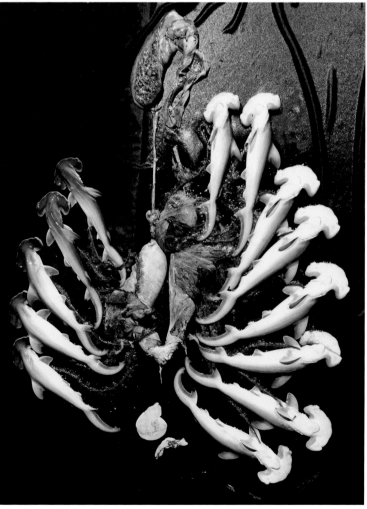

the embryo during birth, and neonates typically feature what could be called a belly button but is more appropriately called a chest button or *umbilical scar*. Immediately after birth, the scar is in fact not entirely healed, a process that can take a month or more, and which can be used (along with other characteristics, e.g., size) to estimate the age of the pup.

Placental viviparity occurs in about 18% of shark species. All of these are in the order Carcharhiniformes, but only in the families Carcharhinidae (requiem sharks), Triakidae (houndsharks), Hemigaleidae (weasel sharks), Leptochariidae (barbeled houndshark), and Sphyrnidae (hammerheads). Interestingly, placental viviparity can co-occur with forms of yolk-sac viviparity in the same family and, in some cases, the same genus (e.g., *Mustelus*). It has been estimated that placental viviparity evolved independently (i.e., in different lineages with as many as 20 times in elasmobranchs), which may account for the high degree of variability in the morphology of the placental connection in sharks using placental viviparity.

Placental viviparity is energetically costly to pregnant females. Little energy is put into vitellogenesis, which occurs early in development and from which recovery of the female is relatively quick (like in lecithotrophic sharks; see above). Instead, the developing embryos require constant nutrition, and the pregnant female must expend considerable energy swimming around carrying the embryos (fig. 6.19). Thus, many sharks in this category (e.g., Sandbar and Blacktip Sharks [*Carcharhinus plumbeus* and *C. limbatus*, respectively]) take a year off after giving

Figure 6.19. Pregnant (A) and postpartum (B) Sandbar Sharks.

A

B

birth, and the Dusky Shark (*C. obscurus*) has a three-year cycle. The reproductive cycle in some species is also condition-dependent. In years when environmental conditions are more favorable (i.e., food is abundant), individuals may not need to wait a year or more before becoming pregnant.

Placental viviparity also entails a potential problem—the potential for umbilical cords to become tangled or disconnected (fig. 6.20A). The likelihood of entanglement is minimized (but not eliminated) by sequestering the embryos in crypts (fig. 6.20B). This may protect the embryos by compartmentalizing them, but this method also comes with a potential danger—the lack of a way to abort or eject an embryo that may be injured or dead, respectively. However, if an embryo dies within a crypt, only other embryos in the same crypt might be affected, not the entire assembly in that uterus.

Which Came First, Oviparity or Viviparity?

We end this chapter with a brief chicken-egg which-came-first discussion. Among vertebrates, live-bearing is more the exception than the rule. In addition to being characteristic of all mammals (except the monotremes; i.e., the platypus and echidnas), live-bearing is found in many reptiles and a lesser number of amphibians and fishes. The remaining vertebrates, including all birds and most reptiles, amphibians, and fish, lay eggs.

Figure 6.20. (A) Sandbar Shark pup strangled by its umbilical cord *in utero*. (B) Blacktip Shark embryo in crypts.

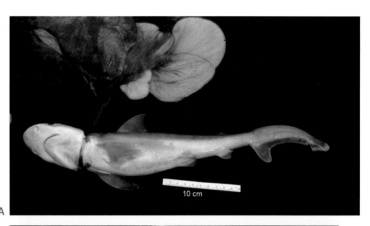

A

B

In elasmobranchs, viviparity (60% of species) is more widespread than oviparity (40%). Reconstructions of trends of reproduction in evolution show that more species have moved from egg-laying to live-bearing than the reverse. This and other evidence have led to the conclusion that egg-laying is the ancestral condition and that live-bearing in this group is thus a modern condition, a conclusion implicit in the order in which we presented the modes of development.

What other evidence supports this conclusion? First, recall that no specimens of the earliest animal in the fossil record that paleontologists consider a shark, *Cladoselache*, had claspers. If indeed *Cladoselache* lacked claspers and this was not a case of sexual segregation, then fertilization could not have been internal, and thus *Cladoselache* was an egg-layer and not a live-bearer. Also, all nine species of the Galeomorph order Heterodontiformes are oviparous, and the members of this order are most similar to the carboniferous hybodont sharks, which are considered ancestors of modern sharks. Scyliorhinids (catsharks) are also oviparous and are considered phylogenetically old. Finally, among vertebrates, the most ancient group are the bony fishes, and 98% of them are egg-layers. The most advanced forms, the mammals, are viviparous. Together, these imply that viviparity arose after oviparity.

Not so fast, however! Shark biologist Jack Musick has pioneered the idea that yolk-sac viviparity is in fact the ancestral form, and the evidence is compelling. This idea suggests that oviparity in the modern sharks (the neoselachians) is an adaptation for increasing the fecundity of smaller species of sharks and batoids.

Consider this supporting evidence:

- Among batoids, the only egg-layers are members of the order Rajiformes (the skates). If in fact oviparity is ancestral, then the rajiforms, which are the smallest group of batoids in physical size, should be the most ancient group of batoids, which they are not. The order Torpediniformes is considered the most ancient, and they exhibit yolk-sac live birth, as do all of the rhinopristiforms (guitarfish and sawfish). The order Myliobatiformes (stingrays) is a recently derived group, and they are all trophonematic (i.e., they secrete histotroph [uterine milk]). Thus, it is reasonable to conclude that the ancestral form of reproduction in batoids is yolk-sac live birth.

- The earliest genus of all living sharks and rays, dating back approximately 190 million years, is the squalomorph shark *Hexanchus*, which has yolk-sac live birth. In fact, there are no oviparous squalomorph sharks.

- Although the galeomorph heterodontiform sharks are egg-layers and their lineage traces back to the Carboniferous, the earliest galeomorph fossil belongs to the order Orectolobiformes (carpet sharks), and the most ancestral family in the brachaelurids (blind sharks). Both of these are yolk-sac live-bearers.

- Scyliorhinids (catsharks) are egg-layers, but the most ancient member of the order Carcharhiniformes are the proscyllids (finback catsharks), which are yolk-sac live-bearers.

Does the idea that oviparity is an adaptation in smaller sharks and rays to increase fecundity hold? Consider that sharks of the family Scyliorhinidae (cat-sharks) are oviparous and small (< 1 m, or 3.3 ft), and their average annual fecundity is 60 eggs. Squalomorph sharks (dogfish) are much larger on average, are more ancient, and give birth to an average of only 4.6 pups.

Skates (order Rajiformes) are also oviparous and small, with similar fecundity, whereas Stingrays (order Myliobatiformes) are live-bearers with a mean fecundity of 4.4. Likewise, guitarfish (order Rhinopristiformes) exhibit yolk-sac viviparity and their mean fecundity is 6.7.

The above examples represent evidence supporting Musick's theory that oviparity is the advanced state, although there is still room for debate.

Concluding Comments

We did not mislead you at the outset of this chapter when we marveled at the diversity of reproductive modes of elasmobranchs. Who knew that being predatory extended into the womb for some species? The diversity is even more remarkable when you consider the relatively low number of species, about 1250, many of which (e.g., deep-sea species) are still poorly known. Stay tuned for more variations!

NOTES

1. Still one of the most authoritative references on the subject: Hamlett, W.C. 2005. *Sharks, Batoids, and Chimaeras.* CRC Marine Biology Series. CRC Press. 562 pp.

2. Pratt Jr., H.L. 1988. *Copeia* 1988: 719–729.

3. The terminology becomes confusing because some authors define the oviduct as the genital tube leading from the ovaries to the nidamental (oviducal, or shell gland) and the nidamental glands and uteri as components separate from the oviducts.

4. Maruska, K.P. and Gelsleichter, J. 2011. In: Norris, D.O. and Lopez, K.H. (eds.). Hormones and reproduction of vertebrates. Fishes 1: 209–237. Academic Press.

5. Yano, K. and Tanaka, S. 1989. Jap. J. Ichthyol. 36: 338–345.

6. Chapman, D.D. et al. 2007. Biol. Let. 3(4): 425–427.

7. Castro, J.I. 2000. Env. Biol. Fishes 58: 1–22.

8. Castro, J.I. et al. 2016. Mar. Biol. Res. 12: 200–205.

9. Sato, K. et al. 2016. Biol. Open 5: 1211–1215.

7 / Circulation, Respiration, and Metabolism

Introduction

In the early 1980s, the following question appeared on a TV game show: *According to marine biologists, who has a bigger heart, a human or a Great White Shark?* The marine biologist in this case was co-author Daniel Abel, when he was a graduate student at Scripps Institution of Oceanography.

If you were the contestant, how would you have answered the question? The answer that was accepted as correct was the Great White Shark (*Carcharodon carcharias*; also known as the White Shark). However, to be accurate, before answering the question we first must qualify that we mean a human and White Shark *of the same size.* A White Shark weighing 900 kg (2000 lb) would have a substantially larger heart than even a large person. But an adult human would have a heart weighing nearly 2.4 times more than that of the White Shark of the same weight as the human.

What if the question was *Who has a bigger heart, a human or a Horn Shark?* The human would, regardless of the size, since Horn Sharks (*Heterodontus francisci*) are substantially smaller. If the human and Horn Shark were of equal weight, the human heart would weigh nearly six times more than that of the Horn Shark.

Even though the game show writer thought he was asking a trivial question, what he was really asking was more meaningful: *What is the heart adapted for and how does this differ in taxonomically and ecologically distinct organisms like humans and White Sharks?* In this chapter we will answer these and other questions about shark physiology, starting with the circulatory system.

Circulation in Sharks

Anatomy of the Heart and Circulatory System

Vertebrates, including elasmobranchs, possess a closed circulatory system, one in which blood is contained in blood vessels. A typical closed circulatory system consists of a propulsive organ (the heart), a high-pressure distributive system (the arterial system), an exchange system (the capillaries), and a low-pressure reservoir (the venous system). The blood in the circulatory system transports oxygen, carbon dioxide, nutrients, waste products, hormones, and other substances.

Unlike mammals, fishes (both osteichthyans and chondrichthyans) have a single circuit system; that is, one main loop and some secondary loops in which the blood goes from the heart to the gills to the body (head and trunk) then back to the heart (fig. 7.1).

Blood vessels leaving the heart make up the arterial system and these vessels get progressively smaller in diameter and more numerous until they reach their destinations as capillaries, where exchange of some of the transport products listed above occurs.

In chondrichthyans and osteichthyans, even though there is only a single major loop, the arterial circulation is given different designations. First is the *bran-*

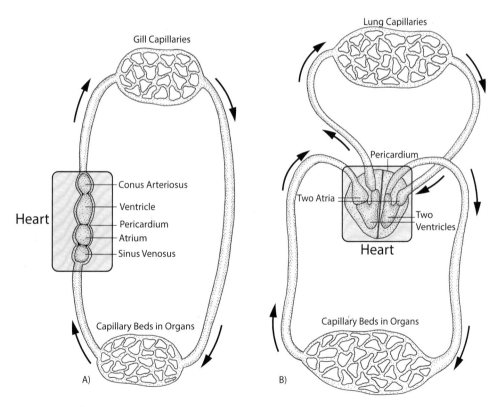

Figure 7.1. Generalized diagram of the closed circulatory system of chondrichthyan fish and mammals. (Redrawn from fig. 9, Furst, B. 2015. J. Cardiothor. Vasc. Anesth. 29: 1688–1701)

chial, or gill circulation (figs. 7.2 and 7.4), that part of the circulatory system from the heart through the gills. In all fishes, the artery leaving the heart is the *ventral aorta*. Once it exits the heart, it almost immediately branches into pairs of *afferent*[1] *branchial arteries*, which carry the blood to the gills, where these subdivide ultimately into capillaries and gas and ion exchange occur. After the gills, these vessels coalesce into *efferent*[2] *branchial arteries* before the now-oxygenated blood is sent to the head and trunk (muscles and viscera).

A subdivision of the branchial circulation is the *coronary* circulation, which arose first evolutionarily in chondrichthyans. The coronary circulation of chondrichthyans branches off the gills and supplies the entire myocardium (heart muscle) with oxygen, nutrients, and so on (fig. 7.3).

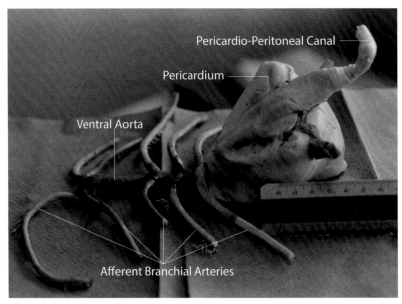

A

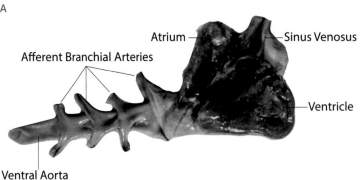

B

Figure 7.2. (A) Silicone rubber cast of the pericardium, ventral aorta, pericardio-peritoneal canal, and afferent branchial arteries from a 2 m (6.6 ft) Blue Shark. The heart is contained within the pericardial space and thus is not shown. (B) Heart, ventral aorta, and afferent branchial arteries removed from a Blacknose Shark.

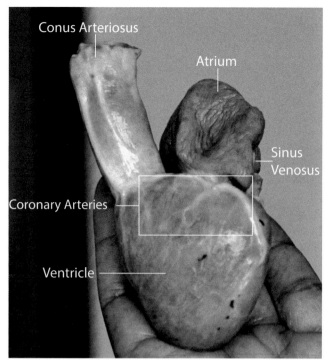

A

Figure 7.3. Photo of the heart of a Blue Shark (A) and drawing of the heart of a Lemon Shark (B), both showing coronary vasculature. Sinus venosus not shown in B. Figures do not depict the orientation on hearts as they would be in the shark (see fig. 7.7A, *top panel*).

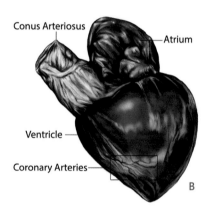

B

The other major subdivision of the arterial system is the *systemic* circulation, in which blood goes to the tissues in the head and trunk (fig. 7.4). The efferent branchial arteries of the gills deliver blood to a *dorsal aorta*, the largest blood vessel and the main distributive channel in all fishes. The dorsal aorta sends blood into either the *cranial* circulation (to the head) or *caudal* circulation (posteriorly, to the trunk). Finally, the caudal circulation divides into *visceral* and *somatic* branches. The visceral branches supply the internal organs whereas the somatic branches supply some muscles, the kidneys, and body wall.

The venous system (fig. 7.4) serves as a low-pressure reservoir of blood. The venous system of chondrichthyans differs from most other vertebrates in that the blood is not always in discrete blood vessels but prior to entering the heart is contained in large sinuses[3] (fig. 7.5). A sinus is a cavity, in contrast to the tubular

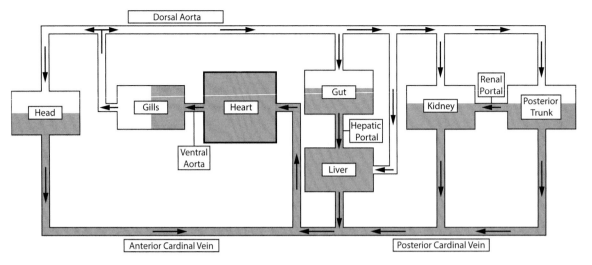

Figure 7.4. Systemic circulation of the shark. Darkened parts represent the venous system. Portal systems carry blood from the capillary system of one organ to another before returning to the heart. Portal systems allow some additional degree of processing of blood or delivering blood components in the second capillary bed. (Redrawn from fig. 6.3, Gordon, M.S. 1982. *Animal Function: Principles and Adaptations*. MacMillan)

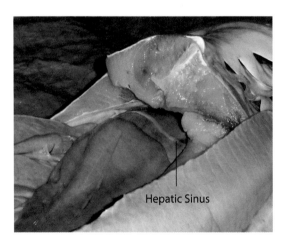

Figure 7.5. Large hepatic blood sinus of the venous system embedded in the liver of a Lemon Shark. The hepatic sinus serves as a reservoir of blood immediately upstream of the heart that could be mobilized when needed. The shark is on its back with its head at the upper right (note the gills).

architecture of blood vessels. Sinuses can accommodate more blood and thus are a ready reservoir when metabolic demands require that the heart pump more blood (e.g., when in pursuit of prey or avoiding predators).

In chondrichthyans, there are three sinuses, the *anterior cardinal, posterior cardinal*, and *hepatic*. The anterior cardinal sinus is located in the head and the other two are at the terminus of the venous system immediately before blood enters the *sinus venosus* of the heart (see below).

The anterior cardinal sinus drains the head before leading to the heart. The posterior cardinal sinus drains the posterior body, including the kidneys, through posterior cardinal veins, and the hepatic sinus accommodates blood from the liver before it goes into the heart.

During dissection of recently deceased sharks and batoids, we have observed greatly expanded hepatic sinuses on the verge of rupturing, like overfilled balloons. Explanations for this observation are unclear, and include diminished cardiac performance associated with capture stress (pooling caused by the effects of gravity and unusual postures when the animal was struggling or lying on the deck without the buoying support of the seawater). Regardless of the direct cause, it is possible that rupture of one or more of the venous sinuses could contribute to stress-related or post-release mortality in some species.

Circulatory specializations in warm-bodied (i.e., *endothermic*) sharks are discussed in Chapter 8.

Blood in Chondrichthyans

Information on characteristics of the blood of chondrichthyans is somewhat scarce. Moreover, blood characteristics vary among species.

Whole blood is composed of blood cells in a fluid medium known as plasma, which is mostly water and dissolved substances (gases, proteins, glucose, minerals, and so on). The percentage of red blood cells, the dominant cellular component of blood, in whole blood is known as *hematocrit*. The pigment in the red blood cells that binds to and transports oxygen in vertebrates is *hemoglobin*.

Most information on blood of chondrichthyans (fig. 7.6) comes from two applied areas, animal husbandry[4] and biomedical science. In combination with pure scientific studies,[5] these have provided the following information:

- Chondrichthyans possess the cells characteristic of vertebrate blood, namely erythrocytes (red blood cells, which function in gas exchange), thrombocytes (platelets, important in clotting), and leukocytes (white blood cells, part of the immune response system).

- Chondrichthyans lack bone marrow, the primary tissue responsible for hematopoiesis (blood formation) in mammals, as well as lymph nodes. In their place they have a thymus and spleen, secondary sites of hematopoiesis in mammals. Elasmobranchs also possess two hematopoietic organs unique to them,[6] the epigonal organ (in the gonads) and Leydig's organ (in the lining of the esophagus). In bony fishes, hematopoiesis occurs in a part of the kidney.

- Erythrocytes of elasmobranchs are nucleated and about 2.5 times larger than those of mammals, whose red blood cells lack nuclei, and are 3–5 times larger than those of bony fishes.

- Elasmobranchs have fewer erythrocytes than bony fishes, although each elasmobranch erythrocyte has more hemoglobin than each bony fishes' erythrocyte.

- Bony fishes have higher overall mean hematocrit, hemoglobin concentration, and erythrocyte counts than elasmobranchs, which translates in greater capacity to carry oxygen. Among elasmobranchs, sharks have higher values than batoids.

- Among sharks, hemoglobin and hematocrit levels are higher in endothermic (i.e., warm-bodied) species (e.g., White Shark and Shortfin Mako, *Isurus oxyrinchus*) than in ectothermic species (most elasmobranchs). Likewise, values are higher in active sharks (e.g., Blacktip Shark, *Carcharhinus limbatus*) than in more sedentary species (e.g., Port Jackson Shark, *Heterodontus portusjacksoni*).

- Elasmobranchs typically have higher blood volumes than bony fishes, although there is considerable overlap. Elasmobranch blood volume measurements range from 18 to 80 ml·kg^{-1} body mass compared to 30 to 70 ml·kg^{-1} for bony fishes. In both cases, higher volumes are associated with more active fish. Because of their large blood volume and fragility of the blood sinuses, dissections of sharks can be very bloody.

The Heart of Sharks

Do you know how many chambers the chondrichthyan heart has? The answer is four, although unfortunately many textbooks of biology frequently mistakenly state that chondrichthyans have two-chambered hearts.

The explanation for this widespread, monstrous mistake is *mammalian egocentrism*. Mammals have four contractile chambers, two *atria* and two *ventricles*, that constitute their heart. Chondrichthyans have a single atrium and a single

Figure 7.6. Blood being withdrawn from the caudal artery or vein of a Sandbar Shark. The shark is on its side with the ventral side at left. (Courtesy of Laura Claiborne Stone)

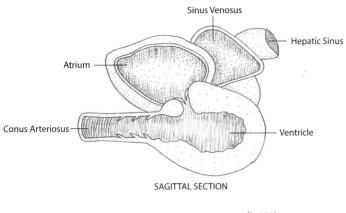

Sinus Venosus

Hepatic Sinus

Atrium

Conus Arteriosus

Ventricle

SAGITTAL SECTION

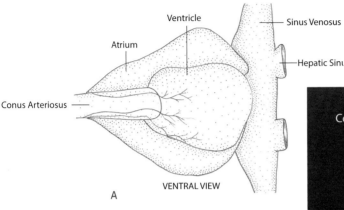

Ventricle

Sinus Venosus

Atrium

Hepatic Sinu

Conus Arteriosus

VENTRAL VIEW

A

Figure 7.7. (A) Anatomy of the shark heart. *Top*: side view. *Bottom*: ventral view. Arrows depict direction of blood flow. (Redrawn from Gilbert, S. 1973. *Pictorial Anatomy of the Dogfish*. University of Washington Press) (B) The heart of a Blue Shark.

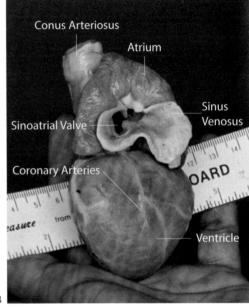

Conus Arteriosus

Atrium

Sinoatrial Valve

Sinus Venosus

Coronary Arteries

Ventricle

B

ventricle. Even though they possess two other chambers composed of contractile, cardiac muscle (the *sinus venosus* and *conus arteriosus*), our human chauvinism resulted in degrading lower vertebrates like sharks and their relatives by not counting their additional chambers as part of the heart, which indeed they are. Thus, the heart of chondrichthyan fishes is a *bona fide* four-chambered heart (fig. 7.7). Unlike the mammalian heart, in which there are two separate atrium-ventricle pairs *in parallel* with each other, and hence two separate circulations, the chondrichthyan heart consists of four chambers *in series* and only a single circulation, as we described earlier (see fig. 7.1).

In the chondrichthyan heart, all venous blood returns to the *sinus venosus*, then it moves to the atrium, ventricle, and the *conus arteriosus* (Figs. 7.4 and 7.7) before being sent to the gills. The chondrichthyan heart thus pumps relatively deoxygenated blood.

How does a muscle that depends on aerobic metabolism (a muscle that requires a continuous source of oxygen) ceaselessly pump if the blood it is pumping has lost most of its oxygen before returning to the heart? First, the blood within the heart, called the *luminal* blood (named because it is found in the lumen or space of a tubular structure), is mostly but not completely deoxygenated. Thus, the small amount of oxygen remaining in the luminal blood supplies the myocardial tissue, albeit minimally. This method of supplying oxygen to the heart likely constitutes the primitive condition among all chordates, with jawless fishes being the only extant chordates utilizing luminal blood as the only source of oxygen for the myocardium.

Second, the chondrichthyan heart has *coronary vasculature* (heart arteries) that carry blood to the heart muscles from arteries that peel off from the circulation after the blood has been oxygenated in the gills (fig. 7.3). This constitutes the major source of oxygen for the elasmobranch heart. A recent study[7] showed 3–4% of the entire *cardiac output* (the volume of blood pumped by the heart per unit time) of Sandbar Sharks (*Carcharhinus plumbeus*) constituted coronary circulation.

All four cardiac chambers of sharks are contractile and thus contain cardiac muscle, but each chamber has a different function and different architecture. The chambers are separated from each other by valves, which are closed, preventing backflow, until pressure develops during contraction of each chamber and opens them. Refer to Figures 7.4 and 7.7 as we describe the chambers.

The first (most upstream) part of the heart, the *sinus venosus*, is a small, very thin-walled chamber that receives the blood from the venous circulation. Do you have a sinus venosus? Not *exactly*. Chordates, of which you are one, possess a sinus venosus at some point in their embryonic development. It persists to adulthood in hagfish, bony fishes, chondrichthyans, and amphibians, but is found in only embryos in other chordates. As the mammalian embryo develops, the sinus venosus is modified and no longer exists as a chamber, but rather as specialized assemblages of cells, including the *sino-atrial node*, the pacemaker of the mammalian heart that initiates the heartbeat.

Some of the blood entering the sinus venosus remains there and some goes directly into the atrium through a *sino-atrial* valve. In contrast to the small size of the sinus venosus, the chondrichthyan atrium is capacious, essentially a large sac, though it is still thin-walled.

Blood moves from the atrium through an *atrio-ventricular* valve into the ventricle. The ventricle is the workhorse of the chondrichthyan heart. It is the heaviest of the chambers, and it contracts with the greatest force. The outer layer of the ventricle consists of *compact myocardium* overlying a thicker later of *spongy myocardium.* The lumen of the chondrichthyan ventricle is small relative to the size of the ventricle (in contrast, the lumen of the atrium is just a fraction smaller than the volume of the atrium). Most of the blood in the ventricle resides in its spongy myocardium, which is said to be *trabeculated*; that is, containing supporting beams of cardiac tissue that create a meshwork, much like a sponge.

Finally, blood moves from the muscular ventricle through a *semilunar* valve into the *conus arteriosus*, another contractile chamber, but more tubular and with a variable number of its own semilunar valves. Unlike the *bulbus arteriosus* in bony fishes, the elasmobranch conus arteriosus is a contractile chamber with myocardial muscle in its walls. Contraction of the conus arteriosus does not greatly increase pressure but rather maintains the pressure developed by contraction of the ventricle for a longer period as blood moves into the ventral aorta, thus protecting the gills from large pressure swings.

A recent paper[8] documented the presence of bulbus arteriosus tissue at the distal end of the elasmobranch heart between the conus arteriosus and the ventral aorta. The function and phylogenetic significance of the bulbous in elasmobranchs is unknown.

The heart of all vertebrates is surrounded by a *pericardium*, also called a pericardial sac (figs. 7.2, 7.8, and 7.9).

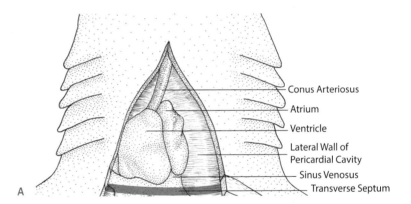

Conus Arteriosus
Atrium
Ventricle
Lateral Wall of Pericardial Cavity
Sinus Venosus
Transverse Septum

A

Figure 7.8. (A) Diagram showing the heart of a shark in the pericardial space, which is extremely spacious and whose walls are connected to surrounding muscle and cartilage (with the exception of the transverse septum). (B) Ventral view of the heart (color) inside the pericardial space of a Blacknose Shark. (Photo by Joshua Bruni)

B

Three features of the elasmobranch pericardium distinguish it from that of other vertebrates. First, the pericardial walls adhere to surrounding tissue, mostly muscle. Because of this, the elasmobranch pericardium is frequently called *semi-rigid* or relatively *non-compliant*. Only the transverse septum (fig. 7.8A), a fibrous sheet that separates the pericardial cavity from the abdominal (peritoneal) cavity and is attached only along its edges to surrounding tissue, is free to move. Second, the pericardial space of elasmobranchs is capacious. Under physiological conditions (normal activity level), it contains around 2.0 ml·kg^{-1} (2 ml in a 1 kg shark) in the Horn Shark compared to about 0.3–0.8 ml·kg^{-1} in humans. Third, elasmobranchs have a duct that connects the pericardial and peritoneal spaces called the *pericardio-peritoneal canal* (PPC; fig. 7.9), which we included in Chapter 1 as an unrecognized distinguishing feature of this group. A connection between these two spaces is a feature in the early embryological development of all vertebrates, but persists only in elasmobranchs, some agnathans (jawless fish), and sturgeon. The significance of these three features to heart function is described below.

Let us see how these pieces work in unison to do what a cardiovascular system does—pump the appropriate amount of blood to where it is needed precisely when it is needed.

We begin with the heart, specifically with a description of a mechanism so simple and elegant that it is included in every textbook in which heart function is discussed. Unfortunately, it is only partially accurate.[9]

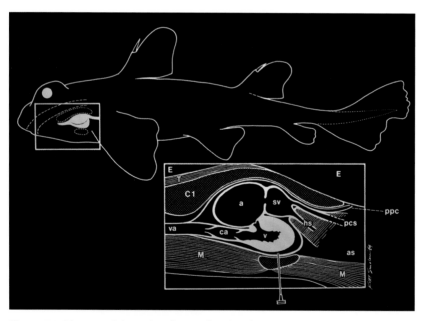

Figure 7.9. Diagram of the pericardium and pericardio-peritoneal canal (labeled ppc). Other labels: a = atrium; as = abdominal space; C1 = basibranchial plate (cartilage); ca = conus arteriosus; E = esophagus; hs = hepatic sinus; M = muscle; pcs = postcardinal sinus; sv = sinus venosus; v = ventricle; va = ventral aorta. (Courtesy of Kurt Smolen)

According to this model, when the muscular ventricle contracts within the semirigid pericardium, two actions occur. First, the blood in the ventricle is propelled to the conus arteriosus and into the branchial (gill) circulation. This much is in fact true, and physiologists call this *vis a tergo* (force from behind).

Second, the pericardial walls are affixed to surrounding muscles, behave as a semirigid box (think thin cardboard, not concrete), and do not move in much as the volume in the pericardial space is reduced as blood leaves the ventricle. Only the transverse septum that separates the pericardial and abdominal spaces is moveable, and it may be pulled inward slightly. This part of the story is also true.

As a result, according to the model prevalent in physiology textbooks, pressure in the pericardial space obeys the laws of physics and declines and becomes very negative, or subambient, like the pressure in a straw the moment you begin to draw up the fluid (fig. 7.10). This diminished pressure pulls open the walls of the thin-walled atrium, immediately upstream of the ventricle, and the suction developed within the atrial space aspirates blood from the upstream sinus venosus and blood sinuses into the atrium. Physiologists refer to this phenomenon as *vis a fronte* (force from the front). This is the part that is not entirely accurate.

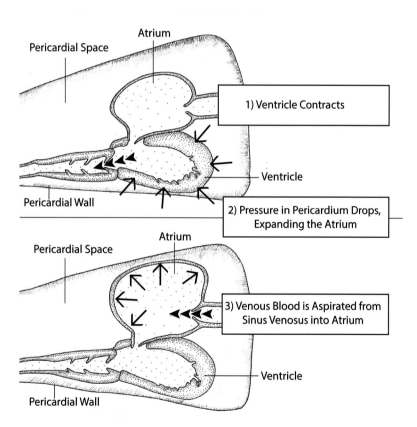

Figure 7.10. Series of drawings showing motions of the heart within the pericardium accompanied by pressure tracings. (Redrawn from fig. 1, Burggren, W.W. 1988. Experientia. 44: 919–930)

In the 1980s, a new conceptual model[10] emerged that showed that pericardial pressures were only *slightly* rather than *very* negative under routine, non-stressed conditions, and that this pressure played a role but worked in concert with muscular pumps in the sharks trunk (similar to those in your legs) to return blood to the heart. There are one-way valves in some of the veins to prevent backflow of blood.

However, as we explain below, under certain factors (e.g., when accelerating to catch prey or repositioning food in their mouths) pericardial pressure drops to the more negative levels assumed in the old model. We will see that the PPC plays a pivotal role in this process, and the new conceptual model is even more elegant than the one it replaced.

Responses of the Heart and Circulation to Activity

As in all vertebrates, the heart plays a major role in supporting increased activity levels and meeting energetic demands that may accompany predation, avoiding being preyed upon, and so on. During these times, muscle and other tissues require additional oxygen and fuel and they must additionally eliminate carbon dioxide and potentially toxic metabolites (end-products) like lactic acid. Thus, the *cardiac output* (the volume of blood pumped per unit time) must be increased under these circumstances. Cardiac output is numerically equal to the volume of blood pumped per beat (the *stroke volume*) multiplied by the number of beats per minute (the *heart rate*), and is typically reported as $ml \cdot min^{-1} \cdot kg^{-1}$ (kg represents body weight and is included to account for differences in body size).

There are only two pathways to achieving increased cardiac output in vertebrates—increasing stroke volume and/or heart rate. In sharks and batoids, faster swimming will push more blood back into the heart, supplementing the suction we previously described. In response to the additional blood, the ventricle will expand and then contract with greater force and increase stroke volume. This is an important physiological principle known as *Starling's Law of the Heart* (or the acronym SLOTH). You could think of this as, once again, the tail wagging the dogfish. Small increases in heart rate can occur, but these are much more limited in scope than in other vertebrates.

Heart rate varies among elasmobranchs. In the Horn Shark, average routine heart rate was about 40 beats per minute (bpm), and in the low- to mid-50s in Lemon (*Negaprion brevirostris*), Sandbar, and Shortfin Mako Sharks. Maximum heart rate found in sharks has not been shown to exceed 60–70 bpm. In contrast, maximum heart rate can be greater than 200 in tuna.[11] Thus, increases in stroke volume constitute the largest contributor to cardiac output in sharks and batoids.

Sharks with high performance cardiovascular systems, like Shortfin Makos, have a larger heart, especially the ventricle, than a Horn Shark, and may have cardiac output several times higher than the Horn Shark's during activity, even if the heart rates are only slightly different. A recent study in our lab documented smaller heart sizes in deep-sea sharks. As we have mentioned, heart size correlates with activity level in elasmobranchs. The Shortfin Mako and other sharks with high metabolisms also have higher blood oxygen capacities than less active sharks, and these are achieved through larger blood volumes, increased hematocrit levels, and higher hemoglobin concentrations.

Role of the Pericardio-peritoneal Canal (PPC)

How does the PPC affect heart function in elasmobranchs? Experiments with Horn, Blue (*Prionace glauca*), and Swell Sharks (*Cephaloscyllium ventriosum*), Shortfin Makos, Dusky Smoothhounds (*Mustelus canis*), and Thornback Fanrays (*Platyrhinoides triseriata*) showed that feeding and movement led to the ejection of pericardial fluid through the PPC.[12] This decompression of the pericardium protected the heart from high pressures that might occur during biting and swallowing food, as well as during swimming as the animal bends and the skin tightens. Ejection of pericardial fluid through the PPC also resulted in decreased pericardial pressure, which translated into increased return of blood to the heart. Recall that this increased blood return can be important for faster swimming through the SLOTH mechanism we discussed above. One structure, two (or more) functions. Evolution is great!

In summary, like that of all vertebrates, the closed circulatory system of sharks and batoids operates under a range of situations to meet the needs of the animal and at the same time participate in whole body homeostasis, which is one of the major goals of physiological systems.

Respiration and Metabolism

In 1982, Scripps Institution of Oceanography physiological ecologist Jeff Graham and colleagues designed and supervised the construction of one of the biggest treadmills on the planet, a behemoth nicknamed the *elasmotunatron* (fig. 7.11) that was built on a 3.3 x 5.3 m (10.8 x 17.4 ft) pallet.

The elasmotunatron, also called a *water tunnel respirometer* for reasons that will become clear below, was not destined for a fish health club but rather the NOAA research vessel *David Starr Jordan*. And you could not walk or run on it, because it was a *water* treadmill, also known as a *flume*. The rather large device was constructed with the goal of swimming a Shortfin Mako or an Albacore (tuna) to determine the amount of oxygen consumed as it swam against currents of different intensity.

The rate of oxygen consumption is a proxy for an animal's metabolic rate, or energy use per unit time. The branch of science concerned with metabolic rates is *energetics*. Understanding how much energy an animal uses is a key ingredient to understanding its *energy budget*, an accounting of all of the energy an animal requires and all of the ways this energy is expended.

Back to the elasmotunatron. Because of the logistical difficulties of capturing a Shortfin Mako, one of the most metabolically charged sharks in the ocean,

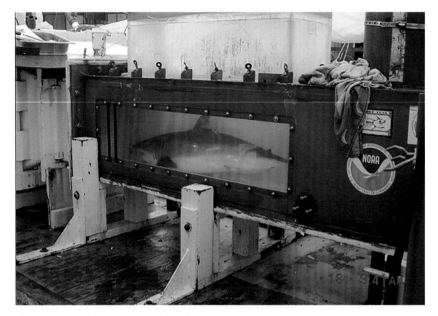

Figure 7.11. Mako swimming in the elasmotunatron. (Courtesy of Chugey Sepulveda)

while minimizing the stresses associated with capturing such a beast, only a handful of Shortfin Makos have been successfully tested. Do you think marine biologists will ever swim a mature White Shark in a specially designed water treadmill? We will revisit this story below.

Why It Is Difficult to Breathe in Water

Acquiring oxygen from water is far more challenging than doing so from air. The first huge advantage goes to air, which contains an almost invariable 20.95% oxygen, or about 30 times the average amount (0.7%) in water. Some aquatic ecosystems are even devoid, or almost devoid, of any oxygen. The second, third, and fourth big advantages also go to air; it is lighter, flows more easily, and oxygen diffuses more readily in it.

The upshot is: to get access to 1 liter of oxygen, it takes a volume of air weighing 0.0062 kg (0.01 lb) but 143 kg (315 lb) of water, assuming the water is 0.7% oxygen. Thus, breathing water requires not only a greater expenditure of energy, but also the evolution of means to efficiently extract the limited oxygen. And these together impose limitations on the upper metabolic rates of bony fishes, sharks, and rays compared to marine mammals, which breathe oxygen-rich, lighter, more flowable air.

Metabolism and Energetics

Recall that conservation of energy is central to the struggle for existence in animals and is a major selective pressure in the evolution of adaptations. How an organism obtains and uses energy underpins all aspects of that organism's life, including its physiology, ecology, behavior, and evolution. For example, to manage a fishery, scientists and policymakers should know growth rates for all of the targeted species and bycatch in the fishery, and this depends in part on energy consumption.

Central to understanding the energetics of any organism is knowing its daily energy budget. An energy budget accounts for how an animal uses, wastes, and eliminates the energy it requires for all of its energy-consuming processes (e.g., pumping the Na^+ and Cl^- ions during osmoregulation; see Chapter 8).

An energy budget for an organism can be represented mathematically as:

$$C = G + M + E$$

Where,
C = consumption of energy (i.e., calories obtained from food)
G = growth in mass (i.e., energy stored in tissue)
M = metabolism (i.e., energy expended in daily life)
E = excretion (i.e., energy lost in urine and feces; these are sometimes separated into two terms, U = urinary excretion and F = egestion)

Sonny Gruber of the University of Miami and his students[13] calculated one of the first shark energy budgets, for a 1 kg (2.2 lb) juvenile Lemon Shark, as:

$$100\% \ C = 22.4\% \ G + 49.4\% \ M + 28.2\% \ E$$

Metabolism, which constitutes a central component of an organism's energy budget, is apportioned among a number of different uses, including cost of locomotion, cost of ventilation, growth and maintenance, reproduction, and specific dynamic action (i.e., the additional energy expended during digestion and protein synthesis).

Measuring an organism's metabolism (its *metabolic rate*) may seem easy in theory but in practice it is very difficult, especially for sharks. Think about why this may be. First, you need a shark that is not stressed, a condition that is extremely difficult to achieve in recently captured or captive animals. Second, there are several levels of metabolism, and ideally a biologist wants to know all of them, from the minimum amount of energy expended during rest to that associated with the maximum level of activity, as well as points in between. Measuring all of these requires a combination of expensive instrumentation, cooperative sharks, and good luck. Finally, seemingly insurmountable logistical problems exist for measuring metabolic rates in rarely encountered sharks (e.g., Goblin Sharks, *Mitsukurina owstoni*), sharks living in extreme environments (the deep-sea, like Gulper Sharks, *Centrophorus granulosus*), big sharks (e.g., Whale Sharks, *Rhincodon typus*), or highly active sharks (e.g., Porbeagles, *Lamna nasus*).

Given these limitations and difficulties, you might be surprised to learn that metabolic rates have been measured or estimated for dozens of species. Before we tell you how these studies have informed scientists about metabolic rates of sharks, we will provide more background.

How is metabolic rate measured? Actually, the true rate at which an organism uses energy is not directly measured, since there is no easy way to measure the conversion of energy; that is, the number of ATP molecules (the biocompounds, frequently called a cell's *energy currency*, that are used at the site of the energy conversion) converted to ADP molecules *in cells in every organ in the body*. Instead, *proxies* (another measurement that is proportional to or is a close estimate of metabolic rate) for metabolic rate are used.

The method most routinely employed involves measuring the amount of oxygen used, a process called *respirometry*. Since most of metabolism is aerobic (oxygen-using), oxygen consumption is one of the most valid estimates of metabolism. It is measured using sharks contained in respirometers, in which sharks either swim in circles (annular/circular respirometers) or they swim against a current (swim tunnels/water treadmills).

Other methods include biotelemetry sensors, which, once implanted, can collect data on heart rate, tail-beat frequency, muscular contractions (electromyograms, or EMGs), muscle temperature (for endotherms only), production of metabolic water, and what is called overall dynamic body acceleration (ODBA).

The last of these, accelerometry, is being used more frequently since the advent of inexpensive accelerometers (found in smartphones, drones, and even Fitbits). Acceleration refers to the rate of change of velocity. As sharks swim, they expend energy for contracting the swimming muscles, and as this happens, there is acceleration of the body and its parts in different planes, and studies have demonstrated that acceleration is proportional to metabolic rate. Accelerometers must first be calibrated using conventional techniques (e.g., swimming a shark with accelerometers in a water treadmill at various speeds).

One final method involves measuring the activity of key regulatory enzymes involved in metabolic processes. These include citrate synthase, lactate dehydrogenase, and gatekeeping enzymes important in cellular respiration.

Early investigations of the metabolic rates of sharks were conducted on smaller species, which were most easily maintained in captivity, the Small-spotted Catshark (*Scyliorhinus canicula*) and the Spiny Dogfish (*Squalus acanthias*). These results suggested that sharks were more metabolically constrained than ecologically similar bony fishes. In other words, bony fishes were thought to have higher metabolic rates than elasmobranchs. However, the two shark species studied are no more representative of typical sharks than a couch potato or Olympic athlete are of typical humans.

Since the metabolic rates of the Small-spotted Catshark and the Spiny Dogfish were measured, similar studies have been conducted on numerous other elasmobranchs. What these studies have shown is that standard metabolic rate (SMR), the minimum energy required for essential biological functions during rest, standardized for body size, was not significantly lower in elasmobranchs and that it correlated positively with environmental temperature and activity. Similarly, maximum metabolic rate was highest in more active species. The most active sharks (i.e., Shortfin Makos) are obligate ram ventilators and, as you would expect, have the highest mass-specific metabolic rate.

A 1993 study[14] tested the hypothesis that sharks and rays have reduced metabolic rates compared to ecologically similar bony fishes. In this study, researchers did not measure oxygen consumption, but rather the speed of chemical reactions of the three enzymes considered biochemical indices of aerobic and anaerobic capacity in tunas and lamnid sharks: *citrate synthase*, *pyruvate kinase*, and *lactate dehydrogenase*. They found that tunas appear to have more and better biochemical specializations, in addition to anatomical and physiological, for high performance than lamnid sharks, even as the two endothermic groups have converged. They also concluded that the metabolic rates of non-lamnid sharks compared favorably to those of bony fishes with similar ecological roles.

Understanding metabolism and energetics is important for managing fisheries and understanding ecological energetics (the sum of all energy flows in a marine ecosystem). Moreover, it is becoming critical to understand energetics of organisms in marine ecosystems as global climate change potentially impacts food supplies, activity levels, and other physiological processes.

Concluding Comments

Sharks are not heartless, though their hearts are smaller than those of similarly sized mammals, and they function differently as well. Remarkably, these relatively small hearts work efficiently to meet the needs of Dwarf Lantern Sharks, Whale Sharks, and all elasmobranchs in between. And despite the difficulty of breathing in water, elasmobranchs have evolved a respiratory system that works in concert with the circulatory system to extract sufficient oxygen and deliver it and other important products to muscles and other tissues as needed and to remove wastes. Elucidating how these functions are accomplished continues to enthrall and amaze us and, we hope, you too.

NOTES

1. In physiology, the term *afferent* is always used for something that is directed toward a tissue, organ, or organ system. In this case, that something is the gills.

2. Similarly, *efferent* refers to something directed away from a tissue, organ, or organ system.

3. Bony fishes and other vertebrates have greatly reduced sinuses embedded in the liver.

4. Walsh, C.J. and Luer, C.A. 2004. In: Smith, M. et al. (eds.). The Elasmobranch husbandry manual: captive care of sharks, rays and their relatives. Special Publication of the Ohio Biological Survey. 307–323 pp. https://sites.google.com/site/elasmobranchhusbandry/manual. (Accessed 9/9/19).

5. Haines, A.N. and Arnold, J.E. 2014. In: Smith, S.L. et al. Immunobiology of the shark. CRC Marine Biology Series. CRC Press. 89 pp.

6. Holocephalans apparently lack both organs, and use the thymus, spleen, and lining of the orbit for hematopoiesis.

7. Cox, G.K. et al. 2017. J. Comp. Physiol. B 187: 315–327.

8. Durán, A.C. et al. 2008. J. Anat. 213: 597–606.

9. See, Abel, D.C. et al. 1986. Fish Physiol. Biochem. 1: 75–83 and Lai, N.C. et al. 1996. Am. J. Physiol.–Heart and Circ. Physiol. 270: H1766–H1771.

10. Abel, D.C. et al. 1986. Fish Physiol. Biochem. 1: 75–83.

11. As cited in Brill, R.W. and Lai, N.C. 2015. Fish Physiol. 34: 1–82.

12. Abel, D.C. et al. 1994. Fish Physiol. Biochem. 13: 263–274.

13. Gruber, S.H. 1984. Ann. Proc. Am. Ass. Zool. Parks and Aq. (AAZPA). 1984: 341–373.

14. Dickson, K.A. et al. 1993. Mar. Biol. 117: 185–193.

8 / Thermal Physiology, Osmoregulation, and Digestion

Introduction

Let us say that you are a juvenile, meter-long (3.3 ft) Sandbar Shark (*Carcharhinus plumbeus*) swimming into Winyah Bay in northeastern South Carolina in late spring, having migrated there from overwintering in Florida as part of an annual ritual. You would not be unlike the people, some call them *snowbirds*, who travel from the northeast United States to Florida to escape the frigid cold and return when the conditions become more clement. Similarly, you will return to your Florida digs when the water temperature cools in late fall.

Being an estuary, Winyah Bay is connected to the ocean on one end, and to several rivers on the other. Thus, the salinity typically ranges from full-strength seawater at the mouth of the bay to full fresh water in the bay's upper reaches. You settle for an area where the salinity is intermediate, about mid-bay, where the ocean and river waters mix.

Your migration and selection of habitat have been driven by your physiological tolerances to and preferences for certain temperatures and salinities, as well as the advantages (e.g., abundant food, protection from predators, and so on) that your seasonal habitats may afford you. In this chapter, we closely examine the effects of temperature and salinity on elasmobranchs, and their often-amazing physiological responses. We also examine the process of digestion. (You can return to being a person again.)

Temperature

Temperature is one of the most important environmental factors for all living organisms and is becoming more so as climate change impacts materialize. It affects not only their diversity, abundance, and distribution, but also their behavior, activity, growth, development, metabolism, heart rate, digestive rate, blood characteristics, and other aspects of their physiology at all levels from the molecular to organ systems.

Physiologically, temperature typically influences the rates of chemical reactions, a phenomenon described by the *temperature coefficient*, or *Q10*, of a reaction or process. Q10 numerically equals the factor by which the reaction rate increases for every 10°C (18°F) increase in temperature. Q10s typically are around 2, that is, the reaction rate *doubles*, or in some cases, triples, for every 10°C temperature increase. For example, the Q10 for metabolic rate for the Horn Shark (*Heterodontus francisci*) is 2.01.

Organisms have both temperature *optima* and *preferences*. Exposure to temperatures outside of this range (beyond an organism's upper and lower thermal limits), or *thermal tolerance scope,* can lead to thermal stress, which may be directly lethal or have sublethal effects that will reduce an organism's fitness (its ability to forage, avoid being eaten, compete, mate, and so on).

The highest temperature an organism can tolerate before it begins a suite of irreversible actions leading to its death is called its *incipient lethal temperature* (ILT) or *critical thermal maximum* (CTmax). Temperatures in this range lead to a cascade of deadly physiological problems, including disrupting cellular function by destroying membranes and denaturing (disrupting the physical structure of) proteins.

The lowest temperature an organism is capable of tolerating is its *critical thermal minimum* (CTmin). Like heat-related death, cold death involves a cascade of physiological impacts; for example, central nervous system and cardiac malfunction, as well as metabolic, blood, and water balance changes.[1]

Both CTmax and CTmin can vary by species, season, life-stage, geography, and other factors. Given the range of factors that affect organisms' critical thermal maxima and minima, it is not surprising that few of these have been determined for elasmobranchs, and some of these are based on observations of animals killed by temperature extremes in their environments more so than in controlled laboratory experiments.

The first CTmax determined for a shark, published in 1924 for the Spiny Dogfish (*Squalus acanthias*), was 29.1°C (84.4°F).[2] In laboratory studies, the CTmax for Atlantic Stingrays (*Hypanus sabina*) from St. Joseph's Bay, Florida, maintained in aquaria at 10.8°, 20.5°, or 35.1°C (51.4°, 68.9°, and 95.2°F) were 35.7°, 39.3°, or 43.2°C (96.3°, 102.7°, and 109.8°F), respectively. In other words, sharks and rays living in warmer waters had higher tolerances than those in cold, and the higher the temperature of acclimatization in captivity, the higher the CTmax and the tolerance to high temperatures.

CTmin for Atlantic Stingrays acclimatized as above were, respectively, 0.8°, 4.8°, or 10.8°C, (33.4°, 40.6°, 51.4°F).[3] Thus, the lower the temperature of acclimatization, the lower the CTmin and the higher tolerance for lower temperature.

Within the range of temperatures in which organisms are found, other variables being equal (e.g., prey availability), there are advantages to living nearer the higher temperatures: physiological processes (e.g., rates of digestion, processing of sensory information, muscle contractions, and rates of chemical reactions) will occur more quickly, along with development *in utero* and growth rate.

For aquatic organisms, however, it is hard to be warm in anything but tropical water because of the thermal properties of water, including its high heat capacity (3000 times that of air) and its thermal conductivity (25 times that of air). These translate into rapidly losing heat in water, about 75,000 times more heat than is lost in air. Some animals (e.g., marine mammals, some fishes) have evolved ways to keep warmer than their surroundings, but most aquatic organisms have a body temperature the same as that of the environment in which they reside.

We call animals whose body temperature is determined primarily by the temperature of the environment *ectotherms* (from *ekto* = out; *therm* = temperature).[4] All invertebrates and most bony fishes and chondrichthyans are ectotherms.

Consider the effect of temperature on just one factor—metabolic rate (how fast an organism uses energy). The metabolic rate of ectotherms decreases in colder temperatures and increases in warmer ones. This translates into being less active with less food required for energy in the former and more active with more food in the latter. Deep-sea sharks, specifically those that do not migrate vertically daily, live in an environment whose temperature is about 4°C (39°F) day and night, across all seasons. These animals have a lower metabolic rate than, say, a Blue Shark (*Prionace glauca*) at the surface. Deep-sea sharks thus expend less energy and require less energy from prey than shallow-water counterparts, which is convenient since the deep-sea is a food-poor environment. This deep-sea lifestyle often leads to relatively delayed life-history characteristics, such as age at maturity or gestation.

While ectothermic sharks do not possess physiological specializations that allow them to maintain body temperatures above those of their environment like the *endothermic* sharks described below, many migrate seasonally to warmer waters and some select higher water temperatures on shorter time scales, presumably as an adaptation that confers upon them one or more of the advantages of the higher temperatures.

In California, female Leopard Sharks (*Triakis semifasciata*) congregate in warm shallow coastal waters during the summer, in the process elevating their core body temperature by as much as 3°C (5.4°F), and in doing so increase their metabolic rate by about 17%.[5] Moreover, the sharks in the warm shallow waters darkened their skin, further increasing their heat gain. These behavioral and morphological adaptations apparently make these sharks better predators, although observational verification is lacking.

The Leopard Sharks leave these same areas in late afternoon as these shallow waters cool faster than deeper ones. Movement to deeper and cooler waters is thought to slow the rates of digestion and defecation, in the process improving the efficiency of extracting nutrients from the food being digested. In other species that feed more frequently, however, the opposite is thought to occur (feeding while cool and more quickly digesting and evacuating while warm).

Similarly, Blacktip Reef Sharks (*Carcharhinus melanopterus*) aggregate in warm shallow waters during daylight and disperse at night.[6] These consistently exhibited core body temperatures of about 1.3°C (2.3°F) above that of their environment. This temperature increase persists for a variable time (depending on the shark's body size and activity as well as the water temperature) even after leaving the warm shallows, a phenomenon called *thermal inertia*.

Bluntnose Sixgill Sharks (*Hexanchus griseus*) migrate vertically in the water column diurnally. In a study conducted in the NW Atlantic, Gulf of Mexico, Central Pacific, and Bahamas, individuals spent days at deeper depths than in evenings, although the depth disparity varied from < 50 m (160 ft) in the NW Atlantic to > 300 m (990 ft) in other locations. It would be tempting to attribute this exclusively to following prey as they migrate vertically, but closer examination showed that these differences in depth in all four locations reflect temperature preference for 5°C (41°F) in the day and 16–17°C (61–63°F) in the evening.

Sandbar Sharks also migrate, at least in part based on water temperature. In a study conducted in Chesapeake Bay,[7] Sandbar Sharks were caught on experimental longlines from late-May until mid-October, with most immigration in mid-June when bottom temperature was above 20.0°C (68°F). Emigration, however, correlated more with day length (photoperiod) than temperature, although the study caught no Sandbar Sharks when bottom temperature was below 15.9°C (60.6°F).

Endothermic Sharks

Only a handful of bony fishes and elasmobranchs have diverged from the mainstream and have evolved to be *endotherms*; that is, their body temperature is determined both by the metabolic heat they produce—their metabolic furnace— and by the environment. Recall that chemical reactions that involve energy conversions are inherently inefficient, about 40% or so in living systems, meaning that as much as about 60% of the energy is lost as heat.

The major problem an aquatic organism faces in evolving endothermy is preventing heat loss. For bony fishes or sharks, most of the heat is lost across the skin and gills. Those same architectural traits of gills that enable them to obtain oxygen and eliminate carbon dioxide, specifically their large surface area and thinness, are huge liabilities for retaining heat since the conductivity of heat is so much higher than that of respiratory gases. As much as 80–90% of the heat of blood is lost in a single passage through the gills, which represents between 20–40% of the total heat loss.

Given that the gills of fish have adapted as exchange organs with large surface areas, there have been no specializations in evolution to conserve heat there. Instead, the focus in evolution has been on preventing heat loss from the body wall. Marine mammals and seabirds (e.g., penguins) have become masters of heat conservation by using thermal mass, blubber, fur, and/or feathers to trap heat inside. It helps that all marine mammals and birds breathe air, and thus do not experience heat loss through the gills (since they lack them), avoiding the 20–40% heat loss of fish.

Members of only seven fish families, Scombridae (14 species of tunas), Lampridae (one species of Opah), Istiophoridae (marlin and other billfishes), Xiphiidae (swordfish), Lamnidae (five mackerel sharks), Alopiidae (three species of threshers), and Mobulidae (at least two species of mantas) have evolved ways of trapping heat and elevating their body (or specific parts of their body) temperatures above that of the environment, a phenomenon known as *regional endothermy*[8] (so named because there are still temperature gradients, i.e., body regions that are warmer than others by a degree or more). Since scombrids, lamprids, istiophorids, and xiphiids on the one hand, and lamnids, alopiids, and mobulids on the other, are also in two different taxonomic classes that diverged as long ago as 450 million years, their regional endothermy is an example of *convergent evolution*, in which distantly related or unrelated organisms evolve similar anatomical, physiological, or behavioral traits (see Chapter 2). We will focus on lamnid sharks, but the general principles *mostly* apply to scombrid bony fishes as well, although there are anatomical differences.

The ability to retain heat in lamnids rests exclusively with the evolution of structures in the circulatory system that allow the engineering principle called *countercurrent exchange* to take place. In countercurrent exchange, there will be virtually complete transfer of some substance or property, in this case heat (but in gills, oxygen as well as heat), from one fluid to another across a permeable membrane if they are juxtaposed and running in opposite directions for a distance long enough for the transfer to occur (see fig. 4.23). In co-current (or concurrent) exchange (when the fluids run in the same direction), only 50% of the heat at best could be transferred.

In sharks, these countercurrent exchangers are known as *retia mirabilia* (singular: *rete mirabile* = wonderful net) and consist of intermingled blood vessels, one set (venules, or small veins) carrying blood warmed near the shark's core, surrounded by another set (arterioles, or small arteries) carrying cooler blood from the periphery (near the skin; fig. 8.1). By the time the warmer blood reaches the periphery, the heat has been transferred to the cooler blood moving toward the core, and thus the core retains the heat.

Lamnid sharks (and tunas as well) are capable of using countercurrent exchange because they have a special arrangement of their main blood distribution system. In most sharks and bony fishes, there is a central blood distribution system, with blood from the gills going to a large dorsal aorta, then to the tissues

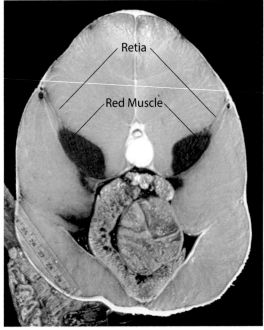

A

B

Figure 8.1. (A) Cross-section through an endothermic shark, a 125 cm (4.1 ft) Shortfin Mako, showing its *retia mirabilia*, centralized red muscle, and white muscle. (Courtesy of Jeff Graham Lab, Physiological Research Laboratory, Scripps Institution of Oceanography) (B) Section through an ectothermic shark, here, a Bonnethead, showing the narrow strip of red muscle confined to the periphery. (Courtesy of Bryan Keller)

(muscles and organs), after which it returns through the post-cardinal vein (fig. 8.2).

The major vessel distributing blood from the gills in lamnids sharks and tunas is not a single dorsal aorta, but, depending on the species, one to four artery-vein pairs just under the skin (fig. 8.2). The lateral artery or arteries carry cooled blood from the gills, and the lateral vein or veins are the main route via which blood returns to the heart. The retia are comprised of the arterioles that branch off of the lateral arteries running next to venules that branch off of veins nearer the backbone (or body core).

In lamnid sharks, the retia are located not only in the locomotory muscles (the *lateral cutaneous* retia) but also in the cranium near the eyes (*orbital* retia) and in the viscera (*suprahepatic* retia). Respectively, these produce varying levels of red muscle endothermy, cranial endothermy, and visceral endothermy.

The main source of heat for the lateral cutaneous retia is *red muscle* (fig. 8.1)— muscle that is red in color due to high vascularization and myoglobin (a relative of hemoglobin that stores oxygen in muscle). Red muscle is primarily used for cruising or slow-speed swimming and is characterized by being slow-twitch, aerobic, and having small-diameter fibers. Interestingly, the amount of red muscle in endothermic and ectothermic sharks is similar, about 2% of body mass. However, the location of the red muscle bundles differs between endothermic and ectothermic sharks. In the former, the red muscle is near the backbone, whereas in the latter it is found near the skin. Thus, in endothermic sharks, contractions of the deeper red muscle generate the heat that the countercurrent exchangers conserve.

The countercurrent exchangers allow lamnid sharks to maintain temperature differentials from 8°C (14°F) over ambient in the Shortfin Mako (*Isurus oxyrinchus*) to 14.3°C (26°F) in the White Shark (*Carcharodon carcharias*) to as high as 21.2°C (38°F) in the Salmon Shark (*Lamna ditropis*).[9]

Two main hypotheses explain the main advantages of red muscle endothermy.[10] The first is the *thermal niche expansion* hypothesis, which asserts that endothermy allows fish to move independently of temperature, both latitudinally and vertically within the water column.

The second hypothesis is that endothermy enables elevated cruising speeds. All endothermic fishes are fast-swimming, highly mobile predators. Thus, advantages relate to their high-performance lifestyle and include more efficient prey capture, greater muscle power, faster swimming, enhanced sensory capacity, and more efficient digestion.

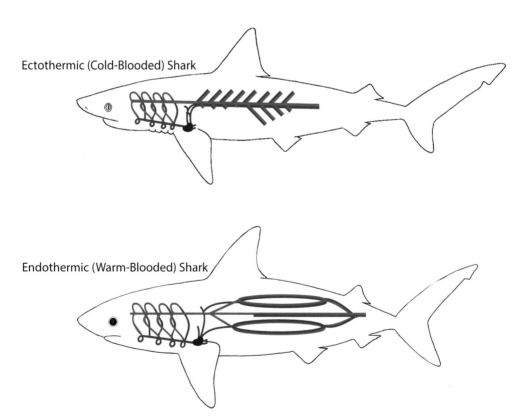

Figure 8.2. Diagram of differences in arterial blood distribution in ectothermic and endothermic sharks. (Redrawn from Carey, F.G. 1974. In: Wessells, N.K. (Ed.). *Vertebrate Structures and Function*. Freeman. 236–244 pp)

Both explanations make sense. A 2015 paper[11] supported the cruising speed hypothesis using modern tagging techniques on free-swimming animals. Their results, that cruising speeds and maximum annual migration ranges of tunas and sharks with red muscle endothermy are 2–3 times greater than ectothermic relatives, also validated the niche expansion hypothesis.

Warming the brain specifically (cranial endothermy) increases sensory acuity and processing, since temperature increases rates of chemical reactions. Thus, endothermic animals likely are superior prey finders and predator avoiders as well.

The cost of being endothermic is that the metabolic furnace generating the heat must be continually fed. An estimate of the food required by an organism of a particular life stage is its *daily ration*, which is typically reported as the mean percentage of an organism's total weight that is consumed over a 24-hour period. Surprisingly, so far these do not clearly demonstrate that endothermic sharks and tunas necessarily consume more food daily on average than their ectothermic relatives.

For ectothermic sharks, daily ration estimates cover a wide range, including 0.2% for adult Broadnose Sevengill Sharks (*Notorhynchus cepedianus*), 2.1% for juvenile Lemon Sharks (*Negaprion brevirostris*), and > 4% for adult and juvenile Bonnetheads (*Sphyrna tiburo*).[12]

For endothermic sharks, daily ration has been studied only in adult Shortfin Makos, with estimates ranging from 2.2–3.0% in one study[13] to 4.1% in another.[14] The latter calculation was based on data not available to the former and is likely more accurate.

The lower than expected daily ration calculations for endothermic sharks and bony fishes could be real and explained by higher digestive efficiencies or more calorically dense prey. Alternatively, they could be artifacts of using inexact data (e.g., metabolic rates, growth rates) in models and a general lack of studies on the subject in a wide variety of sharks and tunas.

One final point—no endothermic sharks are small. The explanation for this is basically the same as why there are no small marine mammals and it invokes a little geometry and physics. Basically, a larger animal stores more heat in its thermal mass than a smaller one, and it also gains and loses heat more slowly. Smaller sharks are not capable of keeping the metabolic heat that is produced, or the heat they gain from behavioral thermoregulation, for very long.

Osmoregulation

Osmoregulation[15] refers to water and ion balance in an organism. Among vertebrates, for proper internal function, both the volume and the solute composition (dissolved components) of an organism's fluid compartments must be regulated so that homeostasis (the constancy of the structure and function of the internal environment) is maintained.

In what ways does the fluid environment affect biological function? Here are just a few:

- The structure of proteins, one of the most important classes of biological molecules, is directly affected by the ionic composition of the internal fluid environment. The function of proteins as enzymes, hormones, energy, in transportation, and as structure are all highly dependent on their complex 3-dimensional structure, which can be disrupted by changes in the fluid environment.

- Ionic composition influences pH, which affects protein structure and function.

- Cells rely on electrical gradients across their membranes for transportation of atoms and molecules and for the development and conduction of nerve impulses. These gradients rely on the proper ionic environment.

The problem that virtually all aquatic vertebrates must face, in fresh water or salt water, is that their internal environments *differ* from their external environments in terms of both *ionic* and *osmotic* characteristics; that is, in terms of the concentration and type of atoms (ions) and molecules as well as the percentage of water. These differences result in *gradients* (differences in the concentration over a distance or across a barrier) of the ions and water inside and outside of the organism. More specifically, these are called *ionic diffusion gradients* and *osmotic gradients*.

According to Fick's Laws of Diffusion, ions and water will each diffuse from where they are in higher concentration to where they are in lower concentration. We do not often think of the concentration of water in a solution, but rather the concentration of salts or ions. But for our purposes, we can assume that if solution 1 is more concentrated in terms of its ionic composition than solution 2, then the opposite will apply to their water concentrations (i.e., solution 2 has a higher water concentration than solution 1).

An important point here is that to achieve homeostasis (i.e., to regulate their internal environments), aquatic vertebrates can regulate their *water balance*, their *salt balance*, or *both*.

Osmoregulation in Bony Fishes

Let us first consider osmoregulation in saltwater bony fishes; for example, a tuna. If you measure the internal concentration of all of the solutes in the blood and other extracellular fluids of a tuna (thus, ignoring the fluid environment *inside* the cells, which is regulated differently), you will find that they are not nearly as concentrated as the salt water in which the tuna resides.

At this point it is necessary to introduce some terminology. We caution you not to be intimidated, because the concepts underlying osmoregulation in both bony fishes and sharks and their relatives are fairly straightforward and rely on only simple physical principles, even if the terminology may take some time to learn.

The salt concentration of seawater is referred to as its *salinity*, and historically this was measured in *parts per thousand* (ppt); that is, the weight of the salts (expressed in grams) per 1000 g (1 kg, or 2.2 lb) of seawater. The average salinity of seawater, expressed as ppt, is 35 (written as 35‰, or 3.5%). Oceanographers continue to use this measurement, but it is measured using conductivity rather than density and is reported as dimensionless ratios (e.g., 35 rather than 35 ppt).

Physiologists do not refer to the internal *salinity* of an organism. Instead, they refer to the *osmolarity* or *osmolality* of the internal and external environments of living organisms, but the concepts are similar. Osmolarity and osmolality are based on the concentration of osmotically active particles (atoms, ions, molecules) in a solution. *Osmolality* is measured as the solute concentration per unit *mass* and is thus for our purposes the same entity as salinity (although in different units). *Osmolarity* refers to the solute concentration per unit *volume* instead of mass. Numerical values of these are close but not identical for the systems we will discuss. We will primarily use osmolality, whose units are milli-osmoles per kg (expressed as mOsm·kg^{-1}).

If the environment on one side or the other (does not matter which) is *more concentrated* (i.e., has more solutes, or osmotically active particles) than the opposite side, the more concentrated side is said to be *hyperosmotic* to the less concentrated side. If the environment on one side or the other (again, does not matter which) is *less concentrated* (i.e., has fewer solutes, or osmotically active particles) than the opposite side, the less concentrated side is said to be *hypoosmotic*[16] to the more concentrated side. Unless the osmotic concentrations are equal (i.e., *isosmotic*), then the more concentrated side will be hyperosmotic to the less concentrated side, which is hypoosmotic to the more concentrated side. Confounding terminology perhaps, but a simple concept.

Unless the animal is totally impervious to the solute (typically salt) and water, which none are, the salt will move (diffuse) from the hyperosmotic (more concentrated) side to the hypoosmotic (less concentrated) side. Water will move in the opposite direction, from where it is more concentrated to where it is less concentrated.

Now, let us return to that saltwater fish, the tuna. The solutes in the tuna's blood and other extracellular fluid, mainly sodium (Na^+) and chloride (Cl^-) ions, are less concentrated than that of the external environment. Thus, the internal environment is hypoosmotic to the external environment. The solutes in seawater (again mainly but not exclusively Na^+ and Cl^-) will diffuse into the tuna, and water will leave. The latter process is referred to as *osmosis* (diffusion of water).

How quickly and how much movement of salts and water occurs depend on several factors (e.g., temperature, permeability of the gills and skin, and so on), but the most important is the magnitude of the gradient (how much difference there is in the salt and water concentrations *inside* and *outside* the tuna) in the seawater. Saltwater bony fishes like our tuna maintain an internal osmotic concentration of between 300–500 mOsm·kg^{-1}. Seawater has an osmotic concentration of around 1000 mOsm·kg^{-1}. Thus, there is a concentration gradient of around 500–700, with the internal environment hypoosmotic to the external environment (i.e., less concentrated than).

Saltwater fish therefore have a *water conservation problem* because water moves from the fish, primarily across the gills, which have a large surface area and are only a thin barrier to seawater. To compensate for this water loss, tuna and other marine bony fishes drink saltwater (three to 10 times as much as a freshwater fish drinks) and water is absorbed across the intestine. Urine flow is low, which also conserves water.

At the same time, Na$^+$ and Cl$^-$ diffuse from the seawater into the bloodstream of the tuna across the gills, and also enter into the body when the tuna eats and drinks. *Monovalent* ions (those with only a single electrical charge), here sodium (Na$^+$) and chloride (Cl$^-$), are eliminated by specialized, mitochondria-rich *chloride cells* at the gills. The process of eliminating the monovalent ions is energy-intensive and involves active transport, which explains the high concentration of mitochondria in the chloride cells. *Divalent* ions (those with two electrical charges), specifically magnesium (Mg^{++}) and sulfate (SO$_4^{--}$), are excreted via the kidneys. The combination of regulating both the ions and water results in osmotic and ionic homeostasis.

This homeostasis is not free—the energetic cost of osmoregulation in marine bony fishes ranges from 8–17% of their resting energy expenditures.

Osmoregulation in Sharks and Rays

Sharks and rays live in a medium that has about the same or slightly lower osmotic concentrations as their interior fluids. We call animals in which this occurs *osmoconformers*, because their internal water concentration conforms to that of the environment and thus water levels do not require regulation. For example, we found the mean osmolality of the blood of 12 Blue Sharks to be 1021, with seawater they resided in at 999 mOsm·kg^{-1}.

Although the absolute value of the osmolality of the internal fluid of sharks and other chondrichthyans is at or, more typically, slightly above that of the seawater in which they live, this does not mean that the two different fluid compartments (i.e., the interior and exterior environments) are composed of the same constituents. There are differences, and the ions responsible for these differences are regulated.

Seawater is mostly Na$^+$ and Cl$^-$ ions. About 50% of the internal fluids of sharks is Na$^+$ and Cl$^-$ ions, a slightly higher level than is characteristic of bony fishes. Why not just elevate the internal Na$^+$ and Cl$^-$ concentrations to that of seawater? Among vertebrates, only the hagfish is isosmotic (same osmolality) to seawater with similar principal ionic constituents (Na$^+$ and Cl$^-$ ions about 500 mOsm·kg^{-1} each). However, this is the exception, and it occurs in a group whose evolutionary success is dubious (perhaps, at least in part, explained by physiological limitations due to being isosmotic with seawater). If elasmobranchs had levels of

Na⁺ and Cl⁻ of the magnitude in hagfish, it would create the kinds of problems we began this section describing (i.e., causing unfolding of proteins, or changing electrical gradients across cell membranes required for normal function).

The remaining components of the internal osmotic concentration are two organic compounds that sharks and rays synthesize from metabolic waste products, *urea* and *trimethyl amine oxide* (TMAO; fig. 8.3). The use of these two osmotically active particles (also called *osmolytes*) was discovered in the late 1920s and early 1930s by renowned physiologist Homer Smith.[17]

Thus, sharks and their relatives utilize two compounds to elevate their internal osmolality to approximately equal that of seawater, their external environment. However, if we ended the story here, we would be negligent by omitting the monumental problems that had to be overcome to allow urea-based osmoregulation. Consider these four major drawbacks:

1 / *Urea is a potent metabolic poison that destabilizes or denatures proteins.*

Proteins are arguably the most diverse biomolecules in organisms. Central to their functioning effectively is that each protein's unique 3-dimensional structure is not altered. Urea severely disrupts the structure of a protein, such that the protein either is unable to fulfill its functions, or it does so less effectively.

Uremia, or urea in the blood, is a horrifying, often terminal, human disease in which urea accumulates because of kidney failure. Symptoms are far-ranging and include fatigue, weakness, nausea, and a long list of other neurological, ocular, respiratory, cardiovascular, and other effects. Urea poisoning

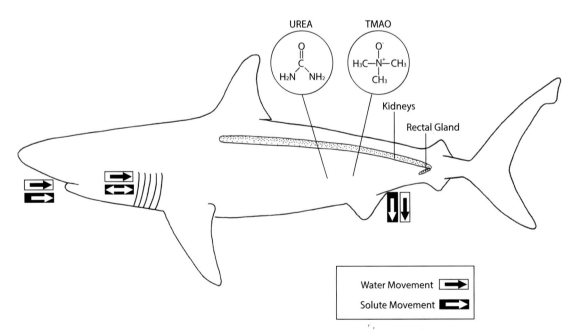

Figure 8.3. Relationships between distribution of solutes and water between a shark and its environment.

sometimes occurs in cows in which there is too much urea in the feed (used as a source of nitrogen) and frequently leads to death.

2 / *Urea is energetically costly to synthesize.*

Recall that the energy currency of the cell is ATP. To synthesize a single urea molecule requires about five molecules of ATP, a significant expenditure of energy.

3 / *Urea is easily lost to the environment because of its high diffusibility.*

Urea and TMAO (i.e., a nitrogen-containing waste product in a class of chemicals called *methylamines*) are small, uncharged, highly soluble molecules that are easily lost to the environment through diffusion and excretion by the kidneys.

4 / *According to Fick's Laws of Diffusion, individual particles (ions, molecules) will diffuse independently of each other.*

Even if the internal and external concentrations of ions are close, each ion will tend to move along *its own* concentration gradient. Thus, there is a large diffusion gradient for Na^+ and Cl^- from outside to inside in sharks and rays, and thus both ions diffuse from seawater into the body. Also, salts enter the shark's body through the food that it eats. And urea and TMAO will tend to diffuse out.

How Do You Solve a Problem Like Urea?

How have these problems been solved such that urea can be retained in the body of sharks and other chondrichthyans?

The solution to the first problem with urea, its toxicity, is the role that TMAO plays in protecting proteins from urea's pernicious effects. Although the mechanism of its action is not entirely clear, the disruptive effects of urea on protein structure and function are prevented in the presence of TMAO. A 1980 study[18] discovered that the ratio of urea to TMAO and related methylamines was 2:1, and that methylamines evolved as osmolytes due to their stabilizing effects on urea-sensitive enzymes. Because of its protective role, TMAO has been called a *chemical chaperone.*

The solution to the second problem, how costly urea is to synthesize, is best explained by the wisdom of evolution. It would seem that two processes being equal, the one that conserves energy should be evolutionarily favored. The process that requires greater expenditures of energy, here, synthesis of urea, should enhance survival in some way to be favored. However, there may be mitigating circumstances or evolutionary constraints of which we may be unaware that dictate what is favored in natural selection. Moreover, becoming superior predators with metabolic systems capable of providing enough ATP early in their evolutionary history may have allowed the use of urea even if it is an energy-hogging osmolyte.

The loss of urea through urine and the gills, the third problem, was solved by resorbing 99% of the urea in the kidneys before it was excreted and blocking urea loss by having tissue impervious to the compound in the gills.

Solving the final problem involves compensating for the influx of Na^+ and Cl^- along their concentration gradients in accordance with Fick's Laws of Diffusion. Marine bony fishes have a similar problem with these and other ions, which they solve with specialized, mitochondria-packed *chloride cells* that expend energy to pump the Na^+ and Cl^- ions back into the environment.

Just as in marine bony fishes, the kidneys of sharks are not able to excrete the excess Na^+ and Cl^-, in part because doing so would require upsetting the animal's water balance by increasing the flow of urine required to eliminate these ions.

What about using chloride cells to excrete the excess Na^+ and Cl^-, like bony fishes do? Sharks and batoids appear to have cells (called *chloride-type* cells) in their gills similar to chloride cells, but these cells lack the complexity of the bony fishes' chloride cell and thus the ability to pump out ions. A study[19] on the Atlantic Stingray (*Hypanus sabina*) suggested that this species, which is *euryhaline* (capable of tolerating a wide range of salinities), has a mechanism for pumping Na^+ and Cl^- ions *in* when in fresh water but not *out* while in seawater. We will discuss elasmobranchs in fresh water below.

The Rectal Gland

The solution to the influx of Na^+ and Cl^- ions is a finger-shaped, bean-shaped, or globular tube called the *rectal gland* (also known as the *caecal* or *digitiform* gland; fig. 8.4), which is also found in the coelacanth. It is also a mitochondria-packed structure (i.e., one capable of expending lots of energy), which selectively accumulates Na^+ and Cl^- in its lumen (fig. 8.4) to twice the levels found in the blood and secretes these ions into the *cloaca*, from where they are dumped into the environment. Thus, the continuous inward diffusion of Na^+ and Cl^- ions from seawater is compensated for by mechanisms primarily in the rectal gland, with limited excretion by the kidneys, whose main responsibility is to eliminate divalent ions like Mg^{2+} and SO_4^{2-}, just as in marine bony fishes.

Sharks and Batoids in Fresh Water

We know that in salt water sharks remain osmotically neutral by elevating their internal osmolality to approximately that of seawater, and they eliminate the Na^+ and Cl^- ions that diffuse in or enter through their food primarily using the rectal gland, with the kidneys assisting with some of the monovalent and all of the divalent ions.

But what about euryhaline and freshwater elasmobranchs? Let us again start by considering bony fishes, freshwater ones this time, which face an entirely opposite suite of problems from those of the saltwater bony fishes we discussed above. For one, body fluids of a freshwater fish are hyperosmotic to (more concentrated than) the environment, since fresh water has few dissolved salts. Thus, the

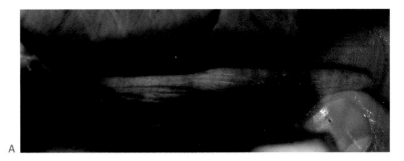

A

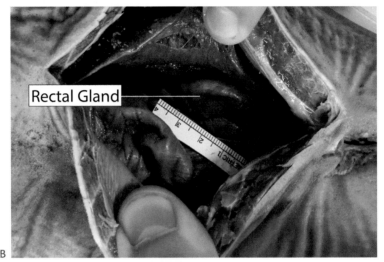

Rectal Gland

B

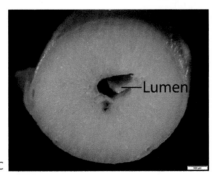

—Lumen

C

Figure 8.4. Rectal glands. (A) Digitiform rectal gland from a Blacktip Shark. (B) Globose rectal gland from a Broadnose Sevengill Shark. (C) Cross-section of a digitiform shark rectal gland showing the central lumen, where the Na$^+$ and Cl$^-$ ions will accumulate before being excreted. (Courtesy of Matthew Larsen)

tendency is for the fish to lose ions, primarily through the gills but also through the urine, because a freshwater fish continually gains water through osmosis from the external environment and urinates frequently to get rid of this excess water. The freshwater bony fishes solve the first problem, loss of ions, by using its chloride cells to pump Na$^+$ and Cl$^-$ back into the bloodstream of the fish from the environment, as well as through eating.

Among elasmobranch species, about 5% enter fresh water and 3–4% are oblig-
atorily freshwater (i.e., they spend their entire lives in fresh water). A 2010 review
paper[20] subdivided freshwater tolerant elasmobranch into three groups based on
the stage of freshwater colonization.

The first group is euryhaline elasmobranchs capable of routinely entering
and leaving fresh water, primarily the Bull Shark (*Carcharhinus leucas*), Atlantic
Stingray (*Hypanus sabina*), and pristids (sawfish) and juveniles of some species
(e.g., Sandbar Sharks [*Carcharhinus plumbeus*]; fig. 8.5). No members of this group
live and reproduce entirely in freshwater systems.

The second group consists of species that live permanently in fresh water but
can tolerate more saline waters. This group includes one population of Atlantic
Stingrays (fig. 8.6) that lives full time and reproduces in fresh water in Florida's
St. Johns River system, the Smooth Freshwater Stingray (*Dasyatis* [= *Fontitrygon*]
garouaensis) from Western Africa, the White-edge Freshwater Whipray (*Fluvit-
rygon signifer*) from Southeast Asia, and the giant Mekong Freshwater Stingray
(*Hemitrygon laosensis*), which is restricted to the Mekong and Chao Phraya Rivers in Laos and Thailand (fig. 8.6).

Figure 8.5 Co-author Abel with
a tagged juvenile Sandbar Shark
caught in low salinity waters
in Winyah Bay, South Carolina,
immediately prior to release.
(Courtesy of George Boneillo)

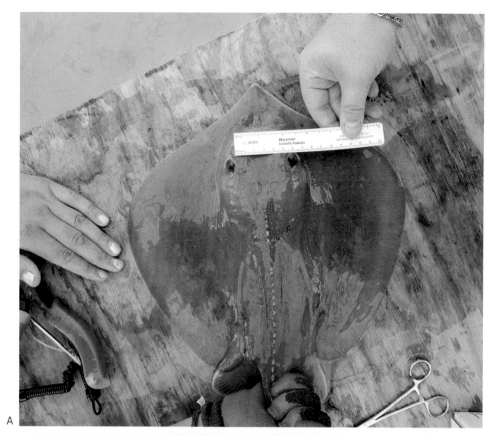

Figure 8.6. Batoids that live in fresh water but can tolerate salt water. (A) Atlantic Stingray. (B) Mekong Freshwater Stingray. (PhumjaiFcNightsky/Shutterstock.com)

The third and final group is found exclusively in fresh water and has lost the ability to tolerate even moderate salinities. These are the rays of the family Potamotrygonidae (fig. 8.7), a mostly landlocked, obligate freshwater assemblage found only in South America.

Stop for a moment and think what problems these elasmobranchs might face in dilute salt water or fresh water and how they might have adapted to these conditions. First, let us consider a Bull Shark, a species known from temperate and tropical shallow marine, estuarine, riverine, and lake ecosystems circumglobally.

If you just caught this shark in the Atlantic Ocean near, say, Recife, Brazil, and *instantaneously transported* it 4000 km (2500 mi) up the Amazon River, where the species has been documented, there would be enormous differences in the osmolality of the fluids inside the Bull Shark, which would measure about 1050 mOsm·kg^{-1}, and that of the river, whose osmolality would be not much higher than zero. Thus, there would be huge gradients for both water inflow and salt and urea loss.

Of course, it would take a lot longer for a Bull Shark to move from a marine environment to a freshwater one, and thus there is more time for physiological ad-

Figure 8.7. Potamotrygonid rays in a public aquarium.

justments to occur, especially since these are mediated principally by hormones, whose action is not immediate. The first order of business, physiologically, is to drop the total osmolality of the internal fluid environment as much as possible so that the gradients for salt and water movement are reduced. There is in fact about a 40% drop. The levels of Na^+ and Cl^- are reduced by about 30% each, but urea levels are reduced by about 50%. The reduction in blood urea levels in freshwater-acclimated Bull Sharks appears to be due to decreases in urea synthesis as well as decreased reabsorption in the kidneys more so than changes in gill permeability to urea.

Even with lowered levels of Na^+ and Cl^-, and thus a lower concentration gradient for diffusion out, these ions are lost to the environment. We know that freshwater bony fishes compensate for this loss (i.e., gain salt) by using chloride cells and eating. But what about our Bull Shark? It would make a nice story to report that the rectal gland simply reverses its function from pumping Na^+ and Cl^- *out* when in full-strength seawater to pumping them *in* while in fresh water. However, this is not the case, since the rectal gland is not in direct contact with the environment like the chloride cells of the gills. All that happens in the rectal gland is that its Na^+/K^+-ATPase activity in fresh water is reduced to near zero (i.e., Na^+ and Cl^- are not removed from the blood and are thus not lost). Some research has shown a reduction in size of the rectal glands of freshwater-acclimated Bull Sharks, which makes sense since the organ is superfluous in fresh water. Loss of ions in other tissues is compensated for by eating, increased uptake at the gills through specialized chloride-type cells, and resorption from the kidneys.

Why not drop urea levels to zero by excreting all of it, especially since urea is so energetically costly to synthesize? The answer is the Bull Shark and other elasmobranchs in this category only secondarily moved into fresh water. Their entire physiology evolved in the marine environment, including the evolution of enzymes and other proteins in the presence of urea and TMAO. Thus, in spite of the toxic nature of urea and the need for TMAO to protect biomolecules, it turns out that enzyme systems of Bull Sharks have become so accustomed to the presence of urea and TMAO over evolutionary time that they now will not work in their absence. We call this *Osmoregulatory Stockholm Syndrome*[21]—in the same manner when hostages become emotionally attached to their captors, enzymes will not work without the presence of the potential toxic compound urea, even though it is not required for osmoregulation under these conditions.

What about elasmobranchs that spend most of their time in fresh water? These sharks and batoids have adaptations similar to those of the Bull Shark, although some differences have been seen. For example, rectal glands of rays in the St. Johns River system were 80% smaller than those of fully marine conspecifics.

Finally, consider the exclusively freshwater potamotrygonids. This group has effectively become as close to osmoregulating like freshwater bony fishes as its constraints as an elasmobranch allow. First, the rectal gland is nonfunctional and vestigial. They reduce their Na^+ and Cl^- levels, and they lower their urea to about

1 mOsm·kg^{-1}. To allow for the loss of urea, their gills are 20 to 50 times leakier than those of the Spiny Dogfish.[22] Additionally, their enzymes and other proteins have lost their urea sensitivity. Moreover, there is an energetic advantage (as much as 14%), since no ATPs must be diverted to synthesize urea.

One final question concerning freshwater elasmobranchs: Why are there so few of them? Notwithstanding the potamotrygonid's success, overcoming the osmotic, sensory, and reproductive constraints of being a freshwater elasmobranch, including the main obstacle of no longer retaining urea, would be extremely difficult to replicate.[23] Additionally, no longer retaining urea would need to co-occur with the evolution of urea-insensitive proteins. Also, fresh water may impose constraints on the electrosensory ability of the ampullae of Lorenzini. The ampullae of potamotrygonids are much smaller than those of marine elasmobranchs and have been called *microampullae*. However, Paddlefish (*Polyodon spathula*) and Sturgeon (*Acipencer*) have ampullae of Lorenzini that they use in fresh water, so this may not be an insurmountable problem. Other problems would include overcoming the reduced viability of sperm in fresh water and living in an environment that might be less stable thermally. Additionally, the freshwater environment may have limited niche spaces for additional elasmobranch species.

Among euryhaline species, other than those previously mentioned, are juveniles of some species (e.g., Sandbar Sharks) who may use habitats of lower salinity than adults, and in doing so, may find refuge from predation from conspecifics and other larger sharks, at least temporarily. Preliminary work that we have done shows that these juveniles inhabit lower salinities during low tide (likely for protection from larger sharks and for food), and that physiologically they lower their solute concentrations (Na$^+$, Cl$^-$, urea) just like Bull Sharks.

Role of the Kidney

If we have neglected the role of the kidney in sharks, it is only because of the enhanced role of the rectal gland and the associated structures that act to eliminate salts and retain urea, one of which is the kidney (fig. 8.8). The other major roles of the elasmobranch kidney are to eliminate divalent ions (e.g., magnesium and sulfate) and to produce dilute urine in response to the intake of water that occurs through the gills and in food. The elasmobranch kidneys are paired, elongated structures located retroperitoneally (i.e., they run lengthwise along the dorsal wall of the abdominal space).

Digestion

When you look at a dead shark, do you wonder what it has eaten?

Previously we have discussed the anatomy of the jaw and the roles of the teeth in feeding. Now, we move to the anatomy and physiology of the remaining processing of food. Refer to Figure 8.9 as we describe the anatomy of the shark digestive tract.

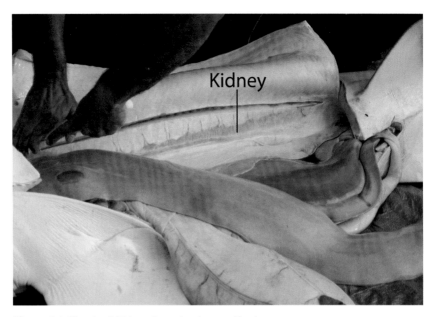

Figure 8.8. The shark kidney, here, in a Lemon Shark.

Food items of sharks and rays, which are protein- and fat-rich, are either ingested whole or are sheared into pieces of varying sizes in the orobranchial chamber. The digestion of these items consists of mechanical, chemical, and microbial processing followed by absorption of the nutrients.

The gastrointestinal, or alimentary, tract consists of a series of organs (fig. 8.9) lined by a *mucosa* containing cells that secrete mucous and digestive enzymes, underlain by supporting connective tissue and, typically, longitudinal and circular smooth muscle.

Food moves from the mouth through the pharynx and into a short but distensible esophagus, which lies dorsal to the moveable basibranchial plate (see fig. 7.9). The esophagus secretes mucous and has striated muscles that facilitate the movement of food into the stomach.

The stomach of sharks varies in form, with the most common being J-shaped (also called *siphonal*). It is divided into an anterior *cardiac stomach* and posterior *pyloric stomach*. It terminates at the *pyloric sphincter*. The stomach stores food and begins digestion of proteins, which is facilitated mechanically by contraction of the stomach muscles. Secretions include acidic (i.e., low pH) gastric juices, mucous that protects the stomach lining, and a precursor that is converted into the proteolytic (protein-digesting) enzyme *pepsin*. The low pH of the stomach also starts to emulsify lipids and helps begin the digestion of carbohydrates and proteins.

An important function of the stomach is to ensure that the partially digested food is not emptied prematurely. The hormone cholecystokinin (CCK), which has been found in Spiny Dogfish (*Squalus acanthias*), Porbeagle (*Lamna nasus*), and Shortfin Mako, is released by the gastric mucosa in the presence of proteins,

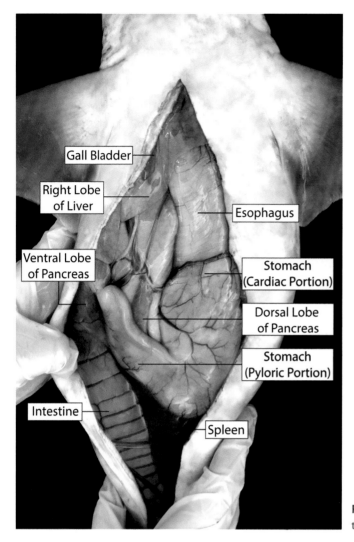

Gall Bladder

Right Lobe of Liver

Esophagus

Ventral Lobe of Pancreas

Stomach (Cardiac Portion)

Dorsal Lobe of Pancreas

Stomach (Pyloric Portion)

Intestine

Spleen

Figure 8.9. Gastrointestinal tract of a female Spiny Dogfish.

and it serves to slow the rate at which the stomach empties. It may also curb the shark's appetite if there is food that is still being digested.[24]

It is generally true in biological systems that surfaces specializing in absorption or exchange of substances (e.g., gills) often have elaborations that increase their surface area, and the stomach is no exception. The stomach contains longitudinal ridges formed by infoldings of the stomach wall known as *rugae* (fig. 8.10). These unfold and allow the stomach to distend to accommodate food that has moved into it. This increased volume occurs without any increase in gastric pressure, a phenomenon known as *receptive relaxation*. The stomach thus stores this food until it begins digesting it.

The pH within the stomach is either continuously low or it varies, depending on the species. A 2005 study[25] showed that during periods of fasting, the stomach pH in Nurse Sharks (*Ginglymostoma cirratum*) rose to > 8, compared to between 2 and 3 when digesting food. Other studies[26] demonstrated persistent

Figure 8.10. Stomach of a shark showing rugae. (Photo by Joshua Bruni)

pH of 1–2 in Leopard Sharks (*Triakis semifasciata*), Blacktip Reef Sharks (*Carcharhinus melanopterus*), and Spiny Dogfish, regardless of whether food was present.

The result of digestion in the stomach is conversion of whole or partial prey into a semi-liquid material called *chyme*. This chyme slurry moves into the small intestine through the pyloric sphincter. The intestines in sharks are not strictly analogous to those of other vertebrates, and a 2017 review[27] suggested that the terms *proximal*, *spiral*, and *distal* intestines be used.

The proximal intestine, sometimes referred to as the duodenum, receives bile from the gallbladder and digestive enzymes, some in an inactive form, from the pancreas. Bile serves to emulsify fats. Pancreatic digestive enzymes may include *proteases* (to digest proteins), *amylases* (to break down large carbohydrates), and *lipases* (to digest fats).

We have already discussed the varied morphologies of the section of the intestine called the *spiral* intestine. Similar to the stomach rugae, the spirals (from 2 to 50), scrolls, and rings serve to increase the surface area for absorption, and thus slow the movement of the chyme (now called *digesta*), although experimental verification is sparse. The intestine also conserves space (it does not receptively relax like the stomach), which is a limited resource given the crowding by a large liver and possibly embryos.[28]

The intestine leads into the more muscular distal intestine, where feces accumulates and dries until released from the rectum, the lower part of the distal intestine, through the anus into the cloaca, and from there, the environment.

Concluding Comments

At this point, we have discussed the physiology of the senses, reproduction, circulation, respiration, metabolism, temperature, osmoregulation, and digestion. These are all fascinating in their own right, and there are still many unanswered questions about the internal function of sharks and their relatives. Recall, however, that all physiological processes arise through natural selection and occur in service to ensuring survival and reproduction. Thus, our coverage of physiology appropriately segues into the relationship between the animal and its living and non-living environment (i.e., ecology) in the following chapter.

NOTES

1. Iwama, G.K. et al. 1999. Am. Zool. 39: 901–909.

2. Huntsman, A.G. and Sparks, M.I. 1924. Contr. Can. Biol. Fish. 2: 95–114.

3. Fangue, N.A. and Bennett W.A. 2003. Copeia 2003: 315–325.

4. Ectotherms have also been called *cold-blooded* and *poikilotherms*. The former is incorrect because some ectotherms live in warm or hot environments and they are not cold at all. Poikilotherms refers to having a variable body temperature. Many ectotherms are poikilotherms, but not all. If you are a fish living on a coral reef, in the tropics, then you are ectothermic but functionally *homeothermic* (with a body temperature that stays the same).

5. Hight, B.V. and Lowe, C.G. 2007. J. Exp. Mar. Biol. Ecol. 352: 114–128.

6. Speed, C. et al. 2012. Ecol. Prog. Ser. 463: 231–244.

7. Grubbs, R.D. et al. 2007. In: McCandless, C.T., Kohler, N.E. and Pratt Jr., H.L. (eds.). American Fisheries Society Symposium. Am. Fish. Soc. 50: 87–108.

8. Some authors differentiate between *systemic* and *regional* endothermy as follows: reserving the former for fishes whose whole body is warmed (those families listed in the text) and the latter for elevated temperatures in certain organs (e.g., the brains or the eyes).

9. Dickson, K.A. and Graham, J.B. 2004. Physiol. Biochem. Zool. 77: 998–1018.

10. Dickson, K.A. and Graham, J.B. 2004. Physiol. Biochem. Zool. 77: 998–1018.

11. Watanabe, Y.Y. et al. 2015. Proc. Natl. Acad. Sci. 112: 6104–6109.

12. As cited in Table 8.1, p. 250 of Wetherbee, B. et al. 2012. In: Carrier, J.C. et al. (eds.). Biology of sharks and their relatives. CRC Marine Biology Series. CRC Press. 3–31 pp.

13. Stillwell, C.E. and Kohler, N.E. 1982. Can. J. Fish. Aq. Sci. 39: 407–414.

14. Wood, A.D. et al. 2009. Fish. Bull. 107: 1–15.

15. Strictly, *osmoregulation* refers to water balance in an organism, but more commonly it refers to both water and solute balance.

16. *Hypoosmotic* is pronounced hy-po-osmotic. If you think the word should have a hyphen (i.e., hypo-osmotic), we agree.

17. Smith, H.W. 1936. Biol. Rev. 11: 49–82.

18. Yancey, P H. and Somero, G.N. 1980. J. Exper. Zool. 212: 205–213.

19. Piermarini, P.M. and Evans, D.H. 2000. J. Exp. Biol. 203: 2957–2966.

20. Ballantyne, J.S. and Robinson, J.W. 2010. J. Comp. Physiol. B 180: 475–493.

21. You will not get any hits in Googling the term, as no *sensible* physiologist has ever used it.

22. Goldstein, L. and Forster, R.P. 1971. Comp. Biochem. Physiol. 39B: 415–421.

23. Ballantyne and Robinson 2010. op. cit.

24. Leigh, S.C. et al. 2017. Rev. Fish Biol. Fisher. 27: 561–585.

25. Papastamatiou, Y. and Lowe, C. 2005. Comp. Biochem. Physiol. A 141: 210–214.

26. Leigh, S.C. et al. op. cit.

27. Leigh, S.C. et al. op. cit.

28. Wetherbee, B.M. et al. 2004. In: Carrier, J.C. et al. (eds.). Biology of sharks and their relatives. CRC Marine Biology Series. CRC Press. 225–246 pp.

Ecology and Behavior

9 / Ecology

Introduction

Ecology is the study of the relationships between organisms and their living and non-living environment. Its subdivisions include:

- *Ecophysiology* (or physiological ecology), which examines physiological adaptations of organisms to their environment (e.g., osmoregulation of Bull Sharks [*Carcharhinus leucas*] in rivers). We have considered physiology previously and will discuss behavioral ecology in Chapter 10.

- *Population ecology* (or autoecology), which studies the dynamics of populations of individual species. This entails *stock assessments*, which are critical to the management of elasmobranchs (e.g., the Sandbar Shark [*Carcharhinus plumbeus*] as the species recovers from overfishing along the US Atlantic Coast).

- *Community ecology* (or synecology), which focuses on the interactions between species within an ecological community (e.g., relative abundances of sharks in an estuary or on a coral reef) and predator-prey interactions.

- *Ecosystem ecology*, which studies the flows of energy and matter through ecosystems (e.g., the trophic, or feeding, structure of an ecosystem).

We begin our discussion of the ecology of sharks with an overview of the important questions asked by shark ecologists.

Big Ecological Questions Related to Sharks and Rays

Studies into the ecology of sharks (i.e., how sharks live their lives and interact with other species and the environment) revolve around searching for answers to a basic suite of questions.

1 / Where do sharks live and what habitats do they require? The answers to these questions vary by species and by life stage. Juveniles often use different habitats than adults and males and females may segregate for much of their lives.

2 / Where do sharks move and what drives these movements? Some species of sharks live in very small home ranges (smaller than a Piggly Wiggly[1]) whereas others migrate across entire ocean basins.

3 / What and how often do sharks eat, and what eats them? Some shark species eat a huge variety of prey whereas others specialize in feeding on a few prey types. Some may eat one big meal that lasts for weeks, whereas others must feed daily. Also, we typically think of sharks only as predators, but it is important to note that most sharks are also potential prey, often to other sharks. In fact, all sharks are subject to being eaten, at least as juveniles.

4 / What life-history characteristics do specific shark species possess? Fundamentally, it is critical to know how long it takes a shark to become sexually mature and how many offspring it can produce over its life, in order to manage and conserve populations. Shark life histories vary widely and are directly a function of the energy available to them and how they use it and thus depends on where they live (e.g., tropics versus the deep-sea), where they fall in the food web (top versus meso-predators), how much energy they spend moving (sedentary versus highly migratory), and which reproductive modes they employ (ranging from laying scores of eggs to live birth of one pup).

5 / What roles do sharks play in the ecosystem? This is perhaps the most difficult ecological question and its solutions rely on the answers to all of the questions above. Simplistically, sharks are often characterized as apex predators that maintain ecosystem health by controlling prey populations and eating the weak and diseased. In reality, empirical evidence to support this notion

has been hard to find and, while some sharks are indeed apex predators, most sharks are not, but instead fill roles in the mid-to-upper tiers of the food web. Sharks play many roles in marine ecosystems that are often complex and not always obvious. In the world of shark research, we are just scratching the surface in answering this fundamental question.

Challenges of Studying Sharks

Unraveling the ecological mysteries of sharks is more complicated with different challenges than studying terrestrial predators like wolves or lions. For one, sharks live underwater, obviously, in a concealing environment rendering observation by air-breathing scientists difficult. Depending on the questions being asked, researchers may need to capture sharks or simply observe them. In some cases, both approaches can be relatively inexpensive. Your authors capture and release thousands of sharks every year in the coastal zone for a variety of ecological studies using small vessels. In clear water, sharks can be observed relatively inexpensively using SCUBA or baited cameras. However, the expense of ships and equipment increases rapidly when studying deep-sea and pelagic species, except in areas such as oceanic islands where the deep-sea slope is very close to shore. Deep-sea research often requires the use of large ships for many days and specialized equipment to capture (e.g., using trawls, baited hooks, or traps) or even to observe (e.g., using submersibles or remotely operated vehicles) the subjects (fig. 9.1).

If we are fortunate enough to overcome the roadblocks above and capture sharks we wish to study, handling them often presents a new suite of problems. Sharks do not want to be handled and many of them have large teeth and will bite. We know many colleagues who have suffered serious injuries from shark bites and even small sharks can be dangerous to the researcher—co-author Grubbs, having tagged thousands of juvenile Sandbar Sharks in Virginia waters, had his thumb nearly removed by a 3-foot-long Sandbar Shark in 1996, and in the summer of 2019 co-author Abel was gnawed by a smaller beast (fig. 9.2). But Grubbs found that studying sawfish can get even more treacherous, as the animals swing a huge tooth-bearing rostrum at the researchers with incredible speed and force. In many cases, researchers seek to release the animals unharmed after all data and sample collection are completed. Many sharks are much more fragile than their reputations would suggest, so researchers try to protect the sharks from damage and release them as quickly as possible while trying to protect themselves from being bitten by a shark, slapped by its tail, or bludgeoned by a sawfish rostrum.

A final challenge to studying the ecology of sharks lies in achieving the meaningful sample sizes. In ecological research, we seek to answer questions or describe patterns within our data that apply beyond our samples or the animals from which they came. In other words, the results allow us to make inferences

Figure 9.1 Co-author Grubbs studied deep-sea sharks from a submersible in the Bahamas as part of an OceanX expedition, during which he photographed this Bluntnose Sixgill Shark. Spearguns used to tag sharks from the submersible are shown.

Figure 9.2. Co-author Abel during an unsettling two-minute period waiting for the small Sandbar Shark to release his gloved finger. The interaction left puncture marks resembling paper cuts. (Courtesy of James Luken)

that apply to whole populations, species, or ecosystems. The sample size needed to answer these questions confidently depends in part on the variability in the measured trait within the population and how well that variability is described in our samples. As an amusing example, years ago we had the opportunity to examine the gut contents of three Blue Sharks (*Prionace glauca*) that were captured together. The only items in their stomachs were green beans—yes, the kind you eat with your family during the holidays. Most likely these were scarfed up after being tossed from a ship. But if we attempted to describe the diet of Blue Sharks based on our available sample size of three, we would conclude that they are herbivores that frequent Piggly Wiggly in search of green beans. In reality, Blue Sharks feed on a wide variety of pelagic fishes and squid and hundreds of individuals are required to describe their diet quantitatively. This is why for biological questions that are amenable to experimentation, model organisms such as fruit flies, zebra fish, and white mice that can be produced in large numbers are used.

And herein lies the challenge in studying sharks. Energy flow in any ecosystem dictates that top predators like large sharks naturally occur in low densities. If we visualize the pelagic food web as the classic trophic pyramid (fig. 9.3), the primary producers (phytoplankton), are the food source for the primary consumers (zooplankton). Zooplankton are eaten by small fishes which are in turn eaten

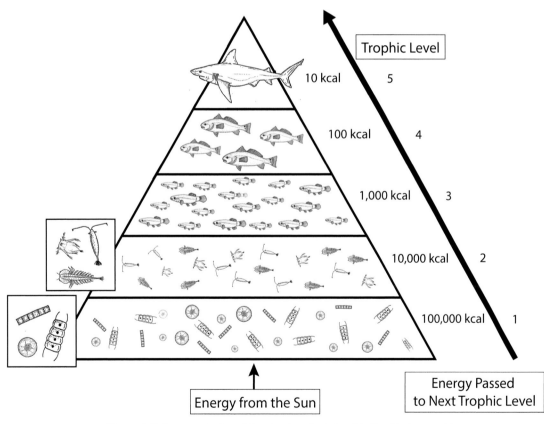

Figure 9.3. Representation of the classic pelagic trophic (feeding) pyramid.

by larger fishes like tunas and sharks. As a rule of thumb, only about 10% of the energy in one trophic level is available to create biomass at the next highest trophic level. This is the *trophic transfer efficiency*. So with our Blue Shark example, 10,000 kilocalories (kcal, a measure of energy more commonly referred to as *calories* in labels on human food) of phytoplankton would support 1000 kcal of copepods, which could support 100 kcal of herring, leading to 10 kcal of Blue Shark tissue. Shark muscle is close to 1 kcal per g, but liver and kidney tissues are much higher. If we assume the overall energy density of a Blue Shark is 2 kcal per g, and that of phytoplankton is roughly 0.6 kcal per g, then an average size Blue Shark weighing 50 kg (110 lb) would require 167,000 kg (370,000 lb) of phytoplankton (or green beans for that matter). That is roughly 12 semi-tractor trailer loads of green beans to produce one adult Blue Shark! It should be obvious now why achieving appropriate sample sizes to answer ecological questions about shark populations is a challenge. This is, of course, compounded when studying species that are truly apex predators or those that are rare, endangered, or live in the deep sea.

Where Do Sharks Live and How Many Are There?

Sharks inhabit nearly all marine environments from tropical coral reefs to cold arctic waters and from estuaries and coasts to the deep sea and open ocean. There are even sharks and rays, though relatively few in number, which live in fresh water. Why does any animal live where it does? The distribution of animals and the habitats they use reflect a combination of congruent and opposing forces.

First, an animal must be physiologically capable of living in a given environment, which requires adaptations to abiotic (non-biological) factors such as temperature, salinity, dissolved oxygen, pH, pressure (depth), and light, to name a few. Temperature tolerances are among the most important factors limiting shark distributions, but also recall the special adaptations required for sharks to live in fresh water (see Chapter 8).

Next, there are a series of ecological constraints that influence not only where sharks live but also the density (how many) of sharks living there. All animals must eat—that is one of their defining characteristics. Thus, the distribution of suitable resources (prey) is a major driver of where sharks live. However, there is often intense competition for those prey from members of the same species or others, so the ability to compete for prey also determines where sharks live. Of course, just as one species of shark knows where to find its prey, its predators (often larger sharks) also know where to find them. This balance between competition for resources and predation risk plays a large role in determining where animals, including sharks, live. Ultimately, an animal's evolutionary fitness is tied to its ability to pass its genes to the next generation (i.e., to reproduce). The probability of finding a mate also influences where a shark will spend a portion of its time.

It may surprise you to learn that over half (~53%) of all species of sharks live their lives in the deep sea (defined as deeper than 200 m, or 656 ft). Most deepwater species of sharks are small (< 100 cm, or 3.3 ft) and this high diversity is likely driven by the relatively low competition for resources and the lower predation risk than in shallower environments. But sharks do not live in the *very* deep sea. The average depth of the world's oceans is 4000 m (13,100 ft), but sharks rarely occur deeper than 3000 m (9800 ft). Explanations for these depth limits are debated, but likely involved both ecological (food limitations for predators) and physiological (e.g., constraints on producing liver oil and synthesizing TMAO under high pressure) factors. About 40% of shark species live in coastal tropical to warm temperate environments, from estuaries to continental shelves. In these regions there is high food biomass and diversity along with very high habitat diversity. Together, these drive niche specialization, which leads to the rise of new species and thus high diversity among coastal sharks. There are very few truly pelagic species of sharks (< 3%), primarily due to limited prey resources in the oligotrophic (low nutrient) open ocean and the lack of barriers to separate them long enough to drive speciation. Thus, most pelagic shark species are found circumglobally. The remaining sharks (~7% of the diversity) are composed of Arctic and freshwater species.

How Do We Know Where Sharks Live? Catching Sharks

Fishery-independent Surveys

In the arsenal of methods used to study the ecology of sharks, tag-and-release fishing has a special place among both the public and scientists. To answer many questions about the biology and ecology of sharks and their relatives, scientists need to handle the specimens so that they can identify the species, make accurate measurements, take tissue samples, and apply tags.

Methods used by shark biologists rely on the same techniques used by commercial and recreational fishers (e.g., longlining, gillnetting, trawling, and hook-and-line fishing; see Chapter 11). Some aspects of shark ecology are learned from data and samples collected directly from commercial and recreational fisheries, which is referred to as fishery-dependent data. However, much of our knowledge comes from *fishery-independent* sources, primarily from scientific surveys. Why is this important? Fishery management as well as understanding fish ecology relies on obtaining estimates of how abundances of sharks vary over space and time. Catch rates (CPUE = catch per unit effort), such as the number of sharks caught on 100 hooks in an hour, number caught in an hour gill net soak, or number caught per hour trawling, are used as the estimate of relative abundance. If we are interested in the distribution of a population of sharks and what habitats they use, we not only must sample where sharks are but also where they are not, regardless of cost or effort required, in order to define the edges of the distribution (i.e., where CPUE goes to zero). To accomplish this, stations are chosen using a random design (e.g., at different depths or geographic locations; scientists would say *stratified by depth* or *spatially balanced*, respectively). If we are interested in changes in the populations over time, we must sample in the same way across the same part of the population repeatedly. These surveys may use a randomized design or fixed stations that are sampled repeatedly for many years. It would be foolish for commercial and recreational fishers to fish in these ways. Fishers weigh the costs and benefits of fishing different areas and target the highest density of fish that is legal and economical.

Fishery-independent shark surveys are often critical to stock assessments for fishery management. Fisheries often use methods that only catch specific portions of the populations of sharks (e.g., longline fisheries tend to catch intermediate-sized sharks because hooks are too large to catch small sharks and the gear is designed to allow large sharks to break free), whereas fishery-independent surveys can be designed to catch sharks that represent the population (e.g., using multiple hook sizes). This provides managers with a more complete picture of the population structure and can highlight critical changes in the stock, such as juvenescence (decreases in the mean age and size), which is a prime indicator of overfishing (fig. 9.4).

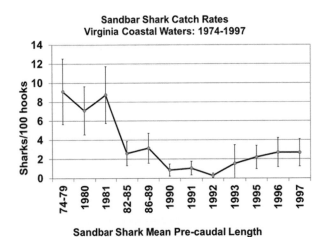

Figure 9.4. Examples of fishery-independent data important to fishery management. *Top*: Sandbar Shark Catch Per Unit Effort (CPUE); that is, sharks per hundred hooks, caught on experimental longlines in Virginia coastal waters from 1974–1997. *Bottom*: Mean Lengths of Sandbar Sharks. Note the period beginning in 1993. Though catch abundance (*top* graph) began to increase after the fishery management plan went into effect in 1993, the mean size of these sharks (*bottom* graph) continued to decline, a classic example of juvenescence from overfishing. (Grubbs and Musick, unpublished data)

Surveys can be used to delineate critical areas for overfished or imperiled species that may be closed to fishing. A survey for juvenile Sandbar Sharks in Chesapeake Bay was used to delineate primary nursery habitat for this species, which led to the federal designation of a *Habitat Area of Particular Concern* (HAPC) for that species in the management plan (fig. 9.5).[2,3]

Fishery-independent surveys can also be used to determine how fishery-induced mortality may change if fisheries changed their methods. For example, the Cuban Dogfish (*Squalus cubensis*) is one of the most common bycatch species in deep snapper and grouper fisheries in the Gulf of Mexico, and 97% are reported as released alive. Fishery-independent surveys were used in conjunction with stress physiology and post-release monitoring to determine that even when handled carefully, about 50% of those released alive die within 24 hours[4] (fig. 9.6). Co-author Grubbs and colleagues are using fishery-independent surveys incorporating hook timers (magnetic devices that begin a clock when a shark bites the hook) to estimate the effect of varying hook soak time on post-release mortality in Great and Scalloped (*Sphyrna lewini*) Hammerheads (fig. 9.7). Shark surveys

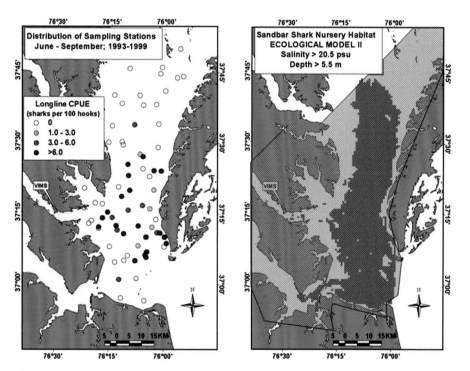

Figure 9.5. Map showing Catch Per Unit Effort (sharks per hundred hooks) for Sandbar Sharks on a randomized longline survey (*left*) used to define primary nursery habitat in lower Chesapeake Bay based on depth, salinity, and catch data (*right*). These catch data informed fishery managers, which led to the designation of the area as a Habitat Area of Particular Concern (polygon in figure on *right*) in the fishery management plan. (Modified from Grubbs, R.D. and Musick, J.A. 2007. In: McCandless, C.T. et al. (eds.). Shark nursery grounds of the Gulf of Mexico and the East Coast waters of the United States. Am. Fish. Soc. Symp. 50: 63–86)

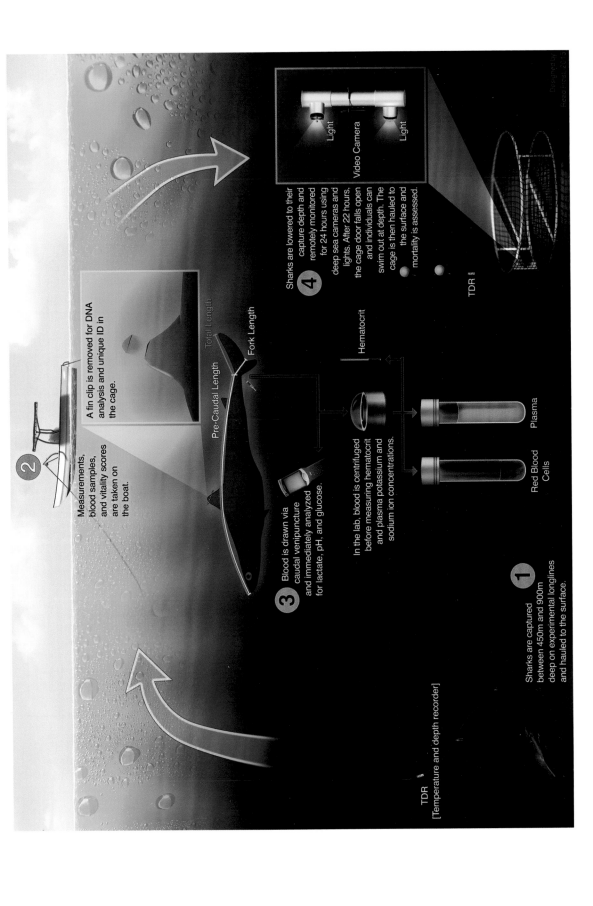

Measurements, blood samples, and vitality scores are taken on the boat.

A fin clip is removed for DNA analysis and unique ID in the cage.

Total Length

Pre-Caudal Length

Fork Length

Blood is drawn via caudal venipuncture and immediately analyzed for lactate, pH, and glucose.

In the lab, blood is centrifuged before measuring hematocrit and plasma potassium and sodium ion concentrations.

Hematocrit

Plasma

Red Blood Cells

Sharks are lowered to their capture depth and remotely monitored for 24 hours using deep sea cameras and lights. After 22 hours, the cage door falls open and individuals can swim out at depth. The cage is then hauled to the surface and mortality is assessed.

Light

Video Camera

Light

TDR

Sharks are captured between 450m and 900m deep on experimental longlines and hauled to the surface.

TDR
[Temperature and depth recorder]

Facing page

Figure 9.6. Graphic showing experimental methods used in a study on the post-release mortality of Cuban Dogfish. (Created by Read Frost and Brendan Talwar)

Figure 9.7. Co-author Grubbs releasing a Great Hammerhead from an experimental longline as part of a study of post-release mortality. (Courtesy of Michael Scholl)

take on many designs depending on purpose, funding, and infrastructure (e.g., vessels, personnel). Here, we briefly describe six surveys we have been involved in or developed that illustrate this variety.

CCU Shark Project (fig. 9.8)—Begun in 2002 at Coastal Carolina University by co-author Abel, the CCU Shark Project is a small-scale, episodic, estuarine survey started with the dual purpose of teaching students about shark research and to understand ecology and physiology of the sharks found in three Northeast South Carolina estuaries: Winyah Bay, Murrell's Inlet, and North Inlet. The initial research goals of the project were to (1) identify sharks inhabiting these systems; (2) describe shark population structure, distribution, and migrations and their environmental influences; (3) determine whether these systems serve as nurseries; and (4) identify human impacts.

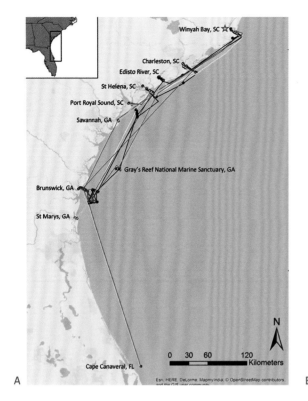

A

B

Figure 9.8. (A) Late fall to spring migration tracks of six acoustically tagged juvenile Sandbar Sharks that exhibited heretofore unknown southern migrations from Winyah Bay, from a study conducted by the Coastal Carolina University Shark Project. (Courtesy of Caroline Collatos) (B) Photo from the Coastal Carolina University Shark Project.

VIMS Shark Survey—The Virginia Institute of Marine Science Shark Survey is a small scale, but long-term survey. Eight fixed stations ranging from inside Chesapeake Bay to offshore Atlantic waters 30 m deep are sampled monthly with bottom longlines from June through September. Started in 1974 to monitor relative abundance of large coastal sharks, the survey's strength lies in its longevity. The VIMS longline survey was the primary fishery-independent data source used in the initial US Federal Fishery Management Plan for Sharks of the Atlantic Ocean implemented in 1993 and remains a primary index used in stock assessments for numerous large coastal and small coastal sharks (fig. 9.9). A second fixed station survey along with a randomized survey using smaller hooks was started at VIMS by co-author Grubbs in 1996 to examine seasonal and spatial use of Chesapeake Bay by juvenile Sandbar Sharks (see fig. 9.4).

FSUCML Big Bend Surveys—Started by co-author Grubbs in 2009 based out of the Florida State University Coastal and Marine Lab, this survey is unique in that it employs multiple gear types. The survey operates in the eastern Gulf of Mexico. Experimental longlines employing four hook sizes and gillnets employing six mesh sizes are used concurrently. Using the dual gears allows the researchers to sample all sharks from the smallest newborn small coastal species to adult large coastal sharks. In addition, the gillnet survey provides an index of prey abundance for large sharks and samples for food web analyses. Fixed stations are sampled one day per month, 12 months per year, providing data on seasonal migration and residency. A larger scale survey in the summer provides an annual index of relative abundance for numerous species. Both surveys and gear types are now used in stock assessments for coastal sharks in the Gulf of Mexico.

FSUCML Deepwater Longline Survey—Very few standardized deep-sea shark surveys exist. Started in 2011, this survey was designed to study the deep demersal (bottom-associated) sharks, bony fish, and invertebrate scavenging

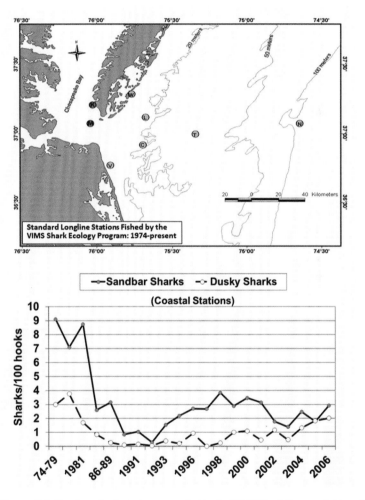

Figure 9.9. *Top*: Standard stations sampled in the VIMS Shark Survey. *Bottom*: Trends in relative abundance of the two most common large coastal sharks, Dusky and Sandbar Sharks, from the project from 1974–2006. Abundance trends reached a minimum in the early 1990s prior to the Federal Management Plan that went into effect in 1993.

community in the northeastern Gulf of Mexico in the wake of the Deepwater Horizon oil spill. The gear employs baited bottom longlines with five hook sizes and two trap types soaked for three hours. The purpose of the survey is to document changes in relative abundance, species assemblages, food web structure, and toxicological responses for sharks and other species as a function of time and space since the oil spill. It is the only such survey of its kind and has become one of the largest surveys of deep-sea sharks ever conducted.

BBFS PIT Study—The Bimini Biological Field Station in the Bahamas started a PIT tag (Passive Integrated Transponder, like the rice-grain-sized microchips implanted in dogs and cats) study of Lemon Sharks (*Negaprion brevirostris*) in 1995 (see fig. 12.8). This is a long-term, fixed station, gillnet survey conducted at night in a semi-enclosed lagoon. It is unusual in being one of few census surveys, where the entire population of small juvenile Lemon Sharks is sampled, counted, and tagged each year. Census sampling is rare among shark surveys due to logistical constraints on capturing all individuals. Over the course of two weeks, sharks are captured, PIT tagged, and placed in holding pens. Sampling continues until no sharks are caught, and all sharks are then safely released. Because all sharks in the population less than about 3—years-old are caught each year, this survey has provided one of the only empirical estimates of natural mortality in a shark population.

National Marine Fisheries Service (NMFS) Pascagoula Longline, Gulf-SPAN, and CoastSPAN Surveys—These are the largest shark surveys. Information can be found on the NMFS website.

Other Methods of "Catching" Sharks

While surveys that actually catch sharks are critical for collection of many of the samples and data needed to understand the ecology of sharks and to manage their populations, there exist numerous non-capture methods of collecting data on shark distributions and habitat associations.

Submersibles

Manned and unmanned submersibles have provided valuable data on the distributions of deep-sea sharks and often provide our only *in situ* observations of these taxa. For example, most of the records for the Caribbean Rough Shark (*Oxynotus caribbaeus*) have come from submersibles[5] (see fig. 3.14). Submersibles may also provide viable methods to even tag deep-sea sharks at depths without exposing them to the trauma of bringing them to the surface. It is even possible to outfit unmanned submersibles with hydrophones and receivers that allow them to track tagged sharks at depths.

Camera Traps

Camera traps have long been used to document the presence of elusive wildlife in terrestrial and underwater environs. Advancements in video camera technology and economical underwater housings have made employing large-scale camera trap surveys possible even in the deep sea. These methods are obviously limited to waters that are relatively clear, but studies using baited remote underwater video (BRUV) surveys specifically to study sharks are now taking place all over the world in coral reef, pelagic, and deep-sea environments.

Environmental DNA

One final method to indirectly investigate the presence of sharks is the use of environmental DNA (eDNA). In concept, every living creature leaves fragments of their DNA in the environment after they are gone (e.g., skin cells, fecal matter, mucous, blood). By analyzing water samples for the presence of these fragments of genetic material, scientists can potentially assess if a species of shark was present in the area. It is important to realize that there are many caveats with these methods. Failure to detect a species does not mean it is absent. DNA degradation rates are usually unknown so if a species is detected in a water sample, there is no way to know if it was there recently or years ago. Also, since the genetic signature is in the water, currents may have carried it long distances from where the animal lived. This emerging technology may prove valuable in detecting species presence but will never replace the need for classic surveys involving fishing since eDNA provides no indication of sex, size, reproductive status, trophic status, or any of the other parameters that require capture.

Where Do Sharks Move and Why?

Before we delve into shark movements and how researchers study them, let us address why they move at all. For that matter, why does any animal move? The answers to this question are fundamentally the same across all species and are related to the reasons fish live where they do. The objectives of most movements are related to gaining access to prey or mates while remaining physiologically comfortable and avoiding potential predators. Patterns of shark movements take place on many temporal scales in reaction to many stimuli ranging from the immediate and patterned reactions to predators, prey, or social cues to regimented seasonal migrations. Within a given day, shark activity patterns are often linked to predictable abiotic patterns, such as tidal cycles (ebb and flood) and time of day (day versus night).

As biologists, we may be interested in what drives animal movements on a daily scale and a lot of work in this arena has been geared toward delineating important nursery grounds. Not all sharks use defined nurseries, but many certainly do. In fact, it has even been hypothesized that sharks in the Paleozoic

Era used nurseries. Use of nurseries may be an evolutionarily stable strategy of size-based habitat selection that minimizes predation risk and limits competition between sharks of different age classes. Some sharks, like juvenile Lemon Sharks in Bimini, Bahamas, establish small, well-defined home range within their nurseries whereas others, such as juvenile Sandbar Sharks in Chesapeake and Winyah Bays are essentially nomadic, ranging widely each day within the estuaries. In both cases, with growth the sharks move to adjacent, more expansive habitats due to different trophic requirements and size-based reductions in mortality risk.

Ontogenetic (i.e., over the course of their growth) expansion in spatial habitat use occurs nearly universally among animals. Activity space increases with body size in most animals, particularly carnivores like sharks. Habitat selection for very young sharks is most often a balance between maximizing accessibility to available resources (e.g., minimizing competition) and minimizing predation risk, which is often lower in shallower habitats typically used as nurseries. Shallower waters are often warmer and more productive than adjacent deeper waters, which facilitates faster growth. Predictably, species that have large birth sizes (e.g., Dusky Sharks, *Carcharhinus obscurus*) and/or fast juvenile growth rates (e.g., Atlantic Sharpnose Sharks, *Rhizoprionodon terraenovae*) often inhabit unprotected coastal areas as juveniles while those that have small birth sizes and slow growth rates typically use more protected areas such as estuaries as nurseries. Predation risk declines with growth while movement rates and foraging abilities increase. Combined with higher energetic demands and intraspecific competition, this drives older individuals to exploit additional habitats, resulting in increased dietary breadth and activity space.

Sharks move on much larger temporal and spatial scales than their daily activity spaces. You often hear that sharks are *highly migratory*, but what does this mean, and is it true? The term *migration* refers to predictable movements between areas, generally related to resource availability, in which the migrants are compelled to return to their place of origin. Migrations occur on tidal, diel (day-night), lunar, and seasonal cycles and over a wide range of horizontal and vertical spatial scales, but it is the larger scale seasonal migrations that draw most attention. Whereas most coastal sharks exhibit some form of seasonal migratory behavior, many deep-sea sharks examined to date do not. Even within those coastal species that do typically migrate, there can be migratory and non-migratory populations and even non-migratory contingents within a migratory population. We classify migrations by the primary driver into four main categories: *climatic*, *alimentary*, *gametic*, and *refuge* migrations.

Climatic migrations are driven by physiological tolerances of individuals to environmental factors such as temperature or salinity. Inshore juvenile habitats in tropical and subtropical regions, such as Lemon Sharks in Bimini, Bahamas, may be occupied year-round, but in temperate climates such as Chesapeake Bay, this is not an option. Juvenile Sandbar Sharks can survive temperatures only down to about 15°C (59°F), but Chesapeake Bay drops to 5°C (41°F) in the winter. Thus, these juvenile sharks must migrate or die.

Alimentary (trophic) migrations are driven by prey availability, allowing sharks access to prey resources that are only present at certain times or only accessible due to abiotic constraints. Juveniles and adults may undergo alimentary migrations.

Gametic migrations are related to reproduction. In sharks, these may be directed toward increasing the survival of offspring by depositing eggs or newborns in habitats where they will be protected from predation and where prey densities are high, or these may be related to locating mates. Migration can facilitate aggregation of reproductively active sharks, thereby increasing the encounter rate with potential mates. It has been shown that the migratory population of Sandbar Sharks off the east coast of the United States had much higher rates of genetic polyandry (i.e., multiple fathers for a single litter) than the nonmigratory population in Hawaii. One theory to explain this posits that gametic migration in the Atlantic leads to a shorter, geographically constrained mating period, with higher encounter rates between mates and thus more multiple matings, than in Hawaii where mating takes place over broad areas for several months.

Refuge migration functions to decrease predation risk. Often, refuge habitats are inaccessible to predators due to physical (e.g., water depth) or physiological (e.g., salinity) constraints. Complex habitats may also serve as refuges even in the presence of predators by allowing concealment. For example, the prop roots of red mangroves provide refuge habitat for newborn Lemon Sharks and Smalltooth Sawfish (*Pristis pectinata*) that is inaccessible to large sharks. Juveniles of many species migrate to distinct nursery habitats that provide refuge but are also highly productive forage areas; therefore, the migration has both refuge and alimentary purposes. Climatic forcings (factors) like temperature are often the proximate causes for migrations that ultimately serve to increase foraging and reproductive success or increase protection. So often migration behaviors fall into multiple categories.

Studying Shark Movements

Determining where sharks move is central to studying their ecology in myriad ways. Movement studies can provide insight into changes in habitat use over scales from days to seasons to lifetimes. They can be used to determine important areas for feeding, mating, and giving birth as well as those habitats most susceptible to human degradation or where the sharks are exposed to fisheries capture. Together, these data are often used to designate Essential Fish Habitat in management plans or Critical Habitats for endangered species such as Smalltooth Sawfish. Shark researchers today have an array of tagging and telemetry tools at their disposal to study the movements and habitat-use patterns of their subjects over a wide range of spatial and temporal scales. These methods range from the very simple to complex.

Conventional Mark and Recapture

Conventional tagging (mark and recapture) methods are the simplest and have been used for nearly a century to study shark and ray movements. The tags employed include discs attached to each side of the dorsal fin using a stainless steel needle, roto-tags attached to the dorsal fin (modeled after livestock ear tags), nylon dart tags anchored through the basal cartilages supporting the dorsal fin (fig. 9.10), stainless steel dart tags anchored in the muscle, as well as PIT tags discussed earlier. Except for PIT tags, which require a reader to detect their presence and identify the tagged shark, all other tags can generally be read by anyone that recaptures the shark. The tag number and the researcher's contact information are printed on the tag. Modern tagging programs often include a website and email address to report recaptures. The advantages of conventional tagging are that the tags are inexpensive, thousands of sharks may be tagged, and the recapture can be reported by anyone. If the appropriate tag is selected and applied correctly, conventional tags can be retained for the life of the shark. While numerous studies have reported recaptures more than 20 years after tagging, the record is for a School Shark (*Galeorhinus galeus*) recaptured in southern Australia after nearly 42 years.[6] The disadvantage of conventional tagging is that obtaining data on a tagged shark requires that it is recaptured and that the person recapturing the shark contacts the scientist. Though recapture rates are low (in the single digits in most cases), conventional tagging has provided a wealth of information on

Figure 9.10. Cuban Dogfish in Eleuthera, Bahamas, with nylon dart tag. (Photo by Lance Jordan)

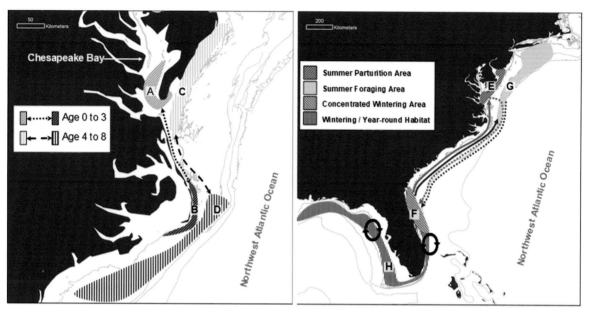

Figure 9.11. Migratory expansions in Sandbar Sharks from Chesapeake Bay based on tag recaptures. *Left*: Younger sharks (age 0 to 3) remain mostly within the Bay, whereas older juveniles (age 4 to 8) expand their range outside of the Bay. *Right*: Seasonal migrations of Sandbar Sharks from their birth and summer foraging areas (*top*) to the overwintering and, in some cases, year-round habitat (*bottom*). (Modified from Grubbs, R.D. and Kraus, R.T. 2019. Fish Migration. In: Choe, J.C. (ed.). *Encyclopedia of Animal Behavior*. (2nd ed.). Elsevier Academic Press. 3: 553–563)

migration patterns, site fidelity, and even estimates of natural mortality rates. For example, tag recaptures from juvenile Sandbar Sharks tagged in Chesapeake Bay and at liberty up to 14 years were used to discern ontogenetic and sex-mediated changes in migratory patterns in this economically important species (fig. 9.11).[7] In addition, since the size of the tagged sharks are typically measured when caught and when recaptured, information on growth rates are obtained. Combining growth rates obtained from conventional tagging with chemical staining of hard parts, such as the vertebrae or fin spines (see below), is an important method in validating age and growth models.

Acoustic Telemetry

Though conventional tagging remains an important tool for studying shark movements, it provides no insight into what the shark is doing and where it is located between the initial tagging and subsequent recapture. The advent of acoustic (sound waves) telemetry greatly improved our capabilities for understanding fish movement and habitat-use patterns. The first use of underwater biotelemetry techniques on sharks dates back to 1963, when researchers tested the use of an ultrasonic (wavelengths higher than human hearing) transmitter for tracking a Scalloped Hammerhead and a Sandbar Shark off the east coast of Florida. Since then, scores of species and sharks and rays have been tracked using acoustic telemetry. The basic approach is to "tag" the shark with a transmitter that frequently emits an aural ping that can be detected using a directional hydrophone attached to a receiver. The frequency of the ping is usually 35–80 kHz, with 69 kHz being the most common frequency in use today. Humans can typically hear no higher than 10 kHz whereas sharks can only hear to around 1 kHz, so these pings are inaudible to us and to sharks. The high frequencies are inaudible to most marine animals, except for some marine mammals that produce high frequency sounds, and some clupeiform fishes (herrings and relatives) that are prey for marine mammals. However, the sound intensity of the transmitters is far too low to be audible even for those species.

Acoustic telemetry is conducted using two major approaches—active and passive tracking. In active tracking, researchers use a directional hydrophone and receiver to detect the shark's position and they physically follow the sharks, recording locations and habitat characteristics continuously. This allows researchers to study the shark's activity space and use of specific habitats and depths at a relatively fine scale and to determine how characteristics such as temperature, salinity, depth, and habitat type influence these movements. Active tracking requires transmitters that emit aural pings between short intervals (~1 s) in order to maintain contact, which also limits their battery lives (usually weeks to a year). Tracking is usually conducted from small boats and the duration of each track is limited by the endurance of the researchers (usually no more than a few days), except in settings where multiple teams are available to rotate on and off of the track.

In passive tracking, individual animals are not followed directly but are instead detected by receivers placed semi-permanently in an array. Each time a shark with an acoustic transmitter passes within the detection range (usually 200–500 m, or 650–1640 ft), the tag number, date, and time are recorded and stored on the receiver. Researchers can then download the data from the receivers periodically. Transmitters used for passive telemetry transmit much less frequently than those used in active tracking (60–180 s intervals instead of 1 s), which permits battery lives of up to 10 years. The design of a receiver array depends on the scientific questions being asked. If interest lies in determining seasonal residency of juvenile sharks in an estuary that serves as a nursery, then a gate of receivers across the entrances to the estuary may suffice for determining the timing of the sharks entering and leaving. However, if interest lies in estimating the amount of space used within that same estuary, then an array that covers the whole estuary is needed and the receivers should be spaced such that all space is within the detection range of a receiver (i.e., ideally there is no place the shark can swim where it would not be detected). Until recently, the scale of receiver arrays was incompatible with the scale of movements of many sharks, particularly adults. In other words, it was logistically impossible and too costly for a researcher to deploy and maintain an array of receivers large enough to track shark migrations or even their seasonal use of large areas. For this reason, most passive telemetry studies until recently were conducted on juveniles or relatively sedentary species. However, in the last few years, teams of researchers have contributed to developing arrays of receivers that cover entire coastlines with hundreds of receivers. A database is maintained of all of the researchers that are tagging animals and their associated tag numbers. When a researcher downloads their receivers and detects tags that are not theirs, they send that detection data to the owner of the tag. Through these large cooperative arrays and long tag battery lives, passive telemetry is becoming a viable tool for studying even the largest of sharks and rays (fig. 9.12).

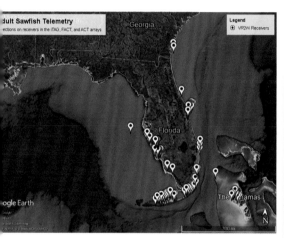

Figure 9.12. (A) Locations of acoustic receivers in Florida and the Bahamas that have detected adult sawfish tagged by co-author Grubbs and colleagues. (B) Co-author Grubbs and graduate student Bianca Prohaska implanting an acoustic tag through a small surgical opening in a Smalltooth Sawfish.

Satellite Telemetry

The primary limitation of acoustic telemetry is that it requires repeated contact between the shark and the receivers. Whenever the shark moves out of the range of the receivers, it can no longer be tracked. The advent of satellite telemetry changed this by allowing animals to be tracked by orbiting satellites independent of the researchers. Essentially, radio transmissions are sent from the tag to satellites, as long as the tag's antenna is above the water surface. The first fish ever tracked by satellite was a Basking Shark (Cetorhinus maximus) tracked for 17 days off the coast of Scotland in 1984.[8] Studies of shark movements using satellite telemetry increased dramatically once the tags became commercially available. Now multiple companies specialize in satellite tag manufacturing. The size of the tags initially limited their use primarily to large species such as White Sharks, Whale Sharks (Rhincodon typus), and Basking Sharks and the results gave us glimpses into shark movements we never thought possible, including tracks of a White Shark moving from South Africa to Australia and a Basking Shark migrating from off Massachusetts to Brazil.[9] Recently, improved design and tag miniaturization have made satellite telemetry amenable to many species, including small sharks like Bonnetheads (Sphyrna tiburo) and Spiny Dogfish (Squalus acanthias), and even skates and rays.

The advantage of satellite telemetry lies primarily in the fact that the researcher needs only to come into contact with the shark one time, to apply the tag. After that, all tracking is done through communications between the tag and satellites. Cost is the major disadvantage of satellite tags, as individual tags cost $1000–$5000 and the satellite service may cost an additional $500 per tag. Though satellite tag technology is continually advancing (e.g., more complex programming, added sensors, improved geolocation capabilities) two basic configurations exist, commonly referred to as SAT (satellite-linked transmitter) and PAT (pop-up archiving transmitter) tags.

SAT tags transmit radio signals that are picked up by orbiting satellites using calculations based on the Doppler Effect to estimate the distance to the tags. The Doppler Effect is the change in the frequency and wavelength of a wave (e.g., sound, radio, light) as the distance between a transmitter object and receiver changes. A simple example of the Doppler Effect is the apparent change in frequency you hear as a motorcycle passes by. The tires and engine of the motorcycle are transmitting sounds that are constant to the driver. But as a bystander, the sound of the motorcycle gets higher pitched as it gets closer to you and lower pitched after it passes. Knowing the radio frequency transmitted from the tag, the distance to a tagged shark can be calculated based on changes in that frequency as the satellite moves closer to or farther from the tag. Each time a signal is detected by the satellite, the change in frequency and time-lag are used to estimate the tag's location. Algorithms that use consecutive location estimates to predict future locations are used to refine the track of the tagged animal.

SPOT (Smart Position and Temperature) tags are the most commonly employed SAT tags for sharks. These have a wet/dry sensor that detects when the shark reaches the water's surface and send up signals, thus avoiding using power when the tag is submerged.

The advantages of SAT tags are the long battery life (one year or more), powerful transmitters that reach satellites 1000 km above the earth, and accurate, near real-time location estimates. This is the type of tag used to track the White Shark mentioned above. You may be able to predict the major disadvantage of SAT tags. Recall that sharks live underwater. SAT tags are only useful for species that come to the surface frequently or live in shallow water. For this reason, these tags are usually attached to the dorsal fin or towed behind the shark with a tether. Figure 9.13 shows an example of a Smalltooth Sawfish tracked using a SPOT tag in the shallow mangrove backcountry of the Bahamas.

The accuracy of the locations produced from SAT tags using the Argos satellites varies considerably and for animals that surface infrequently, few locations may be received. However, making use of GPS satellites, which are far more numerous, can greatly reduce the error. SAT tags are now available that have onboard GPS receivers that can provide location accuracies of 50 m or less when at the surface for about 20 s.

PAT tags may be used with animals that never come near the surface and are therefore used with a wider variety of shark species than SAT tags (fig. 9.14). PAT tags are programmed to remain on a shark for a specific time, which is usually less than a year. While attached to the shark, the tag measures and records (archives) time, temperature, depth, and light data. Once the tag release date is reached, a mechanism such as a burn wire, which predictably corrodes, detaches the tag from the attachment. The tags are positively buoyant on one end and float to the surface such that the antenna breaks the surface vertically. The tag's

Figure 9.13. Example of the track of a Smalltooth Sawfish using a SPOT tag in the shallow mangroves of Andros in the Bahamas. Each red dot represents a fix uploaded to a satellite. Where did this sawfish spend most of its time? (Andrea Kroetz and John Carlson, unpublished data)

current (pop-up) location is transmitted to the satellites along with all the stored data. The light data recorded during the track is used to estimate the animal's location using light-based geolocation. How does this work? Day length varies depending on latitude and time of year. At the equator, there is 12 hours of light and 12 hours of dark all year long. But as you move north or south (increasing latitude) day lengths decrease in the winter and increase in the summer. This is why in Anchorage, Alaska (60 degrees North latitude) there is less than six hours of daylight in the winter and more than 18 hours of daylight in the summer. The time of solar noon varies predictably by longitude. Thus, from the light data stored on the tag, we can determine the time of sunrise and sunset which gives us day length and peak daily light levels provide the time of solar noon. These two pieces of information allow us to estimate the position of the tag over the course of the track. These estimates are far less accurate than those obtained from the surface by the SAT tags. Small errors in day length estimates can cause large errors in latitude estimates, and do not forget about the pesky equinoxes when the sun is directly over the equator in the spring and the fall. During this period, day length is roughly the same all over the globe, causing huge errors in light-based geolocation estimates. However, we do have tricks to reduce these errors. We know exactly when and where the shark was tagged and where the tag popped up. We also know something about the swimming capabilities of our sharks: no shark can swim from Florida to Antarctica in one day (we had PAT locations that suggested this). We also have the temperature data recorded by the tag. By using the tagging and pop-up locations, filtering out erroneous location estimates, and comparing the temperature when the shark is near the surface to sea surface temperature from satellites, we can produce a "most probable track" the shark took. However, due to the geolocation limitations, PATs are better suited for species that travel large distances than those with small home ranges. For example, Figure 9.15 shows the most probable tracks for numerous PAT-tagged Oceanic Whitetip Sharks.[10]

The other great advantage to PAT tags is the archived depth data. The ability to record depth and temperature make PAT tags very useful in studying the depth-use patterns, even in deep-sea sharks where no light data are recorded to allow light-based geolocation. Figure 9.16 is an example of depth data obtained from a PAT tag on a Bluntnose Sixgill Shark. Since the tags are recording temperature as well as depth, the tagged animals can serve as incredibly detailed water samplers. These methods are also very useful in studies of post-release mortality in fisheries. If a released shark goes to the bottom after release and does not move, it likely died. In fact, two manufacturers are now making tags specifically designed to assess mortality. Since most sharks captured in fisheries are bycatch that are discarded, estimating post-release mortality is critical in determining the effects of fishery capture on populations.

Figure 9.14. PAT (or PSAT) tag on a Sandbar Shark in Hawaii.

Figure 9.15. Likely tracks for nine Oceanic Whitetip Sharks tagged with PAT tags. (From Howey-Jordan, L.A. et al. 2013. PloS One 8(2): e56588)

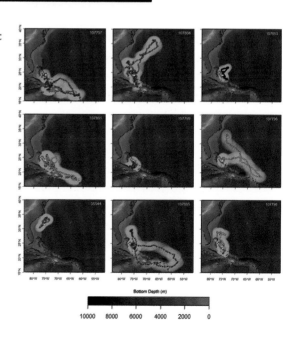

A

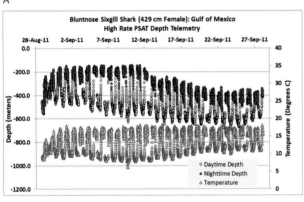

B

Figure 9.16. (A) Co-author Grubbs attaching a PAT tag on a Bluntnose Sixgill Shark in Eleuthera, Bahamas. (Courtesy of Lance Johnson) (B) Depth and water temperature data from a PAT-tagged Bluntnose Sixgill Shark in the Gulf of Mexico, showing nocturnal migrations from colder depths to warmer surface waters. (Grubbs, unpublished data).

Emerging Telemetry Technologies

We will conclude our discussion of methods to study shark movements by briefly mentioning emerging telemetry technologies that are allowing researchers to study shark behavior like never before. Researchers have long wished to gain more insight into what sharks are actually doing while they are being tracked. When are they chasing prey? Avoiding predators? Mating? Feeding? Resting? Some early studies employed tail beat monitors to estimate when sharks were burst swimming versus sustained swimming, but these were relatively cumbersome. Tri-axial accelerometers are taking the technology leaps forward. When incorporated into archival or acoustic transmitters, accelerometers record acceleration in three dimensions (forward-back, side-to-side, up-down) in great detail. Multi-sensor transmitter tracking packages with accelerometers (e.g., daily diary[11]) can provide information on movements in terms of location and animal behavior (e.g., resting, chasing, even mating in Nurse Sharks[12]) and even energetics (see Chapter 8). The accelerometer data can be used to determine overall dynamic body acceleration (ODBA), which has been shown to be correlated with oxygen consumption and thus energy expenditures in a wide range of animals, including sharks.[13] These archival transmitter packages can also include sensors to measure environmental parameters externally (e.g., temperature, depth) and even internal to the shark (e.g., pH). Another emerging frontier in shark movement studies is inter-animal telemetry, using what has been referred to as "business card" tags that are not only transmitters but also include receivers that can detect and archive detections from transmitters on other sharks.[14] These methods hold tremendous promise in studying schooling, aggregation, and social behaviors in sharks.

Using Genetics to Study Shark Movements and Population Structure

Tagging and telemetry provide a wealth of information regarding the movement and habitat-use patterns of sharks. From the standpoint of fisheries management, this is critical because it provides managers with data concerning the likelihood that sharks from a given population will be caught by specific fisheries. However, population genetics offers powerful tools to discriminate whether movements of sharks between two regions are interbreeding or only mixing ephemerally. The techniques employed in population genetics are beyond the scope of this book, but in short involve looking for differences in frequencies of genes and alleles (alternative forms of a gene) between groups of sharks that may reflect isolation and adaptation (local evolution) and indicate whether the groups are members of the same or different populations. Defining the boundaries of a population (or stock) is the critical first step in fisheries management.

Tagging studies may indicate that sharks in two regions rarely if ever mix, but molecular data indicate they are part of the same population. This suggests the sharks crossing from one area to the other happens too infrequently to pick up in the tagging data but frequent enough to genetically mix the populations. For fisheries management, these two groups of sharks may be managed as a single stock. This makes management difficult if a single stock crossed international boundaries, which is common for pelagic sharks. In fact, Blue Sharks have been shown to be part of one single worldwide genetic stock, hence the extreme difficulty in managing them. Conversely, if the genetics data confirm that sharks in the two regions are distinct, they would appropriately be managed as two separate stocks. The peninsula of Florida has been shown to be a barrier to movements and gene flow to many coastal shark species (e.g., Blacktip [*Carcharhinus limbatus*], Blacknose [*C. acronotus*], and Bonnethead Sharks [*Sphyrna tiburo*]) and thus in US Atlantic and Gulf of Mexico waters these are managed as separate stocks.

Life Histories of Sharks: Age and Growth

Sharks are typically slow-growing and long-lived. How slow-growing and how long-lived? Knowing this is central to understanding the ecology of shark species as well as managing them, which differs from managing bony fishes.

Reproduction of sharks as a group is more limited than that of bony fishes. Reproductive limitations are characteristic of animals classified as K-selected (as opposed to r-selected; see Chapter 1). The concept of K- and r-selection is an ecological idea based on diametrically opposed life-history, specifically reproductive, strategies. Although the idea is somewhat outdated, in part because life-history characteristics operate on a continuum and are rarely the extremes depicted in r- and K-selection, it is worth quickly reviewing these as they apply to sharks (fig. 9.17). We will revisit and apply these concepts below.

Before fishery biologists can implement management plans for any species, they must consider the vulnerability of a species or stock (population) to overfishing. To manage a fishery (i.e., to assess this vulnerability), it is crucial to know information about the animal's life history. It is also critical in fisheries with multispecies catches to know life-history characteristics of incidental catch such that

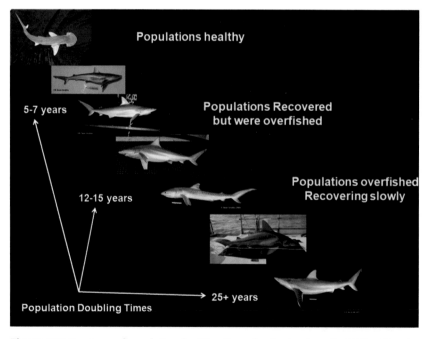

Figure 9.17. Spectrum of population doubling times for sharks along the US East Coast and a generalized view of how they have responded to fisheries and management.

the fishery does not deplete the stocks of bycatch. These characteristics are also important in other areas (e.g., predator-prey studies). Critical life-history characteristics include:

- Size and age at maturity
- Size and age at first reproduction
- Fecundity, both per pregnancy and lifetime
- Frequency of reproduction
- Survival of offspring (the most difficult to accurately determine)
- Maximum size
- Maximum age

How are these characteristics determined? Let us start with age and growth. Access the web, go to Google Scholar, and search for *Age and Growth Sharks*. Surprised at the number of hits? Although not the most exciting research, obtaining these life-history characteristics is the *sine qua non*[15] of fisheries management, which is reflected in the number of studies. Further note that many of these studies are for the same species, but in different locations. We will tell you why below.

How are age and growth studies done? In principle, it is fairly easy and relies on the same idea of trees depositing hard tissue seasonally, annually, or during some environmental events over the course of their entire life. In bony fishes, age is recorded primarily in calcified layers of the *otoliths*, or ear bones. Sharks lack otoliths and instead have *stataconia*, which are like squishy sand grains and which do not lay down concentric rings.

Some bony fishes (e.g., Striped Bass [*Morone saxatilis*]) can be aged by removing a few scales, which record growth. Although sharks also have scales, recall that they are morphologically different from those of bony fishes. Moreover, the dermal denticles of sharks are replaced, perhaps several times over the life of the shark. Thus, the dermal denticles of sharks cannot be used for aging.

The only hard parts that can be used to age chondrichthyan fish with reliability are parts of the vertebral centra (fig. 9.18), the main structural elements of the backbone or, on some species (e.g., some Squaliformes [dogfishes], Heterodontiformes [horn sharks], and Chimaeriformes [chimaeras/ghost sharks]) the spines on one or both dorsal fins (fig. 9.18). For vertebral centra and dorsal spines, until a noninvasive reliable imaging technique is invented, the animal must be killed.

The use of spines to age sharks goes back to the 1930s, for the Spiny Dogfish (*Squalus acanthias*). In some cases (e.g., in deepwater squalids whose vertebral centra are not calcified sufficiently to yield concentric rings), it is the only method available. Using dorsal spines for aging can be problematic in older individuals of some species. First, the bands may become too tightly spaced to accurately infer age. Second, the spines of the first dorsal fin may become too worn down to be of use. Spines are removed in their entirety, including the base below the skin. Figure 9.19 shows ages of spines of a dogfish.

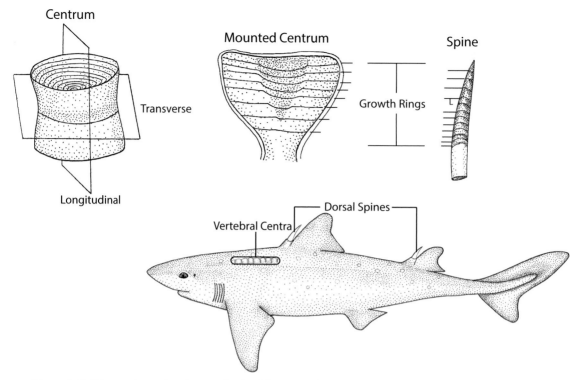

Figure 9.18. Ways to determine the age of a shark from vertebral centra and dorsal spines. Individual centrum (*top*) is pictured rotated 90°. (Redrawn from figs. 6.1 and 6.4, Goldman, K.J. 2005. FAO Fish. Tech. Pao. 474: 76)

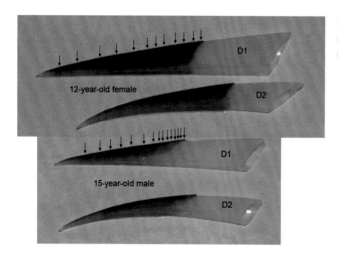

Figure 9.19. Ages from spines of a male and a female Hawaiian Spurdog. (Courtesy of Charles F. Cotton)

Vertebral centra have been used to age sharks for over 90 years. After sacrificing the shark, a section of 10 or more vertebral centra are removed, typically the largest ones in the thoracic (or trunk) region beneath the first dorsal fin. Recall that each vertebral centrum is a disc that is concave on both sides with the remnant of the notochord running through its center. The centra are then cleaned, cut with a diamond blade, and mounted. Figure 9.20 shows the bow tie, or hourglass, shape of a section through the centrum. After sanding the section down until it is sufficiently thin that light passes through it, you can examine rings; for example, the wide, light-colored rings laid down in the summer and the narrower, denser, darker rings deposited during the slower growth winter period.

The aging protocol requires considerable practice and uses multiple readers (i.e., scientists to interpret the number of rings). Additionally, it is crucial to know that the rings that are counted correspond to years and not seasons. There are two main ways to validate the time period of rings:

1 / Catch young-of-year (YOY) or neonates (newborns), measure, and tag large numbers of them, then hope that some will be recaptured three or more years later. Then remeasure and sacrifice recaptured sharks and remove centra and count rings. If the bands in the centra correspond to years at liberty, then you can safely conclude that they lay down bands annually, as well as how much they have grown.

2 / Beginning in the 1970s and 1980s, sharks were injected in their muscles with a chemical, either oxytetracycline (OTC), calcein (a fluorescent dye), crystal violet, or silver nitrate, all of which stained the hard parts. Then the sharks were measured, tagged, and released. Recaptured individuals were measured, sacrificed, and then rings were counted outward from the point of stain and age and growth could then be determined (fig. 9.21).

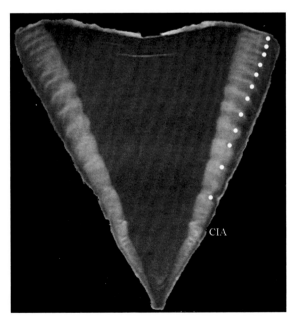

Figure 9.20. Sectioned vertebral centrum from a Sandbar Shark. CIA stands for Change in Angle and depicts the point of birth. If you estimated that the shark from which this centrum came was 12-years-old, you would be correct. (Courtesy of Jason Romine)

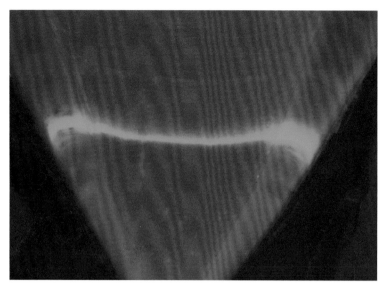

Figure 9.21. Oxytetracycline (OTC)-stained sectioned vertebral centrum from a Sandbar Shark. Green line denotes the time of the OTC injection. (Courtesy of Jason Romine)

AGE AND GROWTH: A CASE STUDY (SANDBAR SHARK)

Sandbar Sharks are circumglobal in distribution. Robust age and growth models from the Virginia Institute of Marine (VIMS)[16] showed that it takes about 15 years to reach sexual maturity and they produce about eight pups every two years. These results further showed that Sandbar Sharks in and near Chesapeake Bay are born at a length of about 45–50 cm (16–20 in) and females reach sexual maturity at about 136 cm (53.5 in). The species has a life span of about 30 to 35 years.

The VIMS data on Sandbar Sharks was critical to the National Marine Fisheries Service's fishery management plan (FMP; see Chapter 11) for this and other coastal shark species, a group that was overfished along the US Atlantic Coast. Among these species, Sandbar Sharks were among the most overfished from the 1980s and early 1990s, primarily for their large fins, which were popular in the shark fin soup industry, as well as for meat. Because of their low reproductive rates, population growth could not exceed 4% per year, and thus the potential for quickly rebounding from overfishing was very low.

In 1993, the FMP that was implemented put heavy quotas on landings of large coastal sharks, including Sandbar Sharks along the US East Coast, based on the life-history characteristics reported by VIMS researchers.

Representatives of the commercial shark fishing industry challenged the validity of the VIMS data, instead suggesting that reproductive rates were in fact much higher. They argued that another scientific study, one on Sandbar Sharks in captivity and published in 1973,[17] suggested that reproduction occurred twice annually, and that the species reached sexual maturity at three years. Thus, the industry concluded that what they interpreted as draconian management policies, were in fact based on bad information. Who was right, the commercial shark fishers or the VIMS scientists?

To answer that question, another age and growth study on Sandbar Sharks was conducted in Hawaii.[18] The reason for repeating the study in Hawaii was twofold. First, it is not unusual for aquarium animals (and those in zoos as well), to grow faster, sometimes *much* faster, than wild counterparts. Second, the ecology of Sandbar Sharks in Hawaii differs significantly from that of those on the US Atlantic Coast. Hawaiian Sandbar Sharks live in deeper water than Atlantic Sandbar Sharks. Moreover, in Hawaii, Sandbar Sharks are year-round, nonmigratory inhabitants, whereas those on the US East Coast are migratory. Additionally, water temperature is high year-round in tropical Hawaii, and the mating season is thus protracted. Finally, Sandbar Sharks are more sparsely distributed in Hawaii than along the US East Coast.

To conduct the study, researchers (including co-author Grubbs) set demersal longlines outside Kaneohe Bay, Hawaii, and caught and sacrificed up to five specimens of each sex and 5 cm size class (between 45 cm and 150 cm, or 18 to 59 in) for a total of 194 sharks. This may seem like a lot of sharks, but it is the minimum required to get robust, accurate estimates of these extremely important life-history parameters. If in fact the estimates of the industry were incorrect and were implemented anyway, which is a real possibility since the fishery management decisions included political and economic dimensions, tens of thousands more sharks would have been killed than the 194 for this study.

The results validated both of the scientific studies. Sharks do in fact mature earlier in Hawaii than in Chesapeake Bay, at 10 years for females and eight for males (compared to 15 combined for both sexes in Chesapeake Bay), though not at three years as the 1973 paper asserted. Moreover, the representatives of the commercial shark fishery ignored another hypothesis from that paper—that an analysis of tooth replacement suggested the age of maturity as 10.2 and 13.1 years for males and females, respectively. Science indeed works, if used as directed.

So, even though Hawaii has no shark fishery and longlining within 80 km (50 mi) of the coast is prohibited, this study was important for Sandbar Sharks about 7900 km (4900 mi) apart. Data from a long-term tagging study of neonate Sandbar Sharks tagged in Chesapeake Bay and recaptured up to 14 years later further confirmed the initial growth models proposed by the VIMS researchers.

Life-history Theory

Life-history theory refers to understanding growth, survival, and the reproduction of species. Understanding the life history of any organism is critical for its management and conservation. The bulk of life-history research on sharks is directed toward estimating five parameters:

- Size or age at first reproduction (i.e., maturity)
- Reproductive output (e.g., fecundity, periodicity)
- Survival of offspring
- Maximum size
- Longevity (maximum age)

All of these parameters are related to *fitness*. By fitness we do not mean the ability of sharks to swim a marathon, but rather their biological or Darwinian fitness, which is defined as success in producing viable offspring that themselves survive to reproduce. The goal of any organism in terms of reproduction and life history is to maximize its fitness. One of the metrics used to estimate how resilient a population may be to fishing is the *intrinsic rate of natural increase* (lowercase *r*), a theoretical maximum value which is a function of two important aspects of its life history—its reproduction rate (uppercase *R*) and the generation time (*G*). Another important estimator of population resiliency is the *doubling time*—the time it takes a population to double in size given a stable growth rate. Doubling time is estimated by 70/% growth rate. For example, a population growing at a rate of 5% per year theoretically doubles in about 70/5 = 14 years.

The *intrinsic rate of natural* increase (or *population growth*), (r) is calculated as follows:

$$r = \log_e(R_0)/G$$

Where,

R_0 = the average number of offspring that a female produces during her lifetime, also called the *net reproductive rate*

G = generation time (i.e., the mean period between when the parents and all their offspring are born)

This intrinsic rate of natural increase, *r* in the above equation, is the "r" in r-selection. r-selected species maximize their rate of population increase. The K in K-selected species stands for *carrying capacity*.

There are three general ways to maximize the intrinsic rate of natural increase r:

- Minimizing generation time (e.g., mature faster)
- Maximizing fecundity (e.g., have more offspring)
- Maximizing survival (e.g., grow faster, have larger offspring)

There are limiting tradeoffs. Maturing fast generally results in small body size. The smaller the young, the higher the mortality. Therefore, mortality is minimized by having larger young. But larger young require higher maternal investments and results in lower fecundity. Small body size also results in fewer offspring unless those offspring develop external to the mother.

Most organisms lie on the continuum of r- and K-selection and not on the extremes. r-selected species are those that have evolved to maximize their rate of population increase versus K-selected species, which have evolved to live near their carrying capacity.

Sharks and their close relatives are on the K-selected side of the spectrum. r-selected species tend to be small, mature quickly, and have very high fecundities (as many as millions of eggs) with very little investment of maternal resources in each egg. An example among fish would be the anchovy. The largest fishery in the world, Peruvian Anchoveta, used for fishmeal and oil, persists because these tiny bony fishes mature in about a year and produce tens of thousands of offspring. Similarly, egg-laying in sharks and rays is likely an adaptation for increasing fecundity in small species. Skates and catsharks can produce dozens of offspring through egg laying, whereas they could produce only a few similarly sized offspring through live birth.

At the other end of the spectrum are K-selected species, which have large body size, take years to decades to reach sexual maturity, and have very low fecundity (just a few offspring at every reproductive event, which may occur every other year, or even every three years) with a large investment of maternal resources in every pup. All large sharks give live birth and again there is a tradeoff. Sand Tigers (*Carcharias taurus*) produce only two pups but they are very large and have a high likelihood of survival. But Tiger Sharks may produce 60 or more pups that are very small and are at high risk of being eaten. Both strategies result in at least enough offspring surviving to replace the mother during each cycle.

Thinking of life histories in terms of fisheries, if a fishery targets tunas, like the Yellowfin Tuna, which matures at age three and produces millions of eggs annually, it takes only two years for the population of this r-selected species to double. So even if the stock is overfished, if management measures are put in place, recovery may be achieved in a few years. Conversely, a Sandbar Shark takes 15 years to reach maturity and produces eight pups every two years (i.e., on average two female pups per year). Their population doubling time is about 25 years. So, an overfished stock of Sandbar Sharks takes many decades to recover, if allowed to.

If you look at the intrinsic growth of increase in all taxa, it is generally inversely proportional to the body weight. In other words, the smaller you are, the faster your population grows or, more appropriately for sharks, the bigger you are, the more slowly your population is going to increase.

This spectrum occurs not only between sharks and other species, but also among different sharks. For example, the Bonnethead takes three years to mature, reproduces every year, has many offspring, and has a population doubling time of six years. Thus, it is important to realize that *within* the sharks and rays, some species are more vulnerable to overexploitation than others. But it is also critical to realize there are notable exceptions to this rule among the sharks. Many of the deep-sea sharks, like the dogfishes and gulper sharks, are around the same size as the small coastal species that grow quickly, but they have among the lowest rebound potentials due to very slow growth, long gestations, and few offspring.

Below, we will discuss the management of sharks along the US East Coast. There, sharks are managed as three major groups—Large Coastal Sharks, Small Coastal Sharks, and Pelagics. The reason for these divisions, in part, is that members of each group tend to have different life histories. For example, in the Small Coastal Shark category, the smoothhounds (e.g., *Mustelus canis*), Bonnethead, and Atlantic Sharpnose Shark (*Rhizoprionodon terraenovae*) all have population doubling times in the 5- to 10-year range. Members of the Large Coastal Shark group, on the other hand, like the Lemon, Sandbar, and Dusky Sharks, have population doubling times in the 20- to 25-year range. Pelagic sharks, including the Blue and Oceanic Whitetip Sharks, are somewhere in the middle.

Large Coastal Sharks are further divided into *ridgeback* sharks and *non-ridgeback* sharks (see Chapter 3), in part because of life-history differences that necessitate that they be managed separately. Historically, the two most-frequently caught Large Coastal Sharks are the Sandbar and Blacktip. The Sandbar, which has a long population doubling time, is a ridgeback. The Blacktip Shark is a non-ridgeback and it matures early and thus has a population doubling time that is half a Sandbar's. Managing them both as Large Coastal Sharks without regard for these life-history differences would most likely be problematic for recovery of Sandbar Sharks.

Complicating management even further, within a species there can be variability. Consider that the age of maturity of Sandbar Sharks is about 16, 15, and 10 years, respectively, for populations in the Indian Ocean, NW Atlantic Ocean, and the Pacific Ocean off Hawaii.

One of the major impediments to manage sharks in the 1980s was trying to do so like bony fishes were managed. In reality, sharks should have been managed more like whales and other cetaceans, or even elephants, because the intrinsic rates of population growth of sharks, like Sandbar Sharks, are more similar to these mammals, or even sea turtles, than they are to bony fishes. The annual rates of population growth for the African elephant, loggerhead sea turtle, and most larger species of sharks are in the ranges of 4.0–7.0%, 2.0–6.0%, and 1.7–6.9%, respectively. Managing sharks as bony fishes, whose annual rates of population growth are much higher, would lead to overfishing in most cases.

Trophic Ecology

Sharks are commonly referred to as opportunistic predators, and prey selection often varies on multiple time and space scales. Most species have diverse diets and can switch prey types in response to changes in prey abundance or species composition. The hypothesis that sharks consume the most abundant prey available has been supported by numerous studies, however, there is also plenty of evidence that some shark species are specialists on certain prey types. For example, Horn Sharks (*Heterodontus francisci*) feed primarily on hard-shelled molluscs and crustaceans, Dusky Smoothhounds feed mostly on a few species of crabs (primarily recently molted crabs), and Frilled Sharks (*Chlamydoselachus anguineus*) eat mostly squids.

How Do We Know What Sharks Eat?

There are multiple approaches to studying the trophic ecology, but two dominate the literature: gut content analyses and analysis of stable isotope ratios. Firstly, a lot of what we know about the trophic ecology of sharks comes from actually looking at a stomach contents and identifying the prey items (fig. 9.22). Typically, the stomach contents are removed in the field, preserved or frozen, and then identified in the lab to the lowest possible taxon. This is a meticulous process that requires a deep knowledge of taxonomy and identification of the species that could be in the potential prey base and, to be done properly, it requires access to scores of stomachs from both sexes across all shark life stages. The advantage is that prey can potentially be identified to species and the relative contribution of each prey species can be quantified. Stomach content analysis remains our most robust method for studying the actual diet of sharks, but it has disadvantages beyond the time, skill, and labor required. Though some studies use stomach lavage with water to flush contents from live sharks that can then be released, most samples are obtained from dead sharks, typically obtained from fisheries or from sharks that have been used in age and growth studies. Many studies are hampered by the prevalence of empty stomachs, since the hungriest sharks take baited hooks. Stomach contents also only tell us what the shark was eating in the hours to days before being caught, so samples are needed over longtime periods to robustly quantify the diet.

Figure 9.22. Gut contents of a Sand Tiger, whose diet includes fishes that may be swallowed whole, like Atlantic Croakers, a Bluefish, and a Clearnose Skate.

Secondly, analysis of stable isotope ratios is based on the adage "you are what you eat." Recall that *isotopes* are multiple forms of a chemical element that vary by the number of neutrons in the nucleus. The more neutrons, the heavier the isotope. These isotopes can be generally considered either *radioactive* (which spontaneously decay over time; that is, they are *unstable*) or *stable*. The element Carbon, for example, has three naturally occurring isotopes, of which two are stable: Carbon-12 (^{12}C or C–12), with six protons and six neutrons, and ^{13}C, with six protons and seven neutrons. The ^{13}C/^{12}C ratio[19] (conventionally reported as heavy isotope/light isotope) is about 1/93.

For analysis of stable isotope ratios, the number in which we are interested would be calculated differently and reported as δ^{13}C (pronounced *delta-13-C*), whose units are parts per thousand (‰, ppt, or per mil). δ^{13}C is calculated by taking the above ratio, ^{13}C/^{12}C, dividing it by a standard, then subtracting 1, and then multiplying by 1000.

Carbon, Nitrogen, and Sulfur isotopes are most commonly used for aquatic diet studies because they are *tracers* of different aspects of trophic ecology. Carbon stable isotopic ratios provide a good idea of which primary producers are at the base of the food chain. Sulfur ratios give insight into whether the animal is feeding in a benthic or pelagic food web. Nitrogen isotopes are most useful for determining a shark's trophic level (how high in the food chain it feeds).

For marine plants at the base of the food web, δ^{13}C values for seagrass, macroalgae, and phytoplankton are, respectively: –11‰, $\cong$ –17‰, and $\cong$ –22‰. These values change, or *fractionate*, at the low rate of only ~1‰ per trophic level. The minus sign indicates that the heavier isotope is *depleted* as it rises through trophic levels.

Nitrogen isotopes fractionate at a rate of 3.0–3.4‰ per trophic transfer, though there is evidence that this fractionation shrinks as you move to higher trophic levels. Increases in the fractionization ratios is known as *enrichment*. One complication is that Nitrogen isotopic ratios also vary by Carbon source, so Carbon isotopes are also needed to interpret the Nitrogen isotopes.

From our own work, a Cownose Ray (*Rhinoptera bonasus*) feeding on bivalves, which are primary consumers of phytoplankton, has a signal of δ^{13}C = –20‰ and δ^{15}N = 10‰. In contrast, an adult Blacktip Shark feeding on fishes and small sharks in a seagrass-dominated system has a signal of δ^{13}C = –13‰ and δ^{15}N = 16‰.

Confusing as it may seem, the advantages of stable isotopes are that the tissues (e.g., muscle, blood, skin) can be collected in relatively safe and noninvasive ways, and the shark can then be released. The results provide a signal of diet that is constrained only by the turnover time of the particular tissue (e.g., days for blood, weeks to months for muscle). The primary limitation of this method is its lack of specificity; that is, the results reflect the signal from all of the prey items eaten and do not provide insight into the actual prey species.

We can also study shark trophic ecology through a functional morphology lens. Variation in jaw and tooth morphology (shape) reflects the varied trophic ecology of sharks (see fig. 4.19). Many familiar sharks like Blacktips, Sandbars, and Lemons have narrow cusped lower teeth for grasping prey, whereas the upper teeth are slightly wider with lateral edges that allow them to slice prey into pieces or grasp and swallow whole prey. Co-author Grubbs saw this when he had his thumb nearly completely removed by a juvenile Sandbar Shark he was tagging. You might say he observed this tooth morphology in action first*hand*. White and Bull Sharks have triangular upper teeth for cutting large chunks from larger prey. Of course, the dalatiid sharks (e.g., Cookiecutter, *Isistius brasiliensis*) take the opposite approach and have small grasping teeth in the upper jaw and triangular cutting teeth on the lower jaw for carving melon-ball chunks out of unwitting prey. Tiger Sharks and Sixgills take flesh removal to extremes, having wide multi-cusped teeth in both jaws and relative weak jaws that are adapted to bending across the body of large prey (e.g., sea turtles, dead whales). They then twist or spin their bodies to carve out huge chunks of flesh. Dogfish sharks actually feed similarly and have oblique bladed teeth in both jaws that lie close together forming a single knifelike blade in both jaws, allowing them to cut small prey in half or cut small pieces from large animals while scavenging.

At the other end, Shortfin Mako and Sand Tigers concentrate on prey they can swallow whole and thus have very narrow cusps on the teeth of both jaws. The Frilled Shark has tri-cusped teeth angled backward to grasp and swallow soft-bodied prey, and like the jaws of a python, once the prey enters the mouth, there is no backing out. Co-author Grubbs also witnessed this when he had to call for assistance when his thumb became stuck in the mouth of a dead Frilled Shark that had been preserved in a museum for decades. (At this point, you would be justified by questioning why co-author Grubbs seems to always be placing his thumb in proximity to the mouths of sharks.)

And then there are many shark species with specialized dentition adapted to some level of durophagy (eating hard prey). The smoothhounds have pebble-like teeth similar to many sting rays for crushing relatively weak shelled prey like shrimp, crabs, and small clams. In the extreme, the Horn Sharks have cusped teeth up front, molariform teeth in the back, and hypertrophied jaw muscles for crushing snails, urchins, and crabs. Nurse Sharks have pavement-like teeth with pointed cusps that are used to crush and tenderize large prey like lobsters until they can be broken into pieces small enough to swallow.

In addition to selecting specific prey taxa, some sharks may also select large or energy-rich prey. For example, one study showed Basking Sharks selected areas to filter feed where there were high densities of large copepod species and low densities of smaller, less energy-rich species. Another study showed that when offshore, Shortfin Makos fed primarily on squid but when they moved inshore, they fed heavily on oily energy-rich bluefish (*Pomatomus saltatrix*) even though there were high abundances of squid prey inshore.

High diversity of sharks in a given region may by supported by opportunistic versus specialized diets. This constitutes a kind of resource partitioning between species (also known as *interspecific specialization*), which refers to dividing limited resources in ways that minimize competition for them.

High densities of sharks can also be supported by ontogenetic niche shifts of the same species (*intraspecific specialization*). Just as habitat use changes with growth and age, sharks undergo changes in diet as they grow. Since mouth gape increases with growth, which allows consumption of larger prey, the trophic position of predators, including sharks, usually increases with age. Larger sharks also eat larger and faster prey than juvenile sharks, but still consume many of the same prey taxa that were consumed as juveniles.

It is often estimated that sharks consume 2–3% of their body weight per day, but this also varies depending on the energy demands of a specific shark species or life stage and the energy content of the prey. The daily ration of Sandbar Sharks was estimated to decrease over the first few years of life, but the energy demand increased, which required consumption of prey with higher caloric content. So as juvenile Sandbar Sharks grew, they consumed larger and more energetic prey.

It is likely that both prey density and accessibility play major roles in habitat selection by sharks. Recall the ontogenetic expansion of habitat use by Sandbar Sharks discussed earlier. Even over the first several years of life, the shift by juvenile Sandbar Sharks to deeper habitats is accompanied by a shift from consuming crustaceans like swimming crabs and mantis shrimp, to bony fishes and skates. The shifts in diet that occur with growth in Tiger and White Sharks are very dramatic. Both species feed mostly on teleost fishes as juveniles. But as adults, Tiger Sharks commonly eat elasmobranchs, sea turtles, birds, and mammals. For adult White Sharks, marine mammals become a major component of their diet. The shape of the teeth of White Sharks even undergo an ontogenetic change that facilitate the dietary shift.

Roles of Sharks in Ecosystems

We end this chapter by addressing the ecological roles of sharks. The first studies to assign trophic levels to sharks (149 species in eight orders and 23 families), skates (60 species), and stingrays (67 species) as groups were published only in 1999,[20] 2007,[21] and 2013,[22] respectively.

These studies validated the perception that most sharks are upper level predators and their food webs are composed of at least four trophic levels. Overall, sharks were considered *tertiary consumers* with a mean trophic level > 4 (see fig. 9.2). Trophic level generally correlated positively with body size, although large planktivorous sharks had low trophic levels (3.2 and 3.4 for Basking and Megamouth Sharks, respectively). Carcharhiniform, hexanchiform, and lamniform sharks exhibited the highest mean trophic levels, as high as 4.7 for the Broadnose Sevengill Shark (*Notorynchus cepedianus*), 4.5 for the White Shark, and

4.4 for the Sand Tiger and Bramble Sharks (*Echinorhinus brucus*). In addition to the planktivores, on the low end of the trophic level scale were the Zebra Shark (*Stegostoma fasciatum*) at 3.1 and the Horn Shark at 3.2.

Thus, large sharks that feed on fish, squid, and mammals are tertiary consumers (trophic level 4) and are at the top of the food chain and in some cases serve as apex predators. Recall, however, that most sharks are not large, but are less than 100 cm (3.3 ft) long as adults. Many of these species as well as large species that feed on invertebrates are secondary consumers (i.e., *mesopredators*, trophic level 3).

It is also critical to understand that large sharks were not always large. Juveniles of sharks that are apex predators as adults are mesopredators as juveniles. Many coastal and deep-sea sharks undergo an ontogenetic shift of more than one trophic level. Since the majority of shark species are small, and all sharks start life small, then by sheer numbers, the vast majority of sharks swimming the world's oceans are not apex predators, or even consumers at trophic level 4. Most sharks are mesopredators.

This means the ecological roles of sharks in marine ecosystems are far more complex than the simplistic notion of them as apex predators influencing marine food webs through top-down forces. They also play critical roles in the middle of the food webs. Even mesopredators can have structuring roles on invertebrate populations and communities. In many cases, sharks that are at trophic level 4 (e.g., reef sharks) function as critical members of an assemblage of shark and teleost species that together occupy the highest trophic level in a system. A recent study[23] found that Bull Sharks and Smalltooth Sawfish share the top predator space in mangrove estuaries. In some specific cases with simple food webs, adults of a single species of shark may be the apex predator. We are tuned to assume the marine food webs should exist in simple linear ways with a single apex predator, as has been depicted in many terrestrial food webs. But marine food webs are often so complex that there is functional redundancy at all trophic levels, including the highest-level predators. The idea that sharks are apex predators is akin to saying all snakes are apex predators because anacondas exist. Sharks are highly connected predators that play complex roles in marine food webs.

Concluding Comments

We began the chapter with a list of the subdivisions of ecology. The overarching goal of ecology is to assemble these reductive views into the big picture of the ecological roles of sharks as some of the questions we raised earlier are answered. Although the gains in our understanding of sharks in their ecosystems are substantial, there remains much to be learned; for example, the ecology of deep-sea sharks, or the impacts of anthropogenic impacts like climate change and overfishing on shark ecology. We discuss the latter in Chapters 11 and 12.

NOTES

1. Your co-authors both hail from the Southern US, where the indigenous grocery store used to be the neighborhood Piggly Wiggly.

2. Grubbs, R.D. and Musick, J.A. 2007. In: McCandless, C.T. et al. (eds.). Shark nursery grounds of the Gulf of Mexico and the East Coast waters of the United States. Am. Fish. Soc. Symp. 50: 63–86.

3. NMFS. 1999. Final fishery management plan for Atlantic tuna, swordfish, and sharks. Silver Spring.

4. Talwar, B. et al. 2017. Mar. Ecol. Prog. Ser. 582: 147–161.

5. Messing, C.G. et al. 2013. Proc. Biol. Soc. Wash. 126: 234–239.

6. Coutin, P. 1992. Austr. Fish. 51: 41–42.

7. Grubbs, R.D. and Kraus, R.T. 2019. In: Choe, J.C. (ed.). Encyclopedia of Animal Behavior 3: 553–563. Elsevier, Academic Press.

8. Priede, I.G. 1984. Fish. Res. 2: 201–216.

9. Bonfil, R. et al. 2005. Science 310: 100–103.

[x] Skomal, G.B. et al. 2009. Cur. Biol. 19: 1019–1022.

10. Howey-Jordan, L.A. et al. 2013. PloS One 8(2): e56588.

11. Wilson, R.P. et al. 2008. Endanger. Species Res. 4: 123–137.

12. Whitney, N.M. et al. 2010. Endanger. Species Res. 10: 71–82.

13. Lear, K.O. et al. 2017. J. Exp. Biol. 220: 397–407.

14. Holland, K.N. et al. 2009. Endanger. Species Res. 10: 287 – 293.

15. *Sine qua non* is Latin for *that without which there is nothing* (i.e., essential element).

16. Sminkey, T.R. and Musick, J.A. 1995. Copeia 4: 871 – 883.

17. Wass, R.C. 1973. Pac. Sci. 27: 305–318.

18. Romine, J.G. et al. 2006. Env. Biol. Fishes 77: 229–239.

19. Note that "$^{13}C/^{12}C$" *is* a ratio and we are being redundant by writing "$^{13}C/^{12}C$ ratio."

20. Cortés, E. 1999. ICES J. Mar. Sci. 56: 707–717.

21. Ebert, D.A. and Bizzarro, J.J. 2007. In: Ebert, D.A. and Sulikowski, J. (eds.). Biology of skates. Springer Science. 115–131 pp.

22. Jacobsen, I.P. and Bennett, M.B. 2013. PLoS One 8(8): p.e71348.

23. Poulakis, G.R. et al. 2017. Endanger. Species Res. 32: 491–506.

10 / Behavior and Cognition

By Tristan Guttridge, PhD

Introduction

Let us be honest—sharks have never had the greatest reputation. Many people today still consider them solitary, mindless, ocean wanderers who exhibit limited behavioral repertoires and cognitive abilities. However, this view could not be further from the truth. Sharks have evolved extraordinary behaviors that set them apart as arguably one of the most successful predators to inhabit the planet. Whether finding food, hiding from predators, or searching for a mate, sharks are always behaving.

In this chapter, we walk you through the behavioral diversity of sharks, revealing insights from their

- *Social lives*—Do Scalloped Hammerheads (*Spyhrna lewini*) displace each other in a school by shaking their heads?

- *Cognitive capacities*—Can reef-dwelling bamboo sharks learn how to navigate a maze?

- *Navigation*—Can Lemon Sharks (*Negaprion brevirostris*) return to where they were born to give birth themselves?

- *Predatory behavior*—Do White Sharks (*Carcharodon carcharias*) use the sun to launch cryptic attacks on unsuspecting seals?

- *Predator avoidance*—Is there a shark that swells up like a balloon to avoid being eaten?

Group Living

There is great diversity in how, when, and why sharks form groups: think resting catsharks in labyrinth-like caves, daytime schooling Scalloped Hammerheads over seamounts (fig. 10.1), or filter-feeding giant Basking Sharks (*Cetorhinus maximus*) congregating in the open ocean.

Like other animals, sharks come together frequently, sometimes in large numbers, but these groups vary in their organization, behavior, and the environmental or ecological conditions under which they form. However, before we begin exploring this phenomenon, it is helpful to distinguish between *aggregations* formed when sharks are drawn together because of a common attractive resource or constraint (e.g., food or shelter) and *social groups* that are formed as a result of individuals being drawn to one another (e.g., influenced by social interaction).

Aggregation

Large numbers of sharks always draw great interest. Show people a photo of a Bull Shark (*Carcharhinus leucas*) and their response is "that's neat," but show them 40 Bull Sharks and their response is more like "Holy cow!, Wow!," and they

Figure 10.1. Scalloped Hammerheads schooling at Cocos Island, Costa Rica. (Courtesy of Shmulik Blum)

sit up and pay much more attention. The spatial distribution of sharks has always fascinated researchers too, and there is widespread evidence of single and multi-species aggregations across the 500+ species that roam our seas. One of the main drivers of these aggregations is food, particularly when it is clumped (as opposed to dispersed) in distribution, as *everyone* comes to exploit the plentiful resource.

One of the most extraordinary examples of sharks aggregating for food is the massive numbers of Whale Sharks (*Rhincodon typus*) that come together during the summer months off Mexico's Yucatan Peninsula (see fig. 3.24). Up to 420 Whale Sharks have been observed in a single aerial survey, all gathered in an approximately 18 km^2 (7 mi^2) elliptical patch of ocean, feeding on dense patches of pelagic fish eggs.[1] This species is regularly seen feeding together throughout the world, and most research focuses on learning about the spatio-temporal dynamics of aggregation sites with a view to conserving them. It is unclear whether such gatherings facilitate some exchange of social information, but putative courtship displays (e.g., follow nose-to-tail and breaching) recorded post-feeding in Basking Sharks, a similar planktivorous ocean wanderer, suggests these displays may serve a secondary reproductive function.

Another incredible sight to behold is the seasonal mass aggregations of migrating Blacktip Sharks (*Carcharhinus limbatus*) off West Palm Beach, Florida (see fig. 1.19). Aerial surveys conducted throughout the year saw huge spikes in Blacktips during the winter months (temperatures below 25°C, or 77°F), with up to 800 sharks per km^2 (2072 per mi^2).[2] It was hypothesized that the sharks gathered to feed on spawning baitfish, however, these Blacktip Sharks also performed coordinated, school-like, evasive maneuvers when attacked by large Great Hammerheads (*Sphyrna mokarran*), suggesting an antipredatory function to the aggregations as well.

The tendency to group is important throughout life stages. Juvenile sharks of many species, particularly in the early stages of their lives, aggregate in nursery areas (often at the same time), presumably to avoid predators and search for food. Newborn Blacktip Sharks, for example, frequently aggregated in specific areas during daylight hours before dispersing at night in a small bay near Tampa, Florida.[3]

Within these nursery areas, more localized responses to environmental factors lead to the formation of aggregations. For example, juvenile Lemon Shark movements are tidally linked in the Bahamas and Brazil, with sharks aggregating at high tide in the back of mangroves or pools, presumably to avoid patrolling predators that have gained access to habitats closer to shore as the tide rises. Lemon Sharks also aggregate in mass off the beaches of Cape Canaveral, Florida, despite this major shift in habitat from mangroves to the high-energy surf zone, providing further evidence that this behavior is adaptive across habitats and regions.[4]

Shifting to the pelagic environment, Blue Sharks (*Prionace glauca*) provide an interesting example of aggregations, with juveniles segregating. Dense aggre-

Figure 10.2. Silky sharks aggregating under an ocean buoy, Andros Island, Bahamas. (Courtesy of Annie Guttridge)

gations of young males (~70 cm, or 28 in, in length) form at oceanic seamounts, while females remain in coastal habitats while they mature.[5]

This segregation does not hold true for other pelagic shark species, such as Silky Sharks (*Carcharhinus falciformis*), that are regularly seen in groups of mixed size and sex aggregating underneath floating aggregatory devices (FADs), buoys, or natural logs in the open ocean (fig. 10.2). Unfortunately, this species' propensity to gather under FADs sees them regularly caught as bycatch in purse seine net fisheries that target tuna (see Chapter 12). Recent statistical modelling suggests an important social component to these aggregations. Thus, studies exploring the mechanisms underpinning this species' aggregatory behavior could prove particularly important for future conservation efforts.

The deep-sea is a challenging habitat to study, for obvious reasons. However, imagine attempting to conduct behavioral trials investigating grouping behavior 500 m (1640 ft) or more below the surface? It probably comes as no surprise to learn that much of our understanding of the associations and aggregations of deepwater sharks is inferred from catch composition in fisheries. A recent study[6] in New Zealand explored these patterns among 14 chondrichthyan species (10 sharks and 4 chimaeras) captured across a 27-year period in 5165 bottom trawls. Results indicated that not all species engaged in aggregative behavior, but those that did showed patterns of sex- and size-specific associations, which varied with catch density. Interestingly, like other sharks that show sex segregation at maturity, adult females were highly associated with other adult females, and adult males were consistently associated with each other.

Some species of sharks have been shown to segregate into small groups or packs in proximity to females. In Whitetip Reef (*Triaenodon obesus*) and Nurse Sharks (*Ginglymostoma cirratum*), multiple males have been observed to pursue and attempt to mate with females. Female sharks are often bigger bodied than males (i.e., they are sexually dimorphic), thus this gang method might increase their chances of successful mating, particularly as larger females can probably outmuscle and deter a single male. Whether this pack-type tactic is common among sharks in general is unknown.

This idea of female sharks aggregating (i.e., sex segregation) to avoid male harassment is not uncommon. Female Leopard Sharks (*Triakis semifasciata*) aggregate in shallow, warm waters during the summer months off the coast of California, and 100+ female Port Jackson Sharks (*Heterodontus portusjacksoni*) were observed in a mass aggregation in waters near Melbourne, Australia. Although it seems that the males are to blame for these large, female only aggregations, some researchers have indicated that the warm water temperatures probably serve a behavioral thermoregulatory function as well, possibly enhancing gestation and embryonic development.[7]

Social Groups

Sharks are social, but how do we know when sharks are socializing? Is it proximity to another shark or physical contact? Or some form of coordinated behavior? Or all the above? Just over 50 years ago, Stewart Springer identified that "some shark populations exhibit complex behavior that constitutes part of their social organization" and that based on anecdotal evidence from aerial surveys "large sharks (and rays) are often in groups and not randomly distributed."[8] A few years later, Samuel Gruber and Arthur Myrberg developed an ethogram (i.e., a catalog of behaviors) of social interactions for captive Bonnetheads (*Spyhrna tiburo*), including definitions of *give-way, following,* and *circling*[9] (fig. 10.3). Using these behaviors, they showed a clear dominance hierarchy existed among 15 captive sharks, with the largest females leading the most groups and deferring the least in head-to-head (also called *give-way*) encounters.

This ethogram of behaviors and results was a tantalizing insight into the social lives of sharks and has since formed the basis of research examining the functions and mechanisms of shark sociality. One study[10] conducted in Bimini, Bahamas over two years documented the grouping behavior of 38 juvenile Lemon Sharks in a shallow-water mangrove inlet. Daily observations of following and circling behavior revealed repeatable social interactions with size-matched

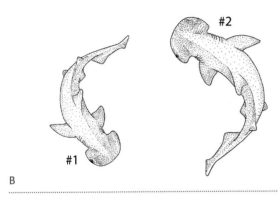

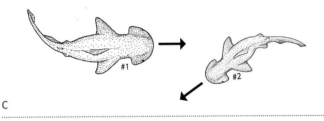

Figure 10.3. Social interactions of Bonnetheads: (A) follow formation. (B) circling. (C) give-way. (Adapted from Myrberg Jr., A.A. and Gruber, S.H. 1974. Copeia 1974: 358–374)

and often related individuals. Groups were typically observed during high tide when larger predators could access the habitats closer to shore, suggesting an antipredatory function. This makes perfect sense, since more sharks equals more senses to detect potential predators and being part of a group means a lower probability of being the one eaten. Through a series of creative semi-captive experiments, the research team went on to explore the mechanisms underpinning group formation, finding that juvenile Lemon Sharks were attracted to other Lemon Sharks (presumably to form groups), preferred to socialize with sharks of similar sizes, and even had preferences for familiar individuals, as well as social personalities.[11]

Concurrent to these discoveries, a research group explored sociality and refuge behavior in the Small-spotted Catshark (*Scyliorhinus canicula*) in the United Kingdom. This benthic species exhibits a very different type of social behavior that involves close or tactile resting in caves or under ledges, almost like roosting insects. Yet despite this obvious difference in sociality between Small-spotted Catsharks and Lemon Sharks, striking similarities existed in the results. For example, like young Lemon Sharks in the Bahamas, juvenile Catsharks showed repeatable associations (e.g., resting with the same individuals), were consistent in their position within the social network (i.e., had a social personality), and familiar sharks (e.g., those kept in the same tank together) formed more groups of greater size than non-familiar groups.[12] Adult Catsharks formed single sex groups, with females resting in labyrinth-like cave systems, and interestingly, males regularly visiting outside, presumably waiting for an opportunity to court a female. Follow-up captive experiments found that the introduction of a male to a tank of females caused disruption to the social network, confirming that sexual harassment can have implications on group behavior in sharks (fig. 10.4).

Arguably the most well-known example of shark grouping behavior is the polarized schools of Scalloped Hammerheads. It is an astonishing sight to witness hundreds of large sharks (approximately 2 to 3 m, or 6.6 to 9.8 ft, in length) showing highly coordinated movements, all within a body length of each other. In the 1970s and early 1980s, shark ethologists Peter Klimley and Don Nelson spent hours free diving among Scalloped Hammerhead schools, recording their behavior in great detail.[13] The results were fascinating! Schools formed during daylight hours that were predominantly female in composition, with larger individuals typically remaining at the top, in more central positions, asserting dominance over smaller sharks through posturing (e.g., body tilting, head shaking, and accelerating) or direct contact (e.g., bumps, hits). They also witnessed a behavior they termed *corkscrewing*, defined as a 360° rotation around the longitudinal

Figure 10.4. Small-spotted Catsharks resting together during lab experiments. (Courtesy of Paul Naylor/marinephoto.co.uk)

axis, used to displace another shark. The function of the schools was difficult to interpret because there are numerous factors that influence school formation. Here are some possible explanations: *Anti-predatory?*—few predators would take on Scalloped Hammerheads; *Foraging?*—no observations of feeding behavior; *Courtship?*—few males present and no mating scars, or attempts observed; *Avoidance/refuging?*—females schooling to avoid costly mating attempts by males; *Social?*—like other animals that are attracted to each other, maybe Scalloped Hammerheads are too.

Evidence of intraspecific communication among sharks (like the above example) suggest more complex social lives and indicates that information transfer could be an important function of their social behavior. Indeed, there is evidence of more coordinated predatory behavior from one of the most archaic species, the Broadnose Sevengill Shark (*Notorynchus cepedianus*; fig. 10.5). These sharks have been observed to feed together, where a group slowly encircles a larger prey item (e.g., a seal), finally converging in on it for the kill.[14] Clearly this pack-type behavior provides the opportunity to access prey items that would not be possible for an individual. Further examples of group attacks have been observed in reef-dwelling Whitetip Reef (*Triaenodon obesus*) and Grey Reef Sharks (*Carcharhinus amblyrhynchos*) trapping and chasing fish among coral heads, Bronze Whalers (*Carcharhinus brachyurus*) and Blacktips blitz-attacking mass sardine schools, and juvenile Lemon and Blacktip Reef Sharks (*Carcharhinus melanopterus*) herding schools of baitfish toward the shore and then feeding on them.[15]

These wild observations provide compelling evidence that sharks are capable of social learning (learning from other individuals); however, only recently has this been demonstrated experimentally (e.g., in juvenile Lemon Sharks).[16] Sharks were trained in a novel food task where they were required to enter a start zone and subsequently make physical contact with a target in order to receive a food reward. Naïve (inexperienced) sharks were then paired with either trained or untrained partners and their performance in the task was compared. The results showed that sharks working with trained partners completed the trials in a quicker time and used more task-related behaviors (e.g., bumping a target).

Understanding the patterns, mechanisms, and functions of shark grouping behavior is particularly important given the conservation status of many elasmobranch fishes, especially as concentrations of sharks are more susceptible to exploitation through spatially focused fisheries.

Learning and Memory

Can you train a shark? Are sharks capable of memory? Sharks live in a watery world where the habitats and occupants are subject to year-by-year, seasonally, day-by-day, or even moment-to-moment variation. Having the ability to retain information about where to find food or learn an association with a stimulus, such as a visual feature of an approaching predator, is adaptive and can ultimately

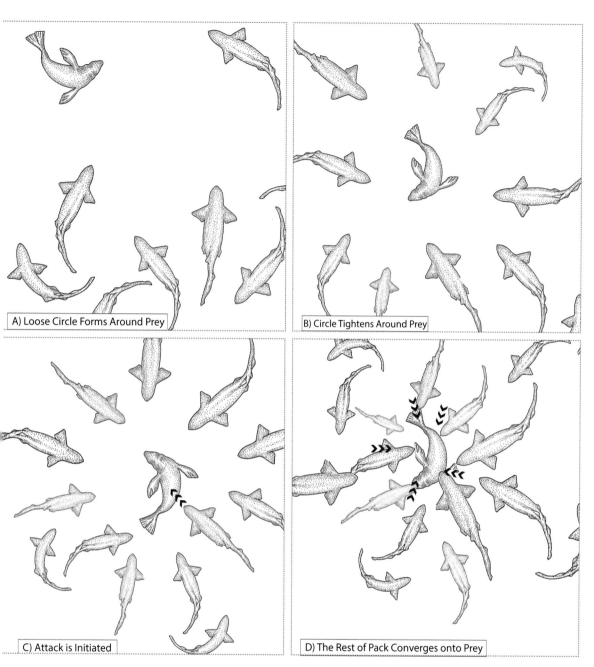

A) Loose Circle Forms Around Prey

B) Circle Tightens Around Prey

C) Attack is Initiated

D) The Rest of Pack Converges onto Prey

Figure 10.5. Sevengill sharks hunting as a pack to feed on a Cape Fur Seal in St Helena Bay, South Africa. (Adapted from Ebert, D.A. 1991. S. Afr. J. Mar. Sci. 11: 455–465)

lead to enhanced survival. The general principles of learning can be described in two forms: *associative learning*, which occurs when an association or relationship between two events is established (e.g., operant, classical conditioning, and observational learning) and *non-associative learning*, which occurs as the result of the presentation of a single stimulus (e.g., habituation and sensitization).[17] These terms, developed by comparative psychologists, have provided a useful conceptual framework for researchers working on learning in a variety of animals and are used here to map out our knowledge of shark learning abilities.

Associative Learning

The capacity of sharks to learn was first demonstrated experimentally by Eugenie Clark and colleagues in the late 1950s, using an operant conditioning regime.[18] Two adult Lemon Sharks were trained to bump an underwater target, on hearing a submerged bell ring, in order to receive a food reward. Training was in accordance with positive reinforcement, where the consequence of bumping a target (correct voluntary behavior) produced a food reward. Sharks were trained in a six-week period and retained a strong response even after a 10-week absence from exposure to the stimuli. Soon after this study, Samuel Gruber and Niel Schneiderman reported the first authoritative account of classical conditioning, again using the Lemon Shark as their study species. Of course, this learning process is most well-known from Ivan Pavlov's work with dogs; however, instead of using food (a positive / reward stimulus), sharks were restrained while exposing them to a light flash that was paired with an electric shock (negative / aversive stimulus), producing an eye-blink conditioned response (not salivation, like in Pavlov's dog experiment; fig. 10.6). After about 60 trials sharks exhibited a reliable conditioned response (e.g., eye blink on light presentation), indicating they had learned the association. This finding was comparable to other vertebrates tested in a similar paradigm, suggesting sharks were able to learn discriminative tasks as rapidly. So, sharks are not just swimming noses after all!

Since these early landmark demonstrations, classical and operant conditioning paradigms have been used effectively to explore a variety of shark cognitive abilities. A team led by Vera Schluessel in Germany, working with captive bred Grey Bamboo Sharks (*Chiloscyllium griseum*), has pioneered research examining

Figure 10.6. Lemon Shark secured during aversive classical conditioning experiments.

recognition and discrimination abilities.[19] Through a series of elegant experiments, her team showed that sharks could discriminate between symmetrical and asymmetrical shapes, contrasts but not colors, stationary 2D-objects, and between moving objects ranging from moving circles to differently moving organisms. She even found that these sharks could perceive optical illusions. For a shark that lives in complex, visually well-structured environments, such as coral reefs, these abilities are probably extremely advantageous. It will be revealing to test other species that occupy different habitats to see how capabilities vary between sharks.

Concurrent to these studies, Culum Brown, an expert in fish cognition, has reported some surprising findings on the Port Jackson Shark, found off the coast of Australia. If you have ever visited an aquarium and seen this species, you would likely not have paid them much attention, as they barely move and show little to no interest in anything. But underneath this rather uninteresting demeanour, there are hidden smarts. Culum's team taught juvenile Port Jackson Sharks a classical conditioning paradigm using an underwater bubbler paired with a food reward. Most sharks reached the learning criterion, biting at the bubbles, within 30 trials, with some remembering the association for 40 days. The researchers also increased the difficulty of the task by adding a 10-second time interval between the bubbles and food reward, and despite this more cognitively demanding task, some of the sharks were still able to learn the association (fig.10.7). His team's latest study[20] investigated whether Port Jackson Sharks could learn about sounds. Low frequency sounds are known to be attractive to sharks (see Chapter 5), and they paired a musical stimulus (jazz) with a food reward. Five of eight sharks learned the association, swimming to the correct corner when presented with the music.

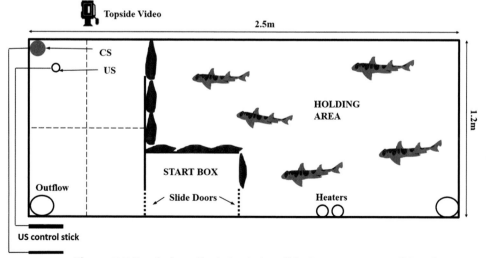

Figure 10.7. Port Jackson Shark classical conditioning setup. CS = Conditioned Stimulus; US = Unconditioned Stimulus.

Non-associative Learning

In the late 1960s and early 1970s, field experiments investigating shark acoustic detection abilities were conducted using artificial low-frequency sounds (see Chapter 5). The behavioral responses of sharks to these sounds were recorded and a simple form of learning known as *habituation*, a decrease in response to stimulus after repeated presentations, became apparent. Fewer sharks were observed to respond to the sound stimulus and their response intensity was seen to decrease within minutes or even seconds after their initial approach to the speaker.

Another context where habituation has been discussed regularly is when people SCUBA or free dive with sharks—this idea was introduced by Don Nelson in the 1970s during his pioneering research on shark behavior.[21] Of course, these days, in addition to repeated exposure to divers, many operators also provide frequent, predictable (e.g., in time and location) food incentives to sharks to enhance the visual spectacle for their guests. As a result, changes in behavior have been documented for various species, from increased shark numbers, to greater residency time and speed of arrival to the dive boat.

More recently, experimental studies have recorded evidence of habituation to electric stimuli. Unrewarded Small-spotted Catsharks exposed to artificial, prey-type electric fields showed reduced responses with eventual cessation, and White Sharks presented with low voltage electric fields (during deterrent studies) showed habituation over short time scales. Amazingly, Clearnose Skates (*Rostroraja eglanteria*), Small-spotted Catsharks, and Bamboo Shark embryos also showed reduced responses to electric stimuli that simulated a passing predator, with Bamboo Sharks taking 30 to 40 minutes to recognize previously presented stimuli versus 10 to 15 minutes for Skates and Catsharks. One study of personality traits in juvenile Lemon Sharks even found evidence for individual differences in the rate of habituation, and a relationship between this and personality, with faster explorers of a novel open-field-test showing more rapid habituation.[22]

Given these types of studies and the growing literature that showcases the diverse cognitive functions that sharks are capable of, we hope the days of people considering sharks as dumb and mindless will disappear very soon.

Orientation and Navigation

Figuring out where to go, or what route to take, is hard enough with a smart phone and an app calling out directions. Now imagine navigating coastal habitats with intricate mangrove systems, homogenous sand flats, tidal ranges, or traveling across endless blue water searching for a profitable food patch. There are changes in temperature to consider, current, light intensity, and sounds coming from all directions. Then, once you have finally figured out how to get from A to D, via C and B, a predator comes out of nowhere and makes a meal of you. The challenges are daunting, and sharks like many other ocean inhabitants have evolved a suite of incredible senses to help them track both abiotic and biotic changes with some considerable accuracy (see Chapter 5).

There are many impressive feats of navigation by sharks. For example, Scalloped Hammerheads rhythmically disperse from islands and seamounts at night to forage in the surrounding pelagic environment; Salmon Sharks (*Lamna ditropis*) make rapid directed migrations between the two most productive ecoregions of the eastern North Pacific; Spiny Dogfish (*Squalus acanthias*) migrations of up to 7000 km (4400 mi) provide evidence for the trans-Pacific connection of stocks; even Greenland Sharks (*Somniosus microcephalus*), a species known for its sluggish, slow moving lifestyle, migrate from Greenland 1000+ km (600+ mi) off the continental shelf in abyssal waters presumably to mate or pup; and Basking Sharks track the daily change in vertical abundance of zooplankton, adjusting their vertical migration depending on prey behavior and ocean habitats.

Some of the early investigations of shark navigation suggested that they used geomagnetic orientation, particularly during directed pelagic movements. Peter Klimley, in his seminal work on Scalloped Hammerheads, mapped the associated magnetic gradients within the region of his study site in Espirito Santo, Baja, Mexico, proposing that Hammerheads were able to orient by comparing the intensity of magnetic stimuli (i.e., *tropotaxis*). Support for this hypothesis has emerged recently from lab-based experiments in Hawaii with Sandbar Sharks (*Carcharhinus plumbeus*), where they were successfully conditioned to magnetic stimuli within the range of values that Klimley previously hypothesized.[23]

Other researchers have explored the contribution of olfactory stimuli (i.e., chemical cues) to navigation by displacing sharks offshore from their typical home ranges. Leopard Sharks rendered anosmic (nares temporarily blocked such that the sense of smell was masked) were displaced 9 km (5.6 mi) from their nearshore habitat and tracked on release. Individuals that could not smell made circular movements on return and ended further from shore than control sharks that exhibited more directed movements.[24]

This idea of being able to return *home* (i.e., homing) was tested with juvenile Lemon Sharks in Bimini, Bahamas, where 31 of 32 individuals that were displaced up to 16 km (9.9 mi) from their capture sites returned to their specific, spatially limited home ranges.[25] Even one shark displaced in the Gulf Stream, a powerful, warm western-boundary current, with 1000+ m (3300 ft) depth, miraculously managed to make it back home. Whether these young sharks imprint (like salmon or turtles) on chemical clues or the geomagnetic signature of their home habitat is unknown, but recent evidence of adult Lemon Sharks returning to their own site of birth to pup themselves, some 14 years later, clearly demonstrates that these homing abilities are retained into adulthood and play an important role in their life history.

Navigation and orientation are both dependent on the formation, storage, and retrieval of spatial memories, which permit repeated visits to fixed points in the environment. In the early 1990s, researchers working with Port Jackson Sharks in Australia speculated that this species was able to remember features of the reef, potentially having some form of mental map. After being displaced 3 km (1.9 mi) away, within 48 hours the sharks were observed to rest in the same reef sites.[26] More recently, scientists analyzing movement tracks of Tiger (*Galeocerdo cuvier*) and Galapagos Sharks (*Carcharhinus galapagensis*) in Hawaii, concluded that the former may use cognitive spatial maps for orientation while moving between different foraging areas.[27]

Using an experimental platform designed for research on cognition, Vera Schluessel and colleagues recently explored the spatial learning and memory capabilities of Bamboo Sharks. Experiments used a two-choice T-maze setup, where sharks were trained to retrieve food from a specialized feeding apparatus and solve spatial tasks. After many weeks of training, all eight sharks were able to remember a feeding location(s), either by memorizing a particular turn response and/or with the help of external visual landmarks (place learning) and were still able to successfully perform the task after various experimental breaks (up to six weeks).[28] These experiments provide an exciting first indication that sharks can form cognitive spatial maps, however, more elaborate experiments should be conducted in the future to explore the phenomenon further.

Predator Behavior

To us, sharks live in a visually concealing world, which makes it extremely challenging to directly observe them. Thus, in comparison to terrestrial predators or smaller-bodied fish, our understanding of their predatory behavior is generally poor. Of course, when most people think about shark predation, they immediately visualize a White Shark at full tilt, launching a seal into the air. And yes, this is a brutal, iconic event to watch, but what makes it even more extraordinary is when you consider the details; that is, how the White Shark got to that moment to launch an attack, resulting in a successful, profitable meal. Beyond the exquisite senses and physical attributes that sharks have evolved, there are tactics and strategies to consider, and split-second decisions that can lead to success or failure. For example, from where should an attack be launched? What depth, habitat, or angle? What time of day or conditions are most effective? Which hunting mode or movement pattern should be used for which prey species or habitat characteristics?

For many sharks, these questions are almost impossible to answer, but the visual nature of White Shark predation events, coupled with their proximity to shore (due to their preferred prey, pinnipeds [seals and sea lions], aggregating on rocky islands) has provided a unique opportunity to study this predator-prey relationship in detail. Early examinations of White Shark predatory behavior were undertaken by Tim Tricas and colleagues in Australia, on the aptly named *Dangerous Reef*.[29] They described behavioral responses to surface and mid-water baits, which included terms such as *surface charge* or *side roll* and, based on these observations and examinations of wounded pinnipeds, suggested a typical White Shark attack scenario. This included an ambush, stealth-type strategy, searching for a silhouette with an attack made from beneath (from depth) or from behind the unsuspecting prey. For large prey items, such as Elephant Seals, they proposed that a *bite-and-spit* strategy might be used to avoid injury, leaving the seal to bleed to death or go into shock after an attack. Interestingly, in other locations White Sharks were not observed to use this tactic for the less dangerous, juvenile Sea Lions, indicating that they alter their predatory tactics depending on the prey species they are targeting.

Since this early investigation, behavioral ethograms of White Shark natural predation events have been developed describing interactions with pinnipeds at Seal Island, South Africa and the Farallon Islands, California.[30] These descriptions generated 20 and 24 distinct behavioral units respectively, and included behaviors such as *polaris attack, surface lunge, lateral head shake,* and *subsurface carry.*[31] At both study sites, the highest rates of foraging success by White Sharks occurred in the hours following dawn. This peak was likely due to a combination of the presence of pinnipeds (the primary prey item) and optimal ambient light conditions for attacking at the surface. Attacks were launched in specific locations (presumably optimal), with larger sharks showing greater success rates and more restricted search areas, indicative of learning. Changing abiotic (i.e., sunlight, wind, currents) and biotic (i.e., presence / absence of competitors, anti-predatory tactics of prey) factors clearly creates a dynamic, complex predator-prey relationship, with White Sharks needing to adapt and learn features of their environment in order to be successful.

The idea that White Sharks attack in low light levels or show behavioral flexibility in their exploitation of environmental features is fascinating and led White Shark researcher Charlie Huveneers to experiment with baited lines at a known aggregation site in Australia. He found that through the course of a sunny day, White Sharks moved their attacks so that the sun was behind them and appeared to reverse their direction of approach along an east-west axis from morning to afternoon. This could be considered a behavioral modification to improve crypsis or a better method for the sharks to track their prey visually.[32] Either way it is impressive!

More recently, a study linked White Shark hunting modes to predation attempts / successes on seals at the surface. Hunting modes were defined as: (1) active search / patrolling when a shark is moving through its environment searching (either directed or random) for prey; and (2) sit-and-wait, or area-restricted-search (ARS), which sees the shark waiting for prey to cross the boundary of its strike space.[33] To undertake such a task, the team actively tracked White Sharks with electronic tags, following them on multiple days throughout a month. Results revealed some interesting patterns, with White Sharks ambushing seals successfully at the surface using both modes, but preferring ARS in the mornings with clear sex and individual differences detected.

Let us move to couch-potato sharks, those that spend most of their day on the seabed. Pacific Angel Sharks (*Squatina californica*) also used ambush tactics by rapidly lunging from the seafloor on oblivious demersal fishes. Resting Angel Sharks were usually located next to reef patches or rock-sand interfaces. These sites serve as refugia for a variety of fishes and no doubt improved encounter rates with potential prey. Sharks showed reuse of ambush sites across the study which was discussed in the context of an ARS hunting mode and models of search behavior.[34]

Traditionally using sharks as models of behavior to test theoretical predictions (e.g., optimal foraging) has proved challenging. However, with the huge improvement in tracking, remote sensing, and data logging technologies, alongside data processing and statistical modelling, it is becoming increasingly possible to conduct hypothesis-driven behavioral studies on elasmobranchs.

Behavioral ecologist David Sims has pioneered this type of approach, with sharks as models of foraging behavior. His team's work over the past two decades has examined pelagic shark movements (e.g., Basking and Blue Sharks) via satellite tracking to examine how their movement patterns conform to optimal search theory (e.g., *Lévy-flight foraging hypothesis,* a type of random movement, typified by many short moves with less frequent long-distant displacements; or *Brownian motion,* which are random movements).[35] Results revealed that Lévy (47%) and Brownian (21%) movement patterns were both important strategies for open ocean sharks, and that environmental context impacted their use. For example, a Blue Shark in the Northeast Atlantic, tracked across shelf (productive) and deep (less productive) waters, displayed dive patterns consistent with Brownian movements when in shelf waters and dive patterns consistent with Lévy flights in open ocean habitats.[36]

Since this pioneering research others have explored the use of *directed walks* (where the shark moves to a known goal) to examine movement patterns of sharks at different spatial scales, and to determine if they were using food patches. Active tracking data for Tiger, Thresher (*Alopias vulpinus*), and Blacktip Reef Sharks, of varying life stages, revealed a variety of movement strategies both between and within species.[37] Tiger Sharks performed directed movements to locations 6 to 8 km (3.7 to 5 mi) away, whereas Blacktips did not show oriented movements. Thresher Sharks also showed directed movement (at scales of 0.4 to 1.9 km, or 0.2 to 1.2 mi), and adult threshers were able to orient at greater scales than juveniles, which may suggest that learning improves the ability to perform directed walks.

Some sharks have impressive weapons at their disposal that aid them in prey capture. Take the Pelagic Thresher (*Alopias pelagicus*) for example, with its huge scythe-shaped tail that is half the length of its body. Recent video recordings taken in Cebu, Philippines showed that Thresher Sharks had remarkable control of the upper caudal lobe, using it to target bait fish in dense schools by slapping and stunning the fish. Amazingly, tail-slaps occurred with such speed (fastest recorded = 21.8 m per s, almost 50 mph) and force that they may have caused dissolved gas to come out of solution-forming bubbles.[38] In addition, there are various other reports suggesting Thresher Sharks hunt in groups, corralling bait fish ready to tail-whip.

Let us switch ends of the body to the head. Probably the most obvious example of sharks with head weaponry are the Sawsharks (order Pristiophoriformes), with their elongated rostrum shaped like an isosceles triangle and teeth lining the edges (fig. 10.8). Although there are no direct observations of Sawsharks using their rostrum in the wild, studies of the microwear surface of the rostral teeth of the Common Sawshark (*Pristiophorus cirratus*) suggests that it uses it to capture prey (though not necessarily to impale them) as well as in defense.[39] This study also examined the rostral teeth of Largetooth Sawfish (*Pristis pristis*), a batoid, finding similarities in the microwear surface of the rostral teeth and ratio of rostrum length to body length. A captive study on Freshwater Sawfish (*Pris-

Figure 10.8. Shortnose Sawshark off Merimbula, NSW, Australia. (Courtesy of Tristan Guttridge)

tis microdon) described 17 different behaviors in which Sawfish used the saw to both sense prey-simulating electric fields and capture prey. Prey encountered in the water column were attacked with lateral swipes of the saw, which can split a fish in half, impale it on the rostral teeth, stun it, or sweep it onto the substrate.[40] Another shark with a weapon-shaped head is the Hammerhead, where the head was used to hit a fleeing stingray and pin it to the bottom ready for consumption (see Chapter 4).

Other sharks may lure prey to themselves. The bearded appearance (branching, freaky-looking dermal lobes around the head) of the Tasseled Wobbegong Shark (*Eucrossorhinus dasypogon*; fig. 10.9), in combination with cryptic color-

Figure 10.9. The bearded appearance of the Tasseled Wobbegong Shark. (Courtesy of Annie Guttridge)

Figure 10.10. Video frame grab of a Sand Tiger shark surrounded by scad school. (Courtesy of Erin Burge/Explore.org)

ation, attracts small unsuspecting fish to refuge. Observations in captivity have also revealed that when it detects a potential prey nearby, this shark slowly waves its tail back and forth, which resembles a small fish, complete with a dark eyespot at the base. Bigger fish come in to investigate and "snap . . . dinner time!" Although not active luring, like the Wobbegong above, the upper jaw of the Megamouth Shark (*Megachasma pelagios*) has luminescent tissue which might attract shrimp and other prey, providing an easy meal for this species.[41]

Ethologist Arthur Myrberg in the early 1990s also suggested that the spot patterns on the fins of the Oceanic Whitetip Sharks (*Carcharhinus longimanus*) were a form of *aggressive mimicry*. From a distance, the white blotches resemble baitfish, which act as a visual lure for fast-moving prey that the shark can more easily capture. Oceanic Whitetips are regularly seen in pairs or small groups, which would likely enhance this visual aid.[42] More recently, observations were made of associative behavior between Sand Tigers (*Carcharias taurus*) and Round Scad (*Decapterus punctatus*) at Frying Pan Shoals, off the North Carolina coast, in which the shoal of scad surrounds the shark (fig. 10.10). This shoal provides camouflage for the shark, and offers up new opportunities for feeding, as the predators of the scad are the prey of the shark. Whether the sharks are actively seeking the scad schools for this predatory camouflage is unknown, but this type of mutualistic behavior (the scad apparently benefit as well) provides yet another interesting predatory tactic.

Predator Avoidance

It may come as a surprise to learn that sharks are regularly prey themselves, in fact most sharks live in constant fear of being eaten by another shark, and even some of the *baddest* sharks on the planet employ antipredatory behaviors to ensure they do not end up as someone's lunch. Of course, having a large body size restricts the list of predators to worry about, but this does not limit the range of behaviors that sharks use—from hiding under crevices, puffing up like a balloon, to counterillumination and even feigning death.

Probably one of the most common (and obvious) strategies to avoid being eaten is to hide or take refuge in a habitat that conceals you or is inaccessible to larger predators. Knowing which habitats are safe and when to use them likely comes from experience, and no doubt many young shark's lives are a series of near misses. Juvenile sharks of many species are known to use shallow-water nursery habitats; for example, the use of coastal bays with low turbidity by juvenile Scalloped Hammerheads (despite low resources), Blacktip Sharks showing strong fidelity to a small bay fringed by mangroves, and Bull Sharks use of some low salinity estuaries. At a finer scale, even within these habitats, sharks use their features to aid in predator avoidance; for example, Nurse Sharks wedge themselves under rock crevices and juvenile Lemon Sharks use the complex mangrove root structure more when in the presence of a predator (fig. 10.11).[43] As discussed earlier in the chapter, often many young sharks will form groups, which likely sees them benefiting from greater vigilance, encounter dilution (lower predation risk), and predator confusion.

Reducing activity levels (i.e., movement speed or direction) is another strategy that can help to avoid being detected by a predator. Although it is difficult to disentangle physiology (i.e., bioenergetics) from behavior, it is likely that resting or slow swimming while refuging is an effective antipredatory tactic employed by many sharks. Indeed, Whitetip Reef Sharks spend much of their days huddled in reef overhangs, and juvenile Port Jackson Sharks rest in seagrass beds for long periods of the day (fig. 10.12).

Some sharks, even as adults, will bury themselves in the substrate to remain undetected (e.g., Angel Sharks, *Squatina squatina*) and many have evolved cryptic coloration to blend into the habitats in which they reside. The Ornate Wobbegong Shark (*Orectolobus ornatus*), or *shaggy beard* in Australian Aboriginal language, uses its incredible camouflage (e.g., mottled appearance and small weed-like whisker lobes) to remain invisible to predators, as well as, selecting daytime resting positions with high complexity and crevice volume. These locations are not good areas for prey availability, and thus were likely chosen to avoid predators.

Amazingly, even as embryos some sharks (e.g., Grey Bamboo and Small-spotted Catsharks) show evidence of antipredatory behavior—they can detect predator-mimicking electric fields and respond by ceasing their respiratory gill movements. Despite being confined in such a small space and vulnerable to pred-

Figure 10.11. A juvenile Lemon Shark patrols the mangroves in Bimini, Bahamas. (Courtesy of Annie Guttridge)

Figure 10.12. Whitetip reef sharks resting during the day under a reef ledge at Cocos Island, Cost Rica. (© Matthew D. Potenski)

ators, these embryonic sharks are still able to recognize dangerous stimuli and react with an innate avoidance response. In addition, Bamboo Sharks, only 10 to 12 cm (4 to 4.7 in) long at birth, are extremely vulnerable to predation and have a distinctive pattern of high contrast banding that mimics the coloration of unpalatable or poisonous prey (e.g., sea snakes), thereby avoiding predation (this is known as *Batesian mimicry*).[44]

Some sharks are actually able to change color to match their background. Experiments conducted on Small-spotted Catsharks and three species of skates demonstrated a darkening or lightening of skin color when placed in either black or all white tanks, effectively background-matching to their environments. Such an ability would likely help them avoid detection by predators.

Another form of predator avoidance comes from the deep-sea, where camouflage by counterillumination is used by species within the kitefin and lantern sharks. These sharks can switch on their luminescence in response to overhead illumination, which makes them invisible to predators. This ability also helps with hiding from prey.

Another fascinating finding that has recently surfaced is that some sharks can glow in the dark (i.e., *biofluoresce*). The Swell Shark (*Cephaloscyllium ventriosum*) and Chain Catshark (*Scyliorhinus retifer*) were found to exhibit bright green fluorescence, and through a series of experiments it was possible to determine that they could visualize their own fluorescent patterns, which could be useful for conspecific communication and possibly avoiding predators that lacked similar visual capabilities.[45]

When a predator is about to strike its prey, the options for escape are limited to either *flight* or *fight*. Some species have evolved morphological characteristics to aid in defense; for example, horn sharks have spines anterior of their dorsal fins that may prevent a predator swallowing them, resulting in ejection and ultimately survival. The eponymous Swell Shark can inflate to almost double its size through taking H_2O into its cardiac stomach. This increase in size likely intimidates a predator or tricks them into thinking the prey is too large for consumption. Of course, stingrays are equipped with a venomous barb that can be used to defend against predators as well as harassment from mates, although a recent study found the venom from the Atlantic Stingray (*Hypanus sabina*) was low in toxicity and metabolic cost compared to other venomous species, suggesting that defense is not the primary purpose of the spine.[46] Other rays may use their electric organ discharges (EODs) in predator defense. Interestingly, the Lesser Electric Ray (*Narcine brasiliensis*) did not use it for foraging, indicating that it was primarily for predator defense or conspecific communication.[47]

Threat displays can also be an effective means of warning off predators. Many sharks have been observed to use agonistic behaviors, such as depressed pectoral fins, arched back, stiff or jerky movements, and jaw gaping. In the 1980s, ethologist Don Nelson examined the intensity and frequency of these behaviors by approaching Grey Reef Sharks in a submersible. He observed that displays were common, often escalating to full-on attacks when sharks were cornered. Bull Sharks, when followed in the shallows, will also exhibit similar behaviors, beginning with flight, then exhibiting the stereotypical pectoral fin depressing, S-shaped swimming followed by the final, last resort defensive yet powerful attack. Most predators prefer the element of surprise during attack and want to avoid injury at all costs, thus the behaviors described above are likely used to inform / signal a predator—"you've been spotted, and if needed I'm ready to attack to defend myself."

Lastly, probably one of the most underreported but fascinating antipredator strategies is death feigning. This bizarre behavior is referred to as tonic immobility (TI) in the shark literature (fig. 10.13), and in other animals is known to be adopted late in the predation sequence, and frequently following physical contact by a predator. A handful of studies have explored the onset and timing of TI in juvenile Lemon Sharks, with one examining stress via blood chemistry parameters.[48] Whether this strategy is used by sharks as a last-ditch attempt to confuse a predator is unknown.

Concluding Comments

It is an incredibly exciting time to be a scientist exploring shark behavior! The very latest studies are now able to use multi-sensor tag packages that can record 3D acceleration, magnetic field strength / direction, swimming speed, depth, and temperature at extraordinary resolution (e.g., 30 times per second), as well as HD

Figure 10.13. Dr. Samuel "Sonny" Gruber placing a juvenile Tiger Shark in tonic immobility. (Courtesy of Bob Crimian)

Figure 10.14. Great Hammerhead quipped with multi-sensor tag package in Bimini, Bahamas. (Courtesy of Eugene Kitsios)

video with 12 hours of record duration (fig. 10.14). At a preprogrammed time, the package releases from the shark and floats to the surface, with embedded acoustic VHF and satellite transmitters to facilitate the retrieval and data download. This detail of data collection is taking behavioral analysis to another level, with scientists now able to know the depth and temperature in which a shark is swimming, as well as its body orientation, compass heading, swim speed, what it is seeing (e.g., prey) and doing (e.g., feeding, socializing, resting).[49]

There are still so many behaviors that we have only just touched the surface of, and these advanced tools, alongside enhanced remote sensing and data processing will pave the way for a better understanding of shark behavior, which will ultimately improve their conservation and survival.

NOTES

1. de la Parra Venegas, R. et al. 2011. PLoS One 6(4): p.e18994.
2. Kajiura, S.M. and Tellman, S.L. 2016. PloS One 11(3): p.e0150911.
3. Heupel, M.R. and Simpfendorfer, C.A. 2005. Mar. Biol. 147: 1239–1249.
4. Reyier, E.A. et al. 2008. Fl. Sci. 2008: 134–148.

5. Litvinov, F.F. 2006. J. Ichth. 46: 613–624.

6. Finucci, B. et al. 2019. ICES J. Mar. Sci. 75: 1613–1626.

7. Hight, B.V. and Lowe, C.G. 2007. J. Exp. Mar. Biol. Ecol. 352: 114–128.

8. Springer, S. 1967. In: Gilbert, P.W. et al. (eds.). Sharks, skates and rays. Johns Hopkins U. Press. 149–174 pp.

9. Myrberg Jr., A.A. and Gruber, S.H. 1974. Copeia 1974: 358–374.

10. Guttridge, T.L. et al. 2011. Mar. Ecol. Prog. Ser. 423: 235–245.

11. Finger, J.S. et al. 2018. Behav. Ecol. Sociobiol. 72: 17 and Keller, B.A. et al. 2017. J. Exp. Mar. Biol. Ecol. 489: 4–31.

12. Jacoby, D.M.P. et al. 2012. J. Fish Biol. 81: 1596–1610.

13. Klimley, A.P. and Nelson, D.R. 1981. Fish. Bull. 79: 356–360.

14. Ebert, D.A. 1991. S. Afr. J. Mar. Sci. 11: 455–465.

15. Jacoby, D.M. et al. 2012. Fish Fish. 13: 399–417.

16. Guttridge, T.L. et al. 2013. Anim. Cogn. 16: 55–64.

17. Lieberman, D.A. 1993. *Learning: behavior and cognition*. Thomson Brooks / Cole Publishing Co.

18. Clark, E. 1959. Science 130: 217–218.

19. Schluessel, V. 2015. Anim. Cogn. 18: 19–37.

20. Vila Pouca, C. and Brown, C. 2018. Anim. Cogn. 21: 481–492.

21. Nelson, D.R. 1977. Am. Zool. 17(2): 501–507.

22. Finger, J.S. et al. 2016. Anim. Behav. 116: 75–82.

23. Anderson, J.M. et al. 2017. Sci. Rep. 7: 11042.

24. Nosal, A.P. et al. 2016. PloS One 11(1): p.e0143758.

25. Edrén, S.M.C. and Gruber, S.H. 2005. Env. Biol. Fish. 72: 267–281.

26. O'Gower, A.K. 1995. Mar. Freshwater Res. 46: 861–871.

27. Meyer, C.G. et al. 2010. Mar. Biol. 157: 1857–1868.

28. Schluessel, V. and Bleckmann, H. 2012. Zool. 115: 346–353.

29. Tricas, T.C. 1985. Mem. S. Calif. Acad. Sci. 9: 81–91.

30. Klimley, A.P. 1994. Am. Sci. 82(2): 122-133.

31. Martin, R.A. et al. 2005. J. Mar. Biol. Assoc. U.K. 85: 1121–1136.

32. Huveneers, C. et al. 2015. Am. Nat. 185: 562–570.

33. Towner, A.V. et al. 2016. Func. Ecol. 30: 1397–1407.

34. Fouts, W.R. and Nelson, D.R. 1999. Copeia 1999: 304–312.

35. Sims, D.W. et al. 2008. Nature 451: 1098.

36. Humphries, N.E. et al. 2010. Nature 465: 1066.

37. Papastamatiou, Y.P. et al. 2011. J. Anim. Ecol. 80: 864–874.

38. Oliver, S.P. et al. 2013. PLoS One 8(7): p.e67380.

39. Nevatte, R.J. et al. 2017. J. Fish Biol. 91: 1582–1602.

40. Wueringer, B.E. et al. 2012. Current Biology 22: R150–R151.

41. Nakaya, K. 2001. Bull. Fish. Sci. Hokkaido Univ. (Japan) 52: 125–129.

42. Myrberg Jr., A.A. 1990. J. Ex. Zool. 256: 156–166.

43. Heupel, M.R. et al. Mar. Freshwater Res. 70: 897–907.

44. Kempster, R.M. et al. 2013. Plos One 8(1): p.e52551.

45. Gruber, D.F. et al. 2016. Sci. Rep. 6: 24751.

46. Enzor, L.A. et al. 2011. J. Exp. Mar. Biol. Ecol. 409: 235–239.

47. Johnson, R.H. and Nelson, D.R. 1973. Copeia 1973: 76–84.

48. Brooks, E.J. et al. 2011. J. Exp. Mar. Biol. Ecol. 409: 351–360.

49. Papastamatiou, Y.P. et al. 2018. Sci. Rep. 8: 551.

Human Impacts

11 / Fisheries

Introduction

Chances are you are far removed from shark fisheries and the products of these fisheries. You may never have caught a shark or eaten one. You likely have not consumed shark fin soup or worn shark-leather products. Most readers will never have seen a commercial shark fishing operation, especially at sea catching sharks. On the other hand, you are very likely aware that fisheries for sharks exist. But for which sharks? And where? And by what methods? And are all shark fisheries unsustainable? Is the shark fin industry responsible for killing 100 million or more sharks annually, as has been widely reported in the media? In what ways are shark fisheries managed differently than fisheries for bony fishes?

Uses for Sharks and Batoids

The use of sharks and batoids by humans is recorded from over 5000 years ago. As this excellent FAO report[1] explains, current demand is reflected in increased exploitation of shark stocks for meat and other shark products (fig. 11.1). Sharks and batoids are currently caught for their fins, meat, liver, cartilage, skin, teeth, and jaws, as well as other body parts. Live sharks are used in aquaria and as ecotourism draws.

Fins

Since the 1980s, shark fins (figs. 11.1 and 11.2) have been the most economically valuable part of the shark and are still one of the most profitable. They are used mostly for shark fin soup, which is considered a delicacy and status symbol in countries in East and Southeast Asia, including China, Hong Kong, Taiwan, Singapore, Malaysia, Vietnam, and Thailand.

It is important that a distinction be made between illegal finning and the legal shark fin trade. The former involves removal of the fins from the carcass,

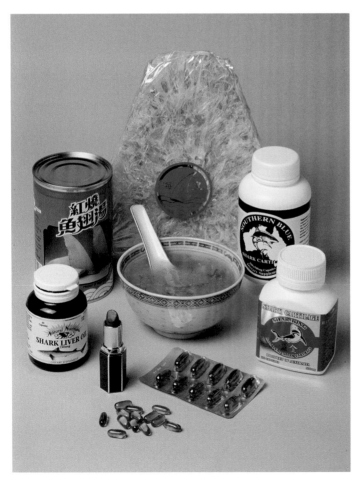

Figure 11.1. Shark Products: shark fin soup, dried shark fin, and shark oil or squalene capsules. (marinethemes.com/ Kelvin Aitken)

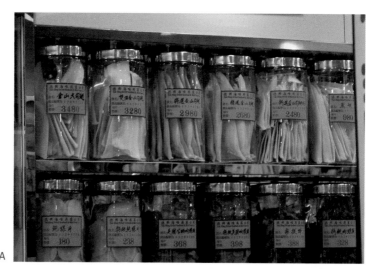

Figure 11.2 (A) Shark fins (*top row*) in a dried seafood shop in Hong Kong. The price per kg for the most expensive fins in the photo, 3480 Hong Kong Dollars per kg, is about $200 US dollars per pound. (B) Finned carcass of Silvertip Shark. (marinethemes.com/ Kelvin Aitkin) (C) Blue Sharks with fins removed in a market in Cadiz, Spain.

most often immediately upon capture when the animal is still alive, and disposal at sea of the less valuable finless carcass. The legal shark fin trade involves fins from legally caught sharks which typically are brought back to port whole and are used for meat and other products. Numerous countries[2] have various kinds of restrictions or bans on finning or on shark fin soup.

All fins on a shark are used except the upper lobe of the caudal (recall that the vertebrae extend all the way to the tip of the tail). Following a week or more of processing, what is left of the fins is the key ingredient, the skeletal elements called *ceratotrichia*. Ceratotrichia, which are composed primarily of an elastic protein similar to the fibrous protein keratin of human hair, taper from a round base to finer strands. In the industry, ceratotrichia are called *fin needles*. The ceratotrichia are boiled and take on the appearance of colorless cellophane noodles, which are then ready to be made into soup (fig. 11.1).

Species of choice in the shark fin industry include carcharhinids, lamnids, sphyrnids, and even pristids. The costs per pound for wet and dried fins varies among species and are also based on how skillfully the fins have been removed and handled. Prices can rise as high as $495/kg ($225/lb) retail, but fishers typically receive only around $16/kg ($7.50/lb) off the boat.

The trade for shark and ray fins reached its maximum of 20,485 tonnes (45,200,000 lb) in 2003, the same years in which landings peaked for sharks and rays. By 2011, it had dropped to 17,899 tonnes (39,500,000 lb).

Peak global value for the shark fin trade was about $300,000,000 US. Hong Kong is still the world's biggest trader of shark fins, with about 40–50% of the global total, followed by Trinidad and Tobago. Commercial fishers from Spain and Indonesia are responsible for the largest shark catches for the shark fin industry.

In the United States, there is a legal shark fin trade, and as long as shark fisheries are allowed, it seems likely that fins will be traded, so that as much of the carcass as possible is used. How do you feel about this policy?

Meat

While shark meat has always been eaten in developing coastal and island countries, it was not until capture fisheries of many bony fishes began to decline in the 1980s that fresh shark meat became more acceptable as an alternative to swordfish, tuna, and so on in wealthier countries.

Shark steaks can be found in seafood markets as both fresh and frozen products. Among the biggest single uses is as a substitute for demersal (bottom) fish cod in the European and British fish and chips industry (fig. 11.3).

Unlike fins, the trade for shark and ray meat is increasing. World exports of shark meat, a category that includes rays, increased from 72,314 tonnes (153,400,000 lb) in 2000 to 117,677 tonnes (259,500,000 lb) in 2011, an average annual increase of 5.23%. The largest importers of shark meat in 2011 were the Republic of Korea, Spain, Italy, Brazil, and Uruguay. Korea's imports are predominantly rays, which are used as substitutes for scallops.

Figure 11.3. Spiny Dogfish in net (A) and on culling table (B). These sharks most likely are destined for the European and British fish and chips market.

The value of shark meat now equals or exceeds the value of shark fins globally, at about $450,000,000 for 120,000 tonnes (2.6 million lb) per year. This value has been increasing over the last 20 or so years, driven by the heavy use of shark meat in fish and chips in Europe and Australia. Batoid landings were not separated in global fisheries statistics until recently, but current global landings for batoids are around 250,000 tonnes per year. Similarly, the value for batoids has increased.

Liver

The largest organ in sharks is the oil-rich liver (see fig. 1.14). It has been found to constitute as high as 19.2% of body weight in the Kitefin (*Dalatias licha*), to as low as 2.9% in the Japanese Topeshark (*Hemitriakis japanika*; fig. 11.4). The weight of the liver of an adult human is about 2–4% of body weight.

In the late 1930s, vitamin A, a fat-soluble compound important to vision, the immune response, reproduction, and cellular communication, was found in high concentrations in sharks. When supplies of cod liver oil, then the major vitamin A source, were cut off during World War II, fisheries for sharks exploded in North America and Australia, particularly for the School (or Tope) Shark (*Galeorhinus*

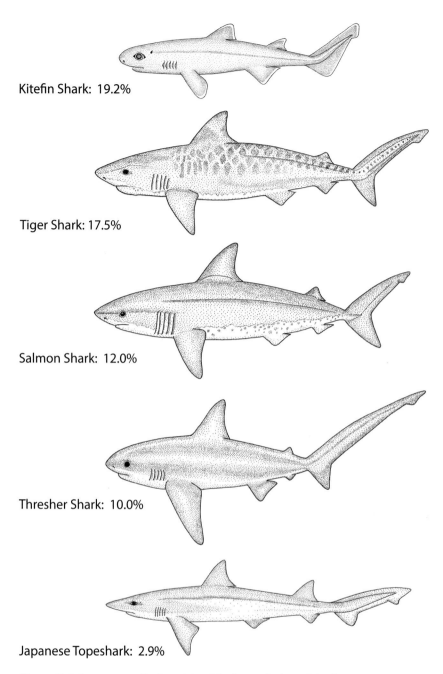

Kitefin Shark: 19.2%

Tiger Shark: 17.5%

Salmon Shark: 12.0%

Thresher Shark: 10.0%

Japanese Topeshark: 2.9%

Figure 11.4. Percentage of body weight of the livers of selected sharks.

galeus) off California, and the Spiny Dogfish (*Squalus acanthias*) off British Columbia. Both of these species were relatively abundant and had high vitamin A levels. To make use of the sharks and rays captured for the livers, the US government instituted a campaign to encourage Americans to eat sharks and rays (fig. 11.5). Synthetic vitamin A was manufactured in the early 1950s, after which the shark-derived form was no longer needed.

Shark liver oil (fig. 11.1) is used in a variety of products, including as a lubricant in the textile industry, cosmetics, health products, fuel oil for lamps, and to protect boat hulls.

World trade in shark liver oil is small. The highest recorded production was 720 tonnes (1,600,000 lb) in 1977, mostly from Japan. By 1997, it had declined to 4 tonnes (8800 lb).

Cartilage

Shark cartilage is widely used as a health supplement, although the health benefits are not always supported by sound, science-based evidence. These so-called benefits include protecting against or curing eczema, ulcers, hemorrhoids, arthritis, and other diseases and disorders, most notably cancer (see Chapter 1).

Shark cartilage is sold as powder, capsules, creams, and tablets, often labeled as *shark cartilage* or *chondroitin*. The latter is used as a dietary supplement for treating osteoarthritis in humans and also in dogs and cats, although there is

Figure 11.5. US Department of Commerce posters encouraging fishing for sharks and rays.

conflicting evidence of its efficacy. Chondroitin is more often sourced from cows (bovine) or pigs (porcine) more so than sharks, of which Blue Sharks (*Prionace glauca*) are especially sought-after.

Accurate records of the shark cartilage trade are lacking.

Other Products

Other products from sharks include skin, gill rakers from mantas, jaws and teeth, rostra from sawfish, and intact, preserved neonate sharks or embryos.

Shark skin is used mostly to make shark leather. The skin can be used for most consumer products where cow leather is used (e.g., furniture, wallets, purses, shoes, book bindings, belts, and so on). Shark skin has also been used as sandpaper and for polishing wood.

Gill rakers from rays of the family *Mobulidae* (mantas) are used in traditional Chinese medicine, purportedly to aid in the immune response as well as to cure or prevent other diseases or disorders.

Shark jaws and teeth, rostra from sawfish, and preserved shark embryos and juveniles are bought and sold at trade shows, markets, and online. All are legal to purchase, except for species like sawfishes and others that are on the US Endangered Species list or are present in Appendix I of CITES (the Convention on International Trade in Endangered Species of Wild Fauna and Flora), which prohibits their trade.

A Primer on Fisheries

The purpose of any capture fishery (as opposed to aquaculture) is to catch targeted species. For any species in a fishery, there is a *maximum sustainable yield* (MSY), which is defined as the highest catch (typically expressed as weight more than numbers) of that species that can be removed annually without irreversibly damaging the population. Note that no attention to impacts on nontarget species or the ecosystem is included in this concept.

Maximum sustainable yield occurs at the point on the growth curve where population growth is highest (fig. 11.6). Mathematically, the MSY should be half of the *carrying capacity*, the maximum number of individuals of a given species that an ecosystem's resources can support in the long term without significantly depleting or degrading those resources. In reality, it is not that simple. For fishes like tunas with very fast-growing populations, MSY may be at half or even less of the carrying capacity. However, for species, like sharks, that take many years to mature, MSY is often far below half the carrying capacity. If the catch of any species is too high (i.e., above the MSY), then resulting population decreases reduce the future yields, or the individuals caught become too small.

Fisheries can be classified by the type of catch, the type of gear, the geographical location as well as position in the water column (surface or bottom, shallow or deep) and the scale (no pun intended) of the fishery.

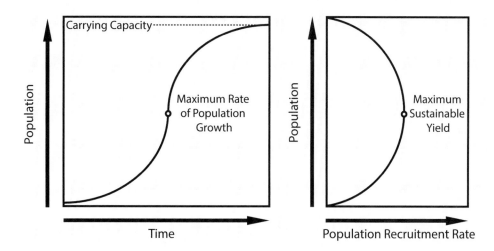

Figure 11.6. Graph of Maximum Sustainable Yield.

Fisheries can be either *subsistence, artisanal,* or *commercial.* Subsistence fisheries are those whose catch is typically for the consumption of the fisher and his or her extended family. Subsistence fishers use traditional techniques and small vessels without advanced technology and are small-scale. Even though the catch of every subsistence fisher is small, there are enough of them globally to contribute to overfishing of some species, although their impact is not fully known since most subsistence fisheries are not monitored and their catches are thus unreported.

Artisanal fisheries are intermediate between subsistence and commercial fisheries in practically every aspect. They provide seafood to their local village or town but usually not for export, and most of their catch is used as caught and is not heavily processed. The gear of artisanal fishers is typically more modern than that of subsistence fishers, but artisanal fishers still use traditional techniques as well. The impact of artisanal fisheries is also difficult to assess but is often substantial.

Finally, the most advanced, largest-scale fishery is commercial/industrial. Vessels may be quite large, and are well-equipped, often with technologically sophisticated instruments for finding fish. The catch may be brought back to port for processing or is processed and frozen at sea. The capital investment is high, as is the impact on target and nontarget species and marine ecosystems. Commercial fisheries are the most closely monitored of the three types, but illegal and undocumented fisheries occur with alarming regularity. It has been estimated that up to 50% of shark and ray landings may go unreported. Most of the shark and ray fisheries we present below are commercial.

Ways to Catch Sharks in Fisheries

There are four major categories of gear on which sharks and batoids (and other fishes as well) are caught, either intentionally or as bycatch (untargeted or unwanted catch; fig. 11.7).

- *Active entrapment* (seines, purse-seines, trawls)—targeting fish like pollack, cod, flounder, tuna, skates, and sharks. Trawl catches are typically multispecies except in some pelagic fisheries for small schooling fish like sardines.

- *Hook and line* (trolling, longlines)—targeting pelagic fish like Mahi-mahi, tuna, grouper, snapper, Spiny Dogfish, and other sharks.

- *Passive entrapment* (trap nets [tonnara], pound nets)—catching tuna and seabass as well as estuarine species.

- *Entanglement gear* (fixed and drift gillnets, trammel nets)—targeting tuna, mackerel, and estuarine fishes, and Spiny Dogfish. A variety of carcharhinid (e.g., Atlantic Sharpnose [*Rhizoprionodon terraenovae*], Blacktip [*Carcharhinus limbatus*], Blacknose [*C. acronotus*], Finetooth [*C. isodon*]) and Bonnetheads (*Sphyrna tiburo*) are targeted in the SE US gillnet fishery. Thresher sharks are caught in the Pacific Ocean using gillnets.

Overview of Shark Fisheries

Data on the annual catches of chondrichthyans as an assemblage (sharks, batoids, and chimaeras) are collated by FAO, based primarily on reports by regional organizations and member nations. Chimaera fisheries represent an insignificant proportion of the chondrichthyan catch, and thus we ignore them in our analyses below. Currently 70% of landings are sharks and about 30% batoids. Finally, the data reported below represent landings only and include only bycatch that was landed and not discarded at sea or unreported.

As a group, elasmobranch fisheries are less than 1% of the total marine capture fishery production. The catch of sharks and batoids reached a maximum of approximately 895,000 tonnes (1,970,000,000 lb) in 2003, representing a tripling of the catch rate of 1950 (fig. 11.8A). This increase was due to several factors, including increased demand and a transition to more modern, industrial-type fishing methods.

One caveat of shark fisheries is that reporting of catches is notoriously bad, in part because identification to species level is low or inaccurate, and about one-third of the elasmobranch catch is lumped together as sharks, rays, skates, and so on.

Facing page

Figure 11.7. Ways to catch sharks as well as sources of shark and ray bycatch.
(Marc Dando. Used with permission from Save Our Seas Foundation © 2016)

SOURCES OF SHARK AND RAY BY-CATCH

Globally, the number of sharks and other elasmobranchs captured as by-catch (species not targeted) is probably greater – perhaps by several times – than the number harvested in fisheries. The greatest biomass of sharks captured as by-catch occurs in pelagic fisheries using long-lines, purse seines and drift gill nets. By-catch in these fisheries is of great concern for the populations affected due to the magnitude of mortality. However, these fisheries affect relatively few species, with catches being dominated by a handful of pelagic sharks like blue sharks and silky sharks. In contrast, fisheries employing gill nets, bottom trawls and long-lines in the coastal zone often affect a great many species of sharks, skates and rays and the biomass of discard mortality is often grossly underestimated. The discard biomass in bottom-trawl fisheries in continental shelf waters is often many times the biomass of the targeted catch. Many deep-sea sharks have life histories that render them much more vulnerable to overfishing than their coastal counterparts, and deep-water trawl and long-line fisheries now operate to the maximum known depth at which sharks occur. Deep-sea edge habitats such as submarine canyons and sea mounts concentrate biomass and biodiversity and often have unique animal communities, including isolated shark populations that can be quickly depleted by relatively small fisheries.

As the human population has grown and coastal marine resources have become fully exploited, and in some cases overfished, fisheries have spread farther from shore and into deeper regions of the world's oceans. Sharks and rays are now subject to harvest and by-catch by fisheries operating throughout their depth range from coastal rivers and estuaries to the deep continental slope more than 3,000 metres deep. The major fisheries responsible for shark and ray by-catch are illustrated and typical species that are exposed to each fishing gear are represented below.

Bottom trawl

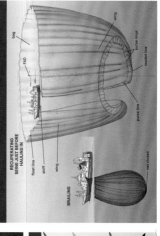

Pelagic long-lines

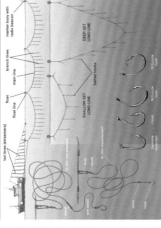

Purse seine

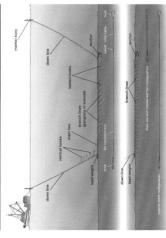

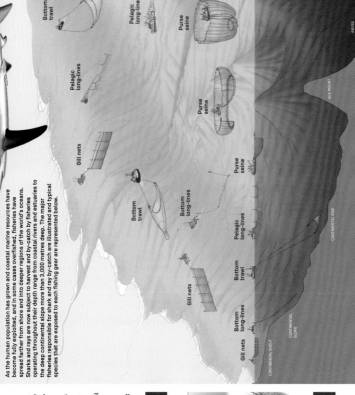

Gill nets

Drift gill net

Set gill net

Ensnared hammerhead

Bottom long-line

Drift gill net	Set gill net	Bottom long-line	Pelagic long-line	Purse seine
Prionace glauca	Carcharhinus obscurus	Carcharhinus plumbeus	Prionace glauca	Carcharhinus falciformis
Lamna nasus	Rhizoprionodon terraenovae	Rhizoprionodon terraenovae	Isurus oxyrinchus	Carcharhinus longimanus
Sphyrna lewini	Squalus acanthias	Etmopterus princeps	Alopias vulpinus	Sphyrna lewini
	Pristis pristis	Centrophorus atromarginatus	Carcharhinus falciformis	Mobula thurstoni

Bottom trawl

Amblyraja radiata
Squatina aquatina
Mustelus mustelus
Apristurus laurussonii

Infographic design by Marc Dando | Content interpretation by Dean Grubbs and Marc Dando

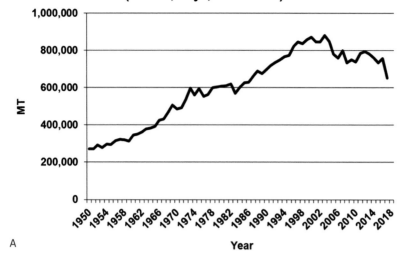

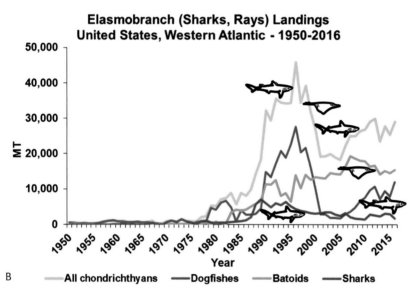

Figure 11.8. (A) Graph of global shark, batoid, and holocephalans (chimaeras) landings combined from 1950–2018. (B) US Landings Western Atlantic landings. *All chondrichthyans* category includes non-dogfish sharks, dogfishes, batoids, and holocephalans (chimaeras). *Sharks* category excludes dogfish.

Since the 2003 peak, reported landings have declined annually. For 2015, the shark and batoid catch was 756,000 tonnes (1.7 billion lb), or about 0.8% of total global fish catch (excluding aquaculture) of 92,630,000 tonnes (2.0 billion lb).

What accounted for the decrease in landings? Globally, the principal explanation is declines of shark and batoid stocks associated with overfishing. Additionally, although to a lesser degree, landings were reduced by the implementation of effective management that included the establishment of marine protected areas or sanctuaries, quotas, and so on.

This contrasts with the explanation for decreased shark and batoid landings in the United States, where catch reductions are best explained predominantly by implementation of fishery management plans (FMPs) and very aggressive enforcement, although some stocks have been and still remain depleted due to overfishing.

In the United States, catch of sharks and rays peaked in 1992 at 54,093 tonnes (119,254,661 lb). By 2016, the landings had declined to 38,820 tonnes (85,583,457 lb). Only 25% of US landings are from Pacific waters, where 90% are skates.

In US Atlantic waters, shark and ray landings increased rapidly in the 1980s and have fluctuated, reflecting changing fisheries and associated management (fig. 11.8B). The major groups landed have changed dramatically as well. For example, over the period 1985–2000, total US Atlantic landings of chondrichthyans were 50% "dogfishes," 16% coastal and pelagic sharks, and 34% skates. However, for the period 2001–2016, these landings shifted to 24% "dogfishes," 11% coastal and pelagic sharks, and 65% skates. These shifts are discussed in more detail under the Contemporary Case Studies below.

Globally, about 40–50% of elasmobranch fisheries are in the Western Central Pacific (>130,000 tonnes, or 286 million lb), and the Eastern and Western Indian Ocean (about 120,000 tonnes each, or 265 million lb each). The elasmobranch fishery in the Northwest Atlantic is only about 40,000 tonnes (88 million lb). The catch for all of Europe is about twice that of the United States.

The top five countries with the highest landings of sharks and batoids, with a combined catch of about 350,000 tonnes (772 million lb) are India, Indonesia, Mexico, Spain, and Taiwan. These five, plus Argentina, the United States, Pakistan, Malaysia, and Japan, make up about 60% of the shark and batoid landings worldwide. The remaining 40% are from small countries, mostly island nations or poorer countries in Africa, and mostly from artisanal fisheries which, as we stated above, are difficult to get good data from. This remains a vexing problem.

In last half of the 20th century, another major problem has been island nations and smaller poor countries selling fishing rights in international waters to big fishing nations like Korea, Japan, India, and Spain. The smaller countries typically receive a one-time payout for these rights. The countries who purchase them thus have no incentive to sustainably fish these stocks. In fact, the incentive is exactly the opposite—come in, exploit, overfish, and wipe out stocks, then move on. More recently, there has been a push in Indo-Pacific countries to fish their own stocks, the idea being that this would serve as an incentive to sustainably fish. To that end, in 2003 the governments of Australia and New Zealand funded a comprehensive manual on longline fishing[3] for pelagic fishes that was distributed by Indo-Pacific countries to teach coastal residents to fish for themselves.

Deep-sea Fisheries

In places where deep-sea fisheries operate, sharks often become quickly depleted. Recall that deep-sea sharks are even more susceptible to overfishing than others, as they have delayed life histories (e.g., slow maturation and low fecundity). A 2014 study assessing the conservation status of chondrichthyan fishes,[4] highlighted three regions of the world where the status of deep-sea chondrichthyans is of most concern—the Eastern Atlantic Ocean from Norway to Southern Africa (including the Mediterranean Sea); the SW Atlantic (primarily off Argentina); and the SE coast of Australia. These areas reflect the places where industrial-scale, deepwater longline, and trawl fisheries have developed.

A study[5] in New South Wales, Australia, revealed declines of more than 90% in a large suite of deep-sea sharks between the late 1970s and the late 1990s as the result of bycatch in a deepwater trawl fishery primarily for redfish and gemfish. In the Maldives during the early 1980s, a deep-sea fishery targeting gulper sharks for liver oil led to steep declines in landings by the early 1990s. Concerns have also been raised over the rapid development of deepwater fisheries targeting sharks in the Arabian Sea off southwestern India.

A 2014 review[6] found that only 5% of the deep-sea shark and ray species they assessed were threatened, a spurious result biased by the fact that deepwater species are severely understudied. Data deficient species (those for which information is too scant to establish the threat status) made up 58% of the deepwater species assessed, but only 39% of the coastal species. Moreover, most of the deep-sea species assessed belong to two large families of small, egg-laying chondrichthyans with high fecundity (softnose skates and catsharks), and these taxa are not representative of deep-sea sharks.

Life-history data are available for only a very small percentage of deep-sea sharks, rays, and chimaeras. What information there is, however, suggests that many of the live-bearing deep-sea taxa (such as gulper sharks, dogfishes, sleeper sharks, and some lantern sharks) have extremely conservative life histories. Females reach sexual maturity slowly, and while interspecific variation exists, the age at maturity can range from 15 to over 100 years of age. Fecundity is limited to fewer than 10 pups (sometimes only one), and gestation periods are among the longest of any vertebrate (e.g., 22–24 months in the Spiny Dogfish [*Squalus acanthias*]). Reproduction often only occurs biennially, but recent research has shown that continuous reproduction can occur. As a result, these species have some of the longest population doubling times of any chondrichthyan, and therefore the lowest rebound potentials if overfished. Unfortunately, most of the species with such conservative life histories have not been assessed for the IUCN Red List due to a lack of information necessary for a reliable assessment.

Concerns over the extremely long recovery trajectories for deep-sea sharks have resulted in some efforts to manage their stocks and limit harvest in certain regions. In the Maldives, for example, all shark fisheries, including the deep-sea demersal fishery targeting gulper sharks, have been closed. However, the closure of targeted shark fisheries is likely to have an insignificant effect on deep-sea chondrichthyan stocks as long as there are deep-trawl fisheries that capture and discard these fish as incidental bycatch.

We know very little about the survival rates of deep-sea sharks that are released after capture, but the evidence suggests, particularly for small species, that very few survive, even if they are released relatively unharmed.[7]

Closing areas of deep sea to trawling may offer the only effective management mechanism for limiting mortality in these inherently vulnerable species.

Unsustainable Fishery Practices

Other issues impact elasmobranch populations and complicate managing them. These include bycatch, multispecies fisheries, varied life histories, and species that may be rare, poorly described, or imperiled.

Bycatch (fig. 11.7) includes juveniles of the target species that are too small to eat, as well as nonfood or nontargeted species. In some circles, bycatch is also known as *bykill* because, while some incidental catch is returned to the water and may survive, many, perhaps most, will die in nets or on longlines or are returned to the water dead or dying.

As many as 25,000 non-recreational fishing boats ply US waters and there are an estimated 4.6 million fishing vessels globally, most less than 24 m (79 ft) in length, and most in Asia (3.5 million) and Africa. Thus, bycatch is a big issue affecting many species, including sharks.

Few fisheries catch only a single species. If the goal of fisheries was overall sustainability of marine ecosystems and their inhabitants, then fisheries should be managed to ensure viability of populations of the most vulnerable taxa encountered, whether bycatch or targeted, rather than the sustainability of the most valuable stock. For example, if a fishery catches tuna, sea turtles, and thresher sharks, then it should be managed, in decreasing order of conservation importance, for thresher sharks, sea turtles, and lastly, the targeted species, tuna. However, as you might have concluded, this rarely happens, since the targeted species is the most economically valuable. Thus, bycatch is often ignored.

A 2008 study[8] estimated the source of the major threats for different at-risk marine vertebrate taxa (cetaceans, sharks, sea turtles, and seabirds). Bycatch was considered the principal current threat for all the groups, followed by directed harvest (i.e., targeted fisheries), habitat degradation and destruction, pollution, and aggregated human disturbance (persecution, noise pollution, and so on).

Most shark and other fisheries are not perfectly *clean* (i.e., the ratio of discards to landings is not zero). A fishery that is perfectly clean catches only the targeted species of marketable size. Fisheries for anchovies, herring, sardines, and menhaden are typically the cleanest, with only 1% by weight bycatch, because the targeted species are found in massive, single-species schools that are easily surrounded and caught by purse-seines. At the other end of the spectrum, the *dirtiest* fisheries are bottom trawls for shrimp and demersal (bottom-associated) fishes like cod. Bottom trawling has been likened to fishing with bulldozers and ocean clear-cutting and all species not capable of outswimming the trawl may be caught.

Even relatively clean fisheries, because of their immense scale, can have a large shark bycatch. The bycatch may be of concern *even if sharks are not caught,* because it may include the forage base (food source) for sharks and other species, and thus the bycatch may have indirect consequences for species not captured, or for the ecosystem.

Not all sharks and batoids are equally impacted when caught as bycatch. Those with high reproductive potential (e.g., early maturity, many offspring, short gestation) may be able to withstand higher levels of bycatch mortality, and rebound more quickly, than those with more conservative life-history characteristics (e.g., lower fecundities, longer gestation periods, later maturities), which make populations of these species more vulnerable.

Consider the sawfishes (family Pristidae), batoids which are considered to be the most imperiled of all chondrichthyans. In the United States, the Smalltooth Sawfish (*Pristis pectinata*) was the first native marine fish to be listed as Endangered under the US Endangered Species Act.

Even though Smalltooth Sawfish are fully protected in US waters, they are still caught incidentally in shrimp trawls along Florida's coastline, and this is believed to be the main source of mortality for adults. The targeted species, shrimp, mature in less than a year and their reproductive potential is high, producing hundreds of thousands of offspring per shrimp annually, whereas the Smalltooth Sawfish takes 8–10 years to mature and produces only about 10 offspring, probably every other year. At this rate, only 10–15% of the sawfish population is replaced annually.

At the other end of the spectrum is the Atlantic Sharpnose Shark (*Rhizoprionodon terraenovae*), which experiences extremely high bycatch mortality in US trawl fisheries. However, because it matures quickly (three years) and has a high reproductive rate, its population has remained strong. Similarly, in the NW Atlantic, the Barndoor Skate (*Dipturus laevis*), whose fecundity is high (about 50 eggs per year), has recovered quickly from high levels of bycatch mortality.

On the other hand, deep-sea squaliform sharks have some of the most conservative and vulnerable life histories. Some gulper sharks (family Centrophoridae) have extremely low reproductive potentials (> 30 years to mature, 1–4 offspring, 2-year gestation period). It takes at least 50 years for their population to double.

Below is a summary of shark and batoid bycatch, based in part on an article by co-author Grubbs in *Save Our Seas* magazine in 2016 (see fig. 11.7).

Trawl Fisheries

Trawls are 10–130 m (33–426 ft) long, funnel-shaped nets that are towed behind one or two fishing boats (fig. 11.9). Trawls collect virtually anything in their paths that cannot actively avoid the net or that do not slip through the mesh. As the trawl moves forward, its contents are pushed to the rear, or *cod end* of the net.

Trawling is practiced worldwide on virtually every different bottom type[9] and in mid-water as well. According to the US National Academy of Sciences,[10] repeated trawling and dredging result in harmful changes in benthic communities, some likely permanent. Saturation trawling, in which the net is repeatedly fished in an area until virtually no fish or shrimp are left, has been compared to clear-cutting a forest. Approximately 100,000 km² (38,000 mi²) of terrestrial forest are clear-cut annually, whereas an area 150 times as large is trawled.

Trawls that are fished mid-water, also known as *pelagic trawls*, catch squid, sardines, pollock, and other schooling fishes and have relatively low bycatch. Some pelagic squid fisheries, however, often catch squid predators like Crocodile

Figure 11.9. Dead Greenland Shark on the deck of a bottom trawler in the North Atlantic. (Juan Vilata / Shutterstock)

Sharks (*Pseudocarcharias kamoharai*) or Sharpnose Sevengill Sharks (*Heptranchias perlo*).

Bottom-trawls are generally dirty, damaging marine habitats and taking excessive bycatch, as much as 90% of the trawl contents, which are sometimes called *trash fish* by fishers.

Mortality often approaches 100% for organisms trapped in trawls because trawls can be pulled for several hours.

Shrimp trawl fisheries in the United States catch numerous species of small coastal sharks, including Atlantic Sharpnose, Bonnethead (*Sphyrna tiburo*), and Blacknose Sharks. Stock assessments for these species indicate shrimp trawl bycatch dwarfs all other sources of fishing mortality.

Bottom-trawl fisheries may be even more harmful to skates and demersal sharks. In the NE Atlantic and Mediterranean, declines in the populations of highly endangered Angel Sharks (family Squatinidae) and guitarfishes (Rhinobatidae) have been attributed to bottom-trawl groundfish fisheries.

Some bottom-trawl fisheries target elasmobranchs and catch other elasmobranchs as bycatch. Two species of skate (Winter Skate, *Leucoraja ocellate*, and Little Skate, *Leucoraja erinacea*) are targeted in the NE US for their wings (pectoral fins) for human consumption and as bait for lobster traps. Five other species are caught as bycatch, one of which (Thorny Skate, *Amblyraja radiata*) has become severely depleted by this fishery.

Squaliform (dogfish) sharks and catsharks (family Scyliorhinidae) constitute significant bycatch in deep-sea bottom-trawl fisheries. The former are often utilized (for meat and livers), but their populations can be quickly depleted. Gulper sharks, dogfishes, as well as sawsharks, angel sharks, and sevengill sharks, all experienced population declines as a result of deep-sea trawling off the coast of New South Wales, Australia.

There has been little research into how the bycatch of sharks in trawl nets can be reduced. Although methods to reduce sea turtle bycatch have been developed successfully—and may also be moderately effective for large batoids such as stingrays—it is unlikely that they could be modified for small coastal and deep-sea sharks. In the case of the former, the effect of limiting tow time on post-release survival should be explored. However, restrictions on the type of trawling gear used and the closure of specific areas to trawling may be the only viable mechanisms for reducing elasmobranch bycatch.

Gillnet Fisheries

Gill nets rely on fishes swimming into them and getting entangled. Their rates of bycatch are very high, especially for nets with smaller mesh.

Like bottom-trawl fisheries, the bycatch in gill nets comprises a wide variety of species (fig. 11.10); some researchers estimate that globally gill nets are responsible for more bycatch mortality of marine mammals, sea turtles, and sharks than any other gear. High seas drift gillnet fisheries, such as those for flying squid and

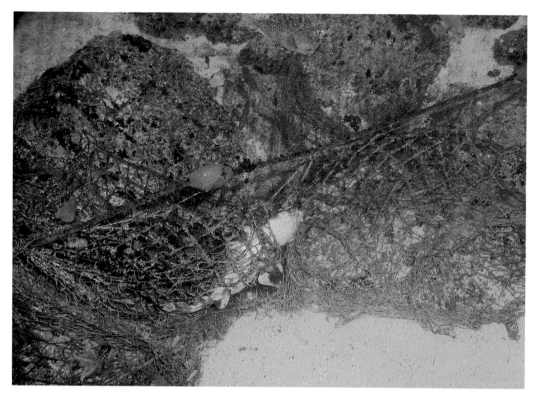

Figure 11.10. Dead shark in an abandoned, or ghost, gill net. (NOAA Fisheries)

salmon in the North Pacific, which have been criticized for their extremely high rates of marine mammal and seabird bycatch, but their rates for shark bycatch are no less extreme. It has been estimated that approximately two million sharks, primarily Blue and Salmon Sharks, were caught in the squid drift-net fishery in the North Pacific in 1990 alone.

Gillnet fisheries for coastal fishes also catch large numbers of stingrays and small coastal sharks such as sharpnose and smoothhounds, as well as juveniles of large coastal species like Blacktip and Bull Sharks (*Carcharhinus leucas*). A recent analysis of bycatch in drift gill nets for Spanish Mackerel, and in sink gill nets for drums and Spanish Mackerel along the SE US, found that more than 20 species of sharks and rays were caught, including prohibited species such as Atlantic Angel (*Squatina dumeril*), Dusky (*Carcharhinus obscurus*), and Sandbar Sharks (*C. plumbeus*), and manta rays. Juvenile Sandbar and Dusky Sharks taken as bycatch in gillnet fisheries has significantly hindered recovery of these species.

A ban on the use of gill nets, particularly in locations where there are vulnerable species or life stages, is clearly the most straightforward way to reduce bycatch. Following the ban on gillnet fisheries in Florida more than 20 years ago, researchers are now reporting signs that the Smalltooth Sawfish population is beginning to recover. An emerging problem is the use of gill nets by artisanal fishers in developing nations, which are difficult to regulate. These can have major conservation implications for species that are endemic or have small regional distributions, such as river sharks (*Glyphis* spp.) and freshwater rays.

Longline Fisheries

Longlines are an effective fishing gear that can be employed across various habitats from rivers to shallow coastal waters, the open ocean and the deep sea (fig. 11.7). Unlike trawls and gill nets, longlines are selective for species that can be lured to take a baited hook, which include most predatory fishes.

A longline consists of a mainline to which a series of branch lines (also called *gangions* or *snoods*) is attached. Each branch line terminates in a baited hook. The mainline generally ranges from hundreds of meters long with only 20 or so hooks in some near-shore and deep-sea fisheries to more than 100 km (60 mi) long with over 1000 hooks in pelagic fisheries.

Pelagic longlines are not anchored but are set adrift and marked with high-fliers (floats with a radar reflector and possibly a radio transmitter) to locate the ends. They employ a combination of floats, weighted branch lines and varied branch line lengths to reach the depths of the targeted species. Bottom-set (demersal) longlines are anchored at both ends and marked by a buoy at one or both ends. The branch lines for pelagic longlines may be as much as 10–20 m (32–64 ft) long, whereas the branch lines for demersal longlines generally range from only 20 cm (8 in) to 3 m (10 ft) long.

Pelagic longline fisheries are often seen in a negative light by environmental groups because of their bycatch, although they are relatively clean when compared to bottom-trawl and gillnet fisheries in terms of their discard-to-catch ratio. In US pelagic longline fisheries, for example, only 3–15% of the catch is discarded. However, as in trawl fisheries, the disparate life histories of target and bycatch species are a concern in longline fisheries. Targeted pelagic fishes, such as tunas and Mahi-mahi, mature early and produce many offspring, resulting in population doubling times in the order of two years, whereas the bycatch often comprises charismatic species, such as pelagic sharks and sea turtles, that have much more conservative life histories and population doubling times that may be much greater than those of the targeted species.

Sharks are often the dominant bycatch in pelagic longline fisheries. In the tuna fishery in the W. Tropical Pacific, shark bycatch has been shown to be relatively high—approximately one shark for every two tunas caught. However, a comparison of many pelagic longline fisheries has suggested that shark bycatch rates were lowest in high seas fisheries that fish deeper, targeting tunas (such as those of Japan, Fiji, and Hawaii), and were highest in fisheries that fish shallower or closer to shore (such as Chile's Mahi-mahi and Hawaii and Chile's swordfish fisheries).

Although the number of individual sharks caught may be very large, relatively few shark species are affected by pelagic longlines in comparison to bottom-trawl fisheries. Globally, the Blue Shark is the dominant species caught in pelagic longline fisheries, followed by the Silky Shark (*Carcharhinus falciformis*) and the Oceanic Whitetip (*C. longimanus*). Most of the sharks that end up in the international fin trade are bycatch in pelagic longline fisheries and there is concern that these shark species may become depleted.

In most pelagic longline fisheries, the Blue, Silky, and Oceanic Whitetip Sharks are reported being discarded alive. Marketable species such as the Shortfin Mako (*Isurus oxyrinchus*) and the three threshers are typically sold and, given their conservative life histories, are of great management concern. It is also important to recognize that some pelagic fisheries retain all the sharks taken as bycatch. For example, the Swordfish fishery off Uruguay, which has an extremely high rate of shark bycatch, typically markets more than 95% of the Blue Sharks caught.

Several mitigation measures to reduce shark bycatch on pelagic longlines have been assessed, though their effectiveness often depends on the species involved. The use of monofilament instead of steel leaders and of squid as bait instead of fish have been shown in some sharks to reduce the number caught, with relatively little—or even a positive—effect on the catch rate of the targeted tunas and swordfish.

The depth of the hooks is often an important consideration. The majority of pelagic sharks, such as Silky and Oceanic Whitetips, spend most of their time in the upper mixed layer shallower than 100 m (330 ft). Increasing the depth of hooks to that depth or to 150 m (500 feet) has been shown to reduce bycatch rates significantly for these and most other shark species. However, Blue Sharks, Bigeye Threshers (*Alopias superciliosus*), and Shortfin Makos make daily excursions to depths of 400 m (1320 ft) and more. For them, factoring in both time of day and depth when deploying hooks may reduce bycatch. More research needs to be undertaken to determine what influences shark bycatch rates in pelagic longline fisheries so that additional mitigation measures can be identified.

At-boat and post-release mortality rates for sharks caught on pelagic longlines vary widely, depending on the species. Some sharks, such as the Smooth (*Sphyrna zygaena*), Scalloped (*S. lewini*), and Great Hammerheads (*S. mokarran*) and all the threshers, suffer at-boat mortality rates of at least 25% and sometimes more than 50%, and it is likely that most of the sharks released alive do not survive. In contrast, the at-boat mortality rate of Blue, Silky, and Oceanic Whitetip Sharks—the three species most often caught on pelagic longlines—as well as of makos, is only 5–20%, and the few data available suggest that post-release survival may be quite high.

Demersal, or bottom-set, longlines are those whose hooks lie directly on the sea floor. Many targeted shark fisheries deploy demersal longlines as their primary gear, but there are also a number of fisheries using the same gear for bony fishes that take significant shark bycatch. In fact, the bycatch on demersal longlines can be 50% or more of the overall catch—a much higher overall bycatch rate than that of pelagic longlines.

Demersal longline fisheries likely take more shark species as bycatch than do all other fisheries combined. In the Gulf of Mexico, the two most common bycatch species taken by the US demersal fishery for groupers and tilefish are often Cuban Dogfish (*Squalus cubensis*) and Blacknose Sharks. It is reported that more than 95% of these sharks are released alive, but a recent study[11] has shown that about half of Cuban Dogfish would likely have soon died, even if they were released in a healthy condition. In addition, on some vessels the hook is ripped from the shark's mouth, breaking the jaw, which probably increases the post-release mortality rate.

The deepest fisheries in the world use demersal longlines, such as those targeting Patagonian Toothfish (marketed as *Chilean Sea Bass*), down to 3000 m (9840 ft) and Greenland Halibut, grenadiers, hake, and ling (2000 m, or 6560 feet). Shark bycatch in these fisheries can be very high for species like Portuguese Dogfish (*Centroscymnus coelolepis*), lantern sharks (Etmopteridae), gulper sharks (Centrophoridae) and their relatives, as well as for numerous species of catsharks (Scyliorhinidae). Some of these species are kept and sold for their livers and meat, but most small sharks are discarded at sea. Mortality is probably 100% in these fisheries as not only are the sharks unable to survive being retrieved from great depth, but their jaws are broken on landing by the auto-line retrieval system that pulls the hooks through a set of steel rollers.

Even in targeted shark fisheries that are reasonably well-managed, bycatch is a major concern. The shark longline fishery in US Atlantic waters targets large coastal sharks and is regarded as one of the best-managed fisheries of its kind in the world. However, more than half of its catch can include small coastal sharks, such as the Atlantic Sharpnose, that are not marketed, and most are discarded dead. Moreover, species that are prohibited because they are overfished (such as Dusky or Sandbar Sharks) or endangered (Smalltooth Sawfish) are also taken as bycatch. Fortunately, many of these are quite resilient and probably survive capture.

However, a challenging current problem in US Atlantic large coastal shark fishery is the capture of hammerhead species that are overfished. Capture mortality is very high in these species, therefore capture prohibitions are ineffective in curbing mortality in a multi-species fishery. A recent study[12] showed that 50% of hammerheads are dead after three hours on the hook in this fishery. Co-author Grubbs is currently studying the survival probability of hammerheads released alive to determine if limiting hook soak time may decrease mortality.

Purse-seine Fisheries

The largest fisheries in the world are purse-seine fisheries that target small pelagic fishes such as anchovies, herring, and menhaden (fig. 11.7). Historically these fisheries have been relatively clean in terms of bycatch as they typically target dense, single-species schools of fish. Industrialized purse-seine fishing for larger pelagic fishes such as tunas has developed relatively recently, and purse-seine nets are now responsible for about 70% of all tuna landings worldwide.

Purse-seine nets are deployed in the open ocean over deepwater. Those used for tunas are typically 1000–2000 m (3280–6560 ft) long and about 300–650 m (985–2130 ft) in diameter, and as much as 200 m (650 ft) in depth. The seine has a float line and a lead line, and it is positioned to encircle a school of fish, with one or two speedboats pulling it from the larger harvest vessel. Once the float line circle is closed, the opening at the bottom of the net is closed, or pursed, by cinching the lead line and thus preventing fish from escaping. The net volume is reduced, and the catch is hauled aboard the harvest vessel, removed from the net, and placed in the ship's hold. Elasmobranch bycatch in purse-seine fisheries is mainly Silky Sharks, Oceanic Whitetips, and mantas, with other species caught episodically (fig. 11.11).

Floating objects that drift around oceanic gyres, such as trees or logs that have been swept into the sea from rivers, or man-made, like lost fishing nets or pieces of wrecked vessel, tend to attract large communities of organisms, as many as 300 fish species, that seek refuge. Pelagic predators such as tunas and sharks in turn are attracted to these floating objects and the potential prey they harbor.

Figure 11.11. Whale Shark caught as bycatch in purse-seine in the Pacific. (© Greenpeace)

Bycatch rates in nets set on these floating objects are much higher than on other sets. The most vulnerable component of this bycatch is pelagic sharks, as well as sea turtles.

Fishers also exploit the tendency for tunas to gather at floating objects by deploying purpose-built fish aggregating devices (FADs) to attract them. These FADs may be floating or anchored and as simple as palm fronds tied together or as complex as large structures with radio or satellite locator beacons and integrated sonars that enable fishers to estimate the biomass of aggregated fish. Pelagic sharks, big and small pelagic fishes, and large amounts of undersize tuna discards make up most of the high bycatch of FADs.

An interesting case study on the complexities of fisheries management, as well as the effects of well-meaning but misguided environmental activism, concerns tuna purse-seine fisheries in the Eastern Tropical Pacific. There, tunas associate with dolphins, which is thought to benefit the tunas, which can take advantage of the dolphins' superior prey-finding abilities. Once these assemblages are located, seines termed *dolphin sets* are then positioned around the combined schools.

The targeting of dolphin-associated tunas generated significant controversy in the 1960s due to the high rate of dolphin mortality in purse-seines and estimates of rapidly declining dolphin populations. However, by the late 1980s, dolphin mortality dropped substantially after the fishery developed its own technique to reduce dolphin bycatch.

Although dolphin mortality was already in a steep decline, the *dolphin-safe* tuna labelling campaign was launched, which led US canneries to adopt dolphin-safe policies in 1990, which required them to buy only tunas that had been caught by methods that did not involve encircling dolphins. By 1993, dolphin mortality had decreased to almost zero, thanks to the procedures developed by the fishery prior to dolphin-safe labelling, and it has remained extremely low for more than 20 years.

In an unintended consequence, however, dolphin-safe labelling led the US purse-seine fleets to develop the use of FADs (fig. 11.12); by 2009, 95% of floating objects used by purse-seiners were FADs. Bycatch increased from 0.5% of the catch in dolphin sets before 2009 to nearly 10% using FADs and included nearly 3500 sharks in the former but > 35,000 sharks in the latter. Capture of one dolphin in dolphin-associated purse-seine nets was equal to the bycatch of approximately 25 sharks and more than 900 bony fishes in FAD-associated purse-seines. FAD-associated purse-seine fisheries now account for more than half the global landings of tuna.

Sharks and mantas are the most vulnerable FAD bycatch because of their life histories. In some regions, more than 20 shark species are caught in purse-seine nets. And of these sharks, Silky and Oceanic Whitetip Sharks dominate worldwide, followed at some distance by Scalloped and Smooth Hammerheads.

In most purse-seine tuna fisheries, sharks caught must be released alive. How-

Figure 11.12. Surface view of a fish aggregating device (FAD) off Hawaii.

ever, these sharks are mostly pelagic and, as obligate ram ventilators, must swim constantly so that sufficient water passes over their gills to oxygenate them. Thus, they tend to be quite fragile and soon die if they are not actively swimming. For this reason, pelagic sharks caught on longlines have a relatively high survival rate, whereas the rate for sharks taken by purse-seine fisheries is likely to be very low. Numerous recent studies[13] examined post-release mortality in Silky Sharks caught in tuna purse-seines set around FADs and all estimated the total mortality 80% or more.

The problem of bycatch in purse-seine fisheries using drifting FADs is compounded by the fact that the structure of the device itself causes additional mortality. In many drifting FADs, old fishing net hangs down from the structure and marine life, including sea turtles and sharks, becomes entangled in it. These deaths are not included in fishery bycatch estimates, but they can be substantial. A recent study[14] conducted in the Indian Ocean estimated that the mortality of Silky Sharks entangled in FADs may be five to 10 times the actual bycatch in the associated purse-seine fishery.

Bycatch and how to reduce it is one of the most difficult and complex issues faced in fisheries management. Relatively few fisheries exist where bycatch is not a major concern, and it is particularly troubling when it includes elasmobranchs whose life histories are more conservative than those of the targeted catch, making them more vulnerable to overfishing. As you may now realize, the real cost of seafood often goes well beyond the monetary cost of the product.[15]

Historical Directed Shark Fisheries

Shark fisheries in the early to mid-20th century have provided historical context to what can be expected if the life-history characteristics of the targeted species are not considered. We give you two examples of classic historical case studies below.

Porbeagle

The most famous, perhaps infamous, historical case is the Porbeagle (*Lamna nasus*) fishery in the Northeast Atlantic.[16] Porbeagles (family Lamnidae) are a cold-temperate shark found across the North Atlantic and in the southern hemisphere. They reach about 3.5 m (11.6 ft). Females mature at about 13 years and have about four pups every other year.

The commercial fishery for Porbeagles was initiated by Norwegian longliners in 1961, prior to which the population was considered virgin (i.e., unexploited). Reported landings for 1961 were 1924 tonnes (4 million lb). By 1964, at its peak, the landings had reached 9360 tonnes (20.6 million lb).

The fishery collapsed in 1967, and by 1970, total reported catch in that fishery was around 1000 tonnes (2.2 million lb), and the average size of the sharks in the fishery had declined. In the 1970s and 1980s, a smaller, more sustainable fishery for the species of about 250 tonnes (550,000 lb) continued in the Faroe Islands of the North Sea.

The species began to very slowly recover until the early 1990s, at which time Canada and the Faroe Islands increased their fishing effort, although not to the 1960s levels. Still, the Porbeagle population in the North Sea could not sustain the fishery and, although overfishing no longer occurs, the population in the Northwest Atlantic remains overfished (i.e., it has not recovered). The IUCN considers the Porbeagle Critically Endangered in the Northeast Atlantic Ocean and Mediterranean Sea, Endangered in the Northwest Atlantic, and Vulnerable globally.

California and Australian Soupfin (Tope) Shark

The fishery for Soupfin Sharks (*Galeorhinus galeus*) in the 1930s and 1940s in California was the biggest US West Coast shark fishery. There was also a fishery for this species at about the same time along the southern coast of Australia. What drove this fishery was the need for a source of vitamin A when cod imports from

Europe stopped during World War II. The livers of this species had among the highest potency vitamin A of any shark species.

Soupfin Sharks were caught using large-mesh, bottom-anchored gill nets or longlines. Prior to the increased demand for vitamin A, the fishery was a small one for fresh fillets, fins for the Asian market, or for rendering for fertilizer. The average annual catch of Soupfin Sharks between 1930 and 1936 was under 272,000 kg (600,000 lb). Data on landings of Soupfin Sharks between 1936 and 1941 have been difficult to locate. In 1939, landings of all sharks in California was about 4170 tonnes (9.2 million lb). Soupfin Sharks comprised about 52% of the catch from 1941–1944. Using that figure, landings for Soupfin Sharks at the peak were over 2040 tonnes (4.5 million lb). The fishery collapsed by 1946, partially due to declining stocks associated with overfishing as well as the development of synthetic vitamin A, which obviated the need for the fishery.

It is worth noting that since this fishery developed so fast and prices for livers was so high, virtually anyone with a boat began fishing for Soupfin Sharks. A story by *The Santa Barbara* (California) *Independent* newspaper[17] quotes locals Ralph Hazard and Red Allen: "'It took about two or two-and-a-half years, but we cleaned them all out,' said Hazard. Allen added, 'And we weren't taking only soupfins.' By 1943, sharks in the Santa Barbara Channel 'were just about all wiped out.'"

CONTEMPORARY CASE STUDIES

US Atlantic Shark and Ray Landings

The development and management of commercial fisheries for sharks and rays in US Atlantic waters offer examples of a suite of contemporary case studies illustrating a pattern of serially depleted stocks, subsequent management plans (often in response to campaigns from environmental advocacy groups), and stock rebuilding or recovery. Prior to the Fishery Conservation and Management Act of 1976 (currently the Magnuson-Stevens Reauthorization Act), foreign fleets frequently exploited shark stocks in US waters but US shark fisheries were very small and the sharks harvested were caught primarily for their livers in the production of vitamin A and liver oil. In the early 1980s, in response to plummeting US stocks of groundfish-like cod off New England, fisheries managers encouraged the development of fisheries targeting Spiny Dogfish to supply *fish and chip* markets in Europe.

Also during the 1980s, directed fisheries for skates, primarily for use as bait in lobster traps, were also developing in the NE US. In addition, the lifting of US trade limits with China in the late 1970s opened up a huge market for US fishery products, including shark fins, leading to the development of directed fisheries for coastal and pelagic sharks. In response to these trends, commercial shark and ray landings in US Atlantic waters rose from less than 900 MT (~20,000 lbs) in 1976 to more than 32,000 MT (70 million lb) by 1990 (fig. 11.8B), an increase of 350,000%! This does not include recreational shark and ray landings, which were on par with commercial landings.

By 1989, 39% of US Atlantic landings were coastal and pelagic sharks caught in unmanaged fisheries to supply the fin trade, whereas 24% were Spiny Dogfish, and 36% were skates. Due to concerns over rapidly declining populations of large coastal shark species such as Dusky and Sandbar Sharks, among the most sought-after species in the shark fin soup market, a fishery management plan for the sharks, the first such effort, was implemented in 1993, with restrictive quotas. By 1996, coastal and pelagic sharks were reduced to 10% of the landings.

However, the fishery for Spiny Dogfish to supply fish and chip markets in Europe continued to expand rapidly and, by 1996, Spiny Dogfish made up 60% of the shark and ray landings, with skates making up the remaining 30%. The Spiny Dogfish stock was soon overfished, and a management plan was implemented in 1998, which began limiting their landings. By 2003, regulated Spiny Dogfish landings were less than 10% of the peak landings and contributed less than 10% to the overall US Atlantic shark and ray landings. Due to a steadily increasing fishery for bait as well as for "wings" for use as food, by 2003 skates constituted 75% of shark and ray landings. Again, evidence of severe overfishing of at least three species of skates in this fishery led to skate management and a rebuilding plan, implemented in 2003.

Currently, all three groups of US Atlantic sharks (large coastal sharks, small coastal sharks, and pelagics) and rays are under management. By 2016, large coastal sharks were managed at a quota that was roughly 8% of their peak landings, and the most vulnerable species remain prohibited. Coastal and pelagic sharks contributed only 6% to the total US Atlantic shark and ray landings, whereas skates continued to be the dominant component at 53%. Due to the apparent recovery of the Spiny Dogfish stock, landings in this fishery increased, making up 39% of the 2016 commercial landings. The astute mathematically minded reader may note these numbers add to 98%. The remaining 2% of 2016 shark and ray landings were Dusky Smoothhounds that were landed in a new targeted fishery, which accelerated in the early 2000s, mostly off North Carolina. All the catch of this species went to Australia and the UK as *flake fish* (i.e., boneless fish fillets). Landings peaked for this single species at about 1200 tonnes (2.6 million lb) in 2012. The science-based advocacy group Shark Advocates International brought this fishery to the attention of NOAA and urged them to act to study and protect the stock. In a rare effort to manage shark fisheries before they are overfished, a 2014 stock assessment found that the population was healthy with no overfishing occurring because it is a small coastal species with about a 5-year doubling time (mature at three years, produce 15 pups annually).

As you can see, the simple examination of landings does not present the whole story. In some cases, declines in landings may be signs of trouble for a stock, but in others they may reflect effective fisheries management.

It Is All in a Name

In our Authors' Note at the start of this book, we mentioned that common names were much less useful than scientific names. Here is a case where the use of common names was in fact *dangerous* for one population of sharks, namely Spiny Dogfish. Until about 2002, Spiny Dogfish and Dusky Smoothhound (also called the Smooth Dogfish) were both heavily fished along the US Atlantic Coast. Since both were small sharks often caught on the same gear, and both had *dogfish* as part of the common names, they were classified for fishery statistics in a single category as *dogfish*.

What harm, you might ask, could that possibly cause? Lots. The two species occupy two different superorders (Spiny Dogfish are squalomorphs, Dusky Smoothhounds are galeomorphs) separated by about 210 million years of evolution and, what is more, have drastically different life-history characteristics. Spiny Dogfish take as long as 20 years to mature and have only 4–6 pups every other year, after a 24-month gestation period. Dusky Smoothhounds mature in three years and have 12 pups annually after a gestation period of 8–9 months. Fortunately, after stocks of the former plummeted, their fisheries were classified separately. Spiny Dogfish have now largely recovered, and, as we established above, the biggest current US shark fishery is for Dusky Smoothhounds.

Save the Bay, Eat a Ray: A Cautionary Tale About Science and Elasmobranchs, with a Barb

We conclude this chapter with an introduction to a case study in which a paper[18] published in the journal *Science* in 2007 led to a wrongheaded movement to eradicate an elasmobranch with one of the lowest known reproductive potentials. The species is the Cownose Ray (*Rhinoptera bonasus*), and the paper, entitled *Cascading effects of the loss of apex predatory sharks from a coastal ocean*, claimed that large coastal sharks (Sandbar, Blacktip, Bull, Dusky, Scalloped Hammerhead, and Tiger Sharks [*Galeocerdo cuvier*]) along the East Coast of the United States had experienced huge population declines which led to what is called *predation release* and a *trophic cascade*.

Predation release here refers to removing predation pressure on smaller sharks and ray mesopredators (mid-level predators; e.g., Atlantic Sharpnose Sharks, Chain Catsharks, Little Skates, Butterfly Rays, and Cownose Rays) by removing their predators (i.e., the larger sharks). The ensuing trophic cascade refers to the reverberations further down the food chain when the population of the mesopredators *skyrocketed*, according to the paper, and the populations of *their* prey dropped.

According to the paper, the largest consequence of the predation release and ensuing trophic cascade was the population explosion of the Cownose Ray and the population crash of its chief prey, the Bay Scallop (*Argopecten irradians*), and the subsequent loss of the North Carolina fishery on which it was based. Another paper[19] by the same research group expanded the list of shellfish affected by the Cownose Ray to include oysters, as well as hard- and soft-shell clams.

A rebuttal to the *Science* paper was published in 2016 in the journal *Scientific Reports.*[20] Here, we summarize parts of that paper, along with additional information, by answering four major questions about the validity of the *Science* paper.

1 / Were the declines in large sharks as severe as reported?

In short, no. The main issue was reliance on a single data set, the University of North Carolina Shark Survey, for estimates of the declines in large sharks of between 87% and 99%. This data set was rigorously collected, but not appropriate for the breadth of the conclusions. Also, sample sizes of three of the seven large shark species were too small to allow the conclusions made by the *Science* paper. For example, the 99% decline of Smooth Hammerheads was based on a sample size of five sharks over a 32-year period. Only 23 Bull Sharks (also 99% decline) were caught by this survey over the 32-year period, and only 39 Tiger Sharks, for which the paper reported a 97% decline. Reliance on small sample sizes is dangerous as small changes in the catch one year can change the perception of both the direction and magnitude of the change of the population of these species.

The *Science* paper reported declines for Sandbar, Blacktip, and Dusky sharks of 87%, 93%, and 99%, respectively. The rebuttal paper published in *Scientific Reports* reanalyzed the UNC data and included the larger VIMS Shark Survey data set and concluded that there was a 50–60%, not 87%, decline in Sandbar Sharks over the time period. Blacktip Sharks showed the opposite trend. While declining in the UNC data set, Blacktip Sharks were increasing in the VIMS data set, likely because of warming coastal waters pushing them north. The important message is that a single survey is insufficient to draw such sweeping conclusions.

2 / Were the trophic links sufficient to elicit this cascade?

The first requirement for a trophic linkage between the large coastal sharks and the mesopredators is spatio-temporal overlap; that is, they must be in the same place at the same time.

Consider two of the mesopredators whose populations were supposed to have exploded after trophic (or predation) release, the Little Skate (*Leucoraja erinacea*) and Chain Catshark (*Scyliorhinus retifer*). Both of these species inhabit only water temperatures of 5–12°C (41–53.6°F). In the *Science* paper, the sharks considered predators of these two species are all found in warmer waters, with little to no overlap, and thus they could not be primary predators for the Little Skate and Chain Catshark because they rarely co-occur.

Let us look closely at the Cownose Rays. First, do the large coastal sharks eat them? Yes, but in small numbers. In fact, Cownose Rays are eaten by more Cobia than by sharks.

Are Cownose Rays significant natural predators of commercial bivalves? What is their natural diet? A variety of studies have shown that Cownose Rays, depending on the location, eat (in no particular order) a large range of prey, including bay scallops, soft-shell clams, other small bivalves, crustaceans, fish, and a number of other invertebrates, including cumaceans (small marine crustaceans), polychaetes, amphipods, and echinoderms.

Do Cownose Rays even eat oysters, as the *Science* paper claims? The diet of Cownose

Rays collected from commercial oyster beds in Chesapeake Bay shows that oysters comprise 5% of their diet. In natural bays, there is no evidence that Cownose Rays eat commercial oysters or clams, but possibly soft clams and bay scallops.

3 / Were increases in smaller sharks and rays credible?

Not necessarily. Recall the case of the Little Skate and Chain Catshark above. Moreover, the Little Skate stock assessments showed either no trend or a declining trend. The Clearnose Skate (*Rostroraja eglanteria*), which is regularly eaten by large coastal sharks, did not show a clear trend in the paper's online supplemental material.

The *Science* paper also showed a clear increase in the Atlantic Sharpnose Shark population in the UNC survey, but the VIMS survey and stock assessment showed no change, which was in line with the federal stock assessment.

Multiple data sets were examined for Cownose Rays and most showed large increases in the population. The problem with assessing Cownose Rays is that they travel in enormous schools of tens of thousands of individuals, and no fishing gear samples them effectively. The sample sizes for all except one of the surveys was small. The survey with the larger sample size showed a moderate increase in the population. Cownose Rays mature at 7–8 years, live to a maximum of 21 years, have a 12-month gestation period, and produce a single pup annually. The population doubling time for Cownose Rays is decades suggesting the population increases reported in the *Science* study are not biologically possible. What probably did occur was shifts in the population similar to what was suspected for Blacktip Sharks in the study.

4 / Was the stated increased Cownose Ray abundance responsible for declines in commercial bivalve populations?

Bay scallops and oysters have been subject to industrial scale fisheries for many decades. One of the more convincing lines of evidence refuting the claim that Cownose Rays were responsible for declines in bay scallops between 1970 and 2005 was that the bay scallops were in steep decline beginning in the 1980s; that is, *before* the purported increases in Cownose Ray populations in the 2000s occurred. One observer commented that the rays would have had to travel back in time to cause the crash. In addition, it was shown that oyster and bay scallop stocks had crashed coast-wide, mostly before 1990, including in regions outside of the range of Cownose Rays. The cause of the collapse of these stocks, including the century-long scallop fishery in North Carolina, was likely a combination of overfishing, loss of seagrass habitat, hurricanes, red tides, wasting disease, pollution, as well as a poorly regulated recreational fishery.

Save the Bay, Eat a Ray—The conflict between shellfish fisheries and Cownose Rays was nothing new. Since the 1970s when stocks of oysters and bay scallops began to collapse in North Carolina and Virginia, there have been calls every few years to implement fisheries for Cownose Rays to eliminate these perceived competitors for the valuable shellfish. These calls were usually dismissed over concerns for the vulnerability of the rays to overfishing, but the *Science* paper was taken by the industry as support for reducing the Cownose Ray population. A lot of effort was expended to develop commercial fisheries and associated markets for the meat and other products (e.g., leather from the skin). The meat was marketed domestically as well as in Europe and Korea. Bow-fishing tournaments were implemented to take place when pregnant female Cownose Rays entered the estuaries in order to kill both the mother and pup at once. These efforts were promoted under the guise of environmental sustainability through the *Save the Bay, Eat a Ray* campaign. Though no stock assessments had been conducted on Cownose Rays, it was known that they have among the most conservative life histories of any shark or ray, rendering their population extremely vulnerable to overfishing. The effects of these fisheries on the Cownose Ray population are unknown.

Concluding Comments

Shark and ray fisheries and bycatch remain complex issues of major concern, involving many species and stakeholders. Fishery managers have the knowledge required to assess the population status of shark and ray stocks, and have used this knowledge to study and help rebuild some overfished stocks, both domestically and globally. However, these gains occur against the overwhelming backdrop of climate change and other human insults, which we will examine in the next chapter.

NOTES

1. Dent, F. and Clarke, S. 2015. State of the global market for shark products. FAO Fisheries and Aquaculture technical paper (590), Rome, FAO, 187 pp.

2. https://awionline.org/content/international-shark-finning-bans-and-policies. (Accessed 8/12/19).

3. Beverly, S. et al. 2003. Horizontal longline fishing methods and techniques: a manual for fishermen. Secretariat of the Pacific Community. 130 pp.

4. Dulvy, N.K. et al. 2014. Elife 3: p.e00590.

5. Graham, K.J. et al. 2001. Mar. Freshwater Res. 52: 549–561.

6. Dulvy, N.K. et al. Op. cit.

7. Talwar, B. et al. 2017. Mar. Ecol. Prog. Ser. 582: 147–161.

8. Finkelstein, M. et al. 2008. PloS One 3: p.e2480.

9. Waitling, L. and Norse, E.A. 1998. Conserv. Biol. 12: 1180–1197.

10. Ocean Studies Board and National Research Council. 2002. Effects of trawling and dredging on seafloor habitat. National Academy Press.

11. Talwar, B. et al. 2017. Mar. Ecol. Prog. Ser. 582: 147–161.

12. Gulak, S.J.B. et al. 2015. Afr. J. Mar. Sci. 37: 267–273.

13. Eddy, C. et al. 2016. Fish. Res. 174: 109–117.

14. Filmalter, J.D. et al. 2013. Front. Ecol. Environ. 11: 291–296.

15. Abel, D.C. and McConnell, R.L. 2012. J. Manage. Contr. 22: 481–485.

16. O'Boyle, R.N. et al. 1996. Atl. Fish. Res. Doc. 1996/024.

17. https://www.independent.com/news/2008/aug/14/world-war-ii-santa-barbaras-waterfront-and-2000-sh/. (Accessed 8/9/19).

18. Myers, R.A. et al. 2007. Science 315: 1846–1850.

19. Baum, J.K. and Worm, B. 2009. J. Anim. Ecol. 78: 699–714.

20. Grubbs, R.D. et al. 2016. Sci. Rep. 6: p.20970.

12 / Climate Change and Other Human Impacts

Introduction

In addition to overfishing (see Chapter 11), sharks must face the same anthropogenic environmental insults as other marine organisms—climate change that causes waters to deoxygenate, warm and acidify; nutrient pollution from agricultural runoff and undertreated human sewage; micro- and macro-plastic pollution; other types of pollution (other chemicals, noise, radioactivity, and so on); habitat destruction; and loss of biodiversity in ecosystems they inhabit. The short- and long-term impacts of these are not all known. They may be minimal, or they can range from sublethal impacts that affect an organism's overall fitness (e.g., its physiological functions and behavior) to mortalities that cause declines in populations.

We lump these impacts together as *pollution*, plus physical habitat loss. We consider *climate change* below separately from pollution, although it is clearly a form of pollution.

As we survey these human impacts one-by-one, you may be frustrated by the unsatisfying lack of clarity about specific impacts on sharks. This lack of clarity is explained by several reasons:

- Many reports of impacts on sharks are anecdotal; for example, opportunistic encounters with one or more dead sharks, such as strandings. Conclusions based on anecdotal occurrences may be useful, but they typically involve low numbers of sharks, which may be at an advanced stage of decomposition, and which also may be temporally and physically distant from the source of the pollution.

- Controlled studies (i.e., exposing captive sharks to or administering doses of potentially harmful substances or biological agents) are logistically difficult and expensive and thus are too infrequently conducted.

- Even if controlled laboratory studies are undertaken, extrapolating from these to ecosystems or generalizing from these studies are both fraught with potential problems because of our limited knowledge of shark ecology.

- Studies on captive animals most often employ species or life stages capable of surviving in experimental systems, and these species do not represent the diversity of forms, physiologies, and habitats of sharks as a group. Examples include juvenile Lemon Sharks (*Negaprion brevirostris*), Spiny Dogfish (*Squalus acanthias*), horn sharks (e.g., *Heterodontus francisci*), and Small-spotted Catshark (*Scyliorhinus canicula*).

- Correlations are easier to establish than cause-effect relationships. In other words, it may be fairly straightforward to relate the presence / absence of sharks or pathological conditions in individuals to levels of pollution, but it is much more difficult to demonstrate that some chemical or combination *caused* the absence or condition.

- Many sharks are migratory, and thus determining the level and duration of exposures to the suite of pollutants encountered during these movements is next to impossible.

A Few Principles of Toxicology

Let us begin with toxicology which, according to the American Society of Toxicologists, is the study of the adverse physicochemical effects of chemical, physical, or biological agents on living organisms and the ecosystem. The agent that causes the adverse effect is known as the *toxicant* or *pollutant*.

Chemicals, including radioactive ones (*radionuclides*), can affect organisms by being:

- *Toxins*—poisonous chemicals of biological origin (e.g., *brevetoxin* [see below], produced by the dinoflagellate *Karenia brevis*).

- *Carcinogens*—cancer-causing agents. An example is the class of chemicals known as polychlorinated biphenyls, more commonly called PCBs, which were widely used as electrical insulators. Radioactive chemicals like Cesium-

137, which is released in nuclear weapon tests and reactor disasters like Chernobyl and Fukushima Daiichi, may also be carcinogenic.

- *Mutagens*—agents that cause changes in DNA (so named because they cause genetic mutations). Polyaromatic hydrocarbons (PAHs), another class of chemicals widespread in many marine and estuarine systems, and metals like arsenic, cadmium, chromium, and nickel, may also be mutagenic.

- *Teratogens*—agents that cause abnormalities or malformations during fetal development. In addition to being carcinogenic, PCBs are teratogenic, as is tributyl tin, a compound used as an antifouling agent on ships.

In all of these cases, toxicologists seek to determine the dose-response relationship; that is, how organisms respond over a range of concentrations or doses of toxicants. Toxicity may be *acute* (acting in the short term and usually fatal) or *chronic* (long term and typically sublethal). Even if the toxicant is sublethal, it could affect the overall *condition* (i.e., health) of an organism and thus reduce its *fitness* (i.e., its ability to grow and develop, forage, swim, sense its surroundings, avoid predation, mate, and so on). Moreover, sublethal toxicants may act on the offspring as mutagens or teratogens.

Toxicants also have different routes by which they enter organisms; for example, across gills, ingested as food, and absorbed across skin. Additionally, they may be systemic (i.e., acting on the entire organism) or they may be more localized, targeting and affecting different organs (e.g., liver, organs of the central nervous system) or tissues (e.g., muscle, nerve).

Delayed mortality and sublethal effects may occur because organisms *bioaccumulate* some harmful substances. For example, mercury and fat-soluble organic compounds bioaccumulate in organisms by *bioconcentration* (i.e., being absorbed from water and accumulating in tissue to levels greater than those found in surrounding water) and *biomagnification* (i.e., increasing in tissue concentrations as larger organisms consume smaller ones [those lower in the food web] and absorb and retain the harmful substances). In some cases, levels of these compounds decrease over time as a result of processes that detoxify, or depurate, the toxic compound.

In some cases, the toxicant may be detected and sidestepped, a phenomenon known as *behavioral avoidance*. While the short-term survival of the organism is enhanced, there may be other impacts on the organism, which may find itself outside of its accustomed surroundings.

Categories of Human Impacts and Their Effects on Sharks

In general, water pollution refers to any hazardous or toxic material released by humans into the aquatic environment. We discuss the nine categories below, as well as an additional human impact not included as pollution—habitat destruction. We conclude with a section on climate change.

Disease-causing Agents

This category includes pathogenic bacteria, viruses, protozoans, and other parasites. Note that not all of these constitute pollution; most in fact are naturally occurring (i.e., not introduced by humans). The main anthropogenic source is untreated or improperly treated human waste, but animal waste from feedlots and meat-packing plants are also sources. Often these infections occur opportunistically after impacts of other stressors (e.g., thermal shock or other types of pollution).

The incidence of diseases in elasmobranch fishes apparently is low. Moreover, distinguishing normal shark pathogens from those considered pollutants is not always possible. Here are a few examples of pathogenic infestations in sharks.

- A 2018 study[1] of Leopard Sharks (*Triakis semifasciata*) that had died after stranding themselves on the shores of San Francisco Bay during the spring of 2017 found that they had severe meningoencephalitis—dangerous inflammation of the brain and parts of the spinal cord, likely but not definitively caused by the ciliated protozoan pathogen *Miamiensis avidus*.

- Nematode worms (also called roundworms) are common endo- and ecto-parasites of sharks (fig. 12.1). A 2002 paper[2] found what the authors characterized as *potentially debilitating lesions* on the pancreases of 10 of 20 Spiny Dogfish (*Squalus acanthias*) caused by the nematode *Pancreatonema americanum*. The authors could not establish a causal relationship between the pathology and pollution.

- Toxins from *harmful algal blooms* (HABs; see Plant Nutrients section below), sometimes also called red or brown tides, have been shown to cause mortality in sharks (fig. 12.2). Increased anthropogenic nutrient input is leading to more frequent or larger scale HABs. In 2000, the cyanobacterium *Karenia brevis*, which produces a potent neurotoxin in the *brevetoxin* family, was responsible for a

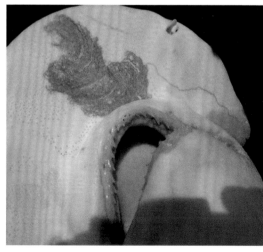

Figure 12.1. Bonnethead skin infected with nematode ectoparasites. (Courtesy of Caroline Collatos)

red tide in NW Florida that caused mass mortality in Blacktip Sharks (*Carcharhinus limbatus*) and Atlantic Sharpnose Sharks (*Rhizoprionodon terraenovae*).[3] Presumably, the toxin entered the sharks from their prey, but the sharks could have also died from the presence of anoxic water that often accompanies HABs.

In 2018, a dead 8 m (26 ft) male Whale Shark (fig. 12.2) washed ashore at Sanibel Island in Florida. Tissues tested positive for brevetoxin, but sufficient knowledge on toxic levels for this species do not exist and thus brevetoxin could not be definitively assigned as the cause of death.

Oxygen-demanding Wastes

One of the ways of assessing the health of aquatic ecosystems is by examining the amount of dissolved oxygen (DO) in the water. Water with DO of > 3 to 8 parts per million (ppm) is said to be *normoxic* and can support virtually all aquatic life. Water with less than 2 to 3 ppm is said to be *hypoxic* and typically will support only anaerobic and relatively inactive fauna. Water devoid of DO is *anoxic* and cannot support vertebrate life except for brief periods.

Oxygen-demanding wastes are those that lead to decreases in the DO level. These include sewage, paper pulp (a byproduct of manufacturing paper prod-

Figure 12.2. 8 m (26 ft) Whale Shark presumably killed by the 2018 Florida red tide. (Courtesy of the Florida Fish and Wildlife Conservation Commission)

ucts), and food processing wastes. These substances can either be decomposed by oxygen-consuming microbes (bacteria or fungi), or can be chemically oxidized, in both cases reducing DO.

Dissolved oxygen levels, particularly in nearshore environments, are known to affect the distribution of some elasmobranch fishes; for example, juvenile Bull (*Carcharhinus leucas*) and Sandbar Sharks (*C. plumbeus*) and adult Gray Smooth-hounds (*Mustelus californicus*), but not Atlantic Sharpnose (*Rhizoprionodon ter-raenovae*) and Spinner Sharks (*C. brevipinna*).[4] The lack of an observed effect on the two latter species does not indicate that they can survive in anoxic conditions, but that they may be less sensitive to deviations in dissolved oxygen levels.

Unfortunately, studies that make the connection between hypoxia caused by oxygen-demanding wastes and sharks and batoids are lacking, even though members of both groups occupy systems subject to hypoxia or anoxia caused by these wastes.

Water-soluble Inorganic Chemicals

This large category includes toxic metals, acids, bases, and salts. Examples include heavy metals like mercury, lead, tin, and cadmium, as well as selenium, arsenic, chlorine, and nitrates. These derive from the weathering of rocks accelerated by humans by mining, manufacturing processes, oil field drilling operations, and agricultural practices. A problem of increasing concern is ocean acidification due to increased CO_2 levels brought about by burning fossil fuels, deforestation, cement manufacturing, and so on.

Let us focus on mercury. The Food and Drug Administration (FDA) has cautioned pregnant women, women who may become pregnant, nursing mothers, and children on the hazard of consuming fish, including sharks, that may contain high levels of *methylmercury*, the positive ion that mercury is converted to by aquatic microbes. Predatory fish like sharks accumulate the highest levels of methylmercury and therefore pose the greatest risk. Unlike *lipophilic* (fat-loving) substances like PCBs and dioxin (see below), which accumulate in fat, mercury accumulates in muscle tissue (the part of the fish we eat), so it cannot be avoided by removing fat and skin of fish.

Since industrialization, mercury in the environment has increased by 2–4 times. Sources include burning coal, burning medical and other waste that contain mercury, and gold and mercury mining.

Mercury causes changes in the function of cells, inhibits protein synthesis, and causes formation of reactive oxygen compounds or radicals, which can damage DNA and disrupt cell division. Mercury interferes with the development of the nervous system and affects brain function.

Methylmercury is rapidly taken up but only slowly eliminated from the body, so each step up in the food chain biomagnifies the concentration from the step below. Generally, sharks whose diets include larger predatory fish and mammals have higher levels of mercury and other metals, but this varies by species and metal. Sandbar Sharks, which are bottom associated in many ecosystems, show higher mercury levels because they are found near sediments containing methylmercury and other contaminants.[5]

Two recent studies[6,7] examined methylmercury levels in 11 species of sharks in a Florida lagoon. As might be expected, there was significant variability in methylmercury levels among the species. In the Blacknose (*Carcharhinus acronotus*), Blacktip, Shortfin Mako (*Isurus oxyrinchus*), Blue (*Prionace glauca*), and Oceanic Whitetip (*C. longimanus*) Sharks, the methylmercury levels increased with body size and were at levels that could be harmful to these sharks. As meaningful as are the reports of high levels of mercury in sharks, not all studies of methylmercury levels in sharks have found high levels.[8]

Plant Nutrients

Plant nutrients are water-soluble inorganic nitrate and phosphate salts. Sources include sewage, runoff from livestock feedlots, fertilized fields and lawns, and industrial wastes. The result of nutrient pollution ranges from plankton blooms to dead zones (e.g., large tracts of ocean with little or no dissolved oxygen).

While nutrients are essential for the growth of marine plants (especially microscopic forms), too many nutrients, a condition known as *eutrophication,* can lead to blooms, which can deoxygenate water when the phytoplankton die and decompose, using oxygen in the process. Hypoxic (low oxygen) or anoxic (no oxygen) conditions can result and any aerobic organism in this area must either leave or potentially die.

Over the last several decades, these hypoxic events have increased. The marked increase in human population in coastal regions worldwide has contributed to greater nutrient and organic matter inputs into coastal watersheds and thus these events.

The Gulf of Mexico Dead Zone covers an area within about 30 m (98 ft) of the bottom as extensive as 20,000 km² (9000 mi²), about the size of New Jersey, seasonally, extending from the mouth of the Mississippi River. It is due to nutrient enrichment from agriculture in the Mississippi River watershed.

When they encounter areas of low dissolved oxygen, sharks tend to emigrate rather than adapt or succumb to the conditions, if the option exists. Anecdotal accounts of increases in sharks near beaches following nearby hypoxic events support their moving *en masse* from areas depleted of oxygen.

On the other hand, dead sharks have been reported among fish kills associated with red tides. In some cases, toxins associated with the fish kills were responsible for the shark mortality, but in others whether the cause was a toxin or hypoxia/anoxia was not determined.

Sharks may also take advantage of kills of bony fishes caused by plankton blooms; for example, in the 1971 red tide in Tampa Bay, Florida, hundreds of sharks of several species were seen *gorging themselves on dead, decomposing, or dying fishes.*[9]

Organic Chemicals

This category includes a wide variety of both natural and synthetic chemicals such as oil, gas, solvents, pesticides, and even pharmaceuticals. Significant sources are improper disposal of industrial, household, and medical waste, and pesticide runoff from farms, forests, roadsides, and golf courses. We will focus on oil spills, the persistent organic pollutants (POPs) dichloro-diphenyl-trichloroethane (DDT) and polychlorinated biphenyls (PCBs), and some pharmaceuticals.

Oil

Oil receives perhaps the most attention in this category. Crude oil is a complex mixture of hydrocarbons. Crude oil is sticky, insoluble, and floats on water, although some components may sink or evaporate.

Polycyclic aromatic hydrocarbons (PAHs) are considered the most toxic components of crude oil and are the common focus of studies on the ecotoxicology of marine oil spills. Sharks, like most fishes, can metabolize PAHs taken in with contaminated food and are capable of ridding themselves of these contaminants through the liver. However, episodic or prolonged PAH exposures that exceed the body's detoxification capabilities can result in cellular damage, mutagenic effects, reproductive impairment, and even death.

Oil exposure on sharks can be estimated by measuring concentrations of PAHs in the shark's tissues and by studying how hard the body is working to rid itself of these toxins. For example, exposure to PAHs in sharks has been studied by examining the activity levels of enzymes in the liver that detoxify it and measuring concentrations of PAH metabolites in the bile (i.e., the byproducts associated with metabolizing the toxins). Research done by co-author Grubbs and colleagues has been conducted on the 2010 Deepwater Horizon oil spill in the Gulf of Mexico. In this study, liver detoxification enzymes in the Little Gulper Shark continued to increase long after the spill was stopped, peaking three years later. Research into the long-term effects of this, the largest oil spill in US history, is ongoing.

Persistent Organic Pollutants (POPs)

Since the early 20th century, humans have been synthesizing chemicals not naturally found on earth, chemicals that bacteria may be unable to render harmless. Concentrations of many of these chemicals have reached levels in the oceans that have begun to threaten the survival of many species of phytoplankton, invertebrates, large mammals, and fish. These chemicals are persistent because they tend to be difficult for organisms to decompose. Moreover, even if organisms can break down the chemicals, these decomposition products are often toxic as well. DDE (dichloro-diphenyl-dichloroethylene), a breakdown product from DDT, is one notorious example.

POPs, or persistent organic pollutants, include a bewildering array of synthetic organic compounds that persist in the marine environment, are easily distributed globally, and lodge and become concentrated in the fatty tissues of animals by bioaccumulation and biomagnification. Many, including DDT, are pesticides, part of the "Green Revolution" that radically increased food production during the 20th century. Dioxins and furans are industrial byproducts of combustion. Polychlorinated biphenyls (PCBs) were used for years in electrical insulation. The manufacture and use of all POPs (except for furans and dioxin, which are combustion by-products) have for years been banned in the United States.

POPs are of global concern because there is firm evidence of *global* transport of these substances, by air and water, to regions where they have never been used or produced, such as the North American Arctic.

They cause a range of harmful effects among humans and animals, including cancer, birth defects, damage to the nervous system, reproductive disorders, disruption of the immune system, and even death. POPs can damage the reproductive and immune systems of exposed individuals as well as their offspring. Some POPs are *endocrine disrupters* (or endocrine-active chemicals). An endocrine disruptor is a chemical that interferes with the function of the endocrine system. It can mimic a hormone, block the effects of a hormone, or stimulate hormone production. Endocrine disrupters have been known to cause feminization of males, masculinization of females, and fertility problems in fishes.

Let us consider PCBs first. PCBs are a group of over 200 synthetic chemicals called *organochlorines*, which are complex compounds made of one or more chlorine (Cl) atoms attached to carbon-based structures such as chains and rings.

PCBs do not exist naturally on earth—they were synthesized during the late 19th century. By 1977, the manufacture of PCBs was banned in the US, the UK, and elsewhere. However, they are still being used in many developing countries. By 1992, scientists estimated that 1.2 million metric tons (264,500,000 lb) of PCB existed worldwide, while scientists estimate that 370,000 metric tons (810 million lb) have been dispersed globally, much of it into the oceans.

PCBs are lipophilic and thus tend to accumulate in the fatty tissues of animals. They enter food chains and webs mainly through the feeding of organisms called *sediment* or deposit feeders. PCBs move up the trophic tiers and become further concentrated.

PCBs have become widespread and are serious pollutants that have contaminated many marine trophic tiers. They are extremely resistant to breakdown (which of course was one of their virtues) and are known to be carcinogenic and probably mutagenic as well. In terms of their specific effect on life, PCBs have been shown to cause liver cancer and harmful genetic mutations in animals. PCBs have been linked to mass mortalities of striped dolphins in the Mediterranean, to declines in orca populations in Puget Sound, and to declines of seal populations in the Baltic.

Although PCBs may threaten the entire ocean, the Northwest Atlantic is believed to be the largest PCB reservoir in the world because of the volume of PCBs produced in countries that border the region.

PCBs have been detected in a number of sharks and batoids at levels across the spectrum of those found in other marine organisms. PCBs have also been shown to transfer from embryo to embryo in some species. Virtually all of these studies report on the occurrence and concentration of PCBs in the liver and other tissues, but none have examined pathologies associated with PCBs in sharks and batoids.

Interpreting the meaning of the studies of PCBs in sharks and batoids is difficult, in part because there are numerous PCB compounds. Moreover, determining threshold levels for all of these compounds at which measurable damage is done is all but impossible because of interspecific differences in rates of accumulation and synergistic effects with each other and with other chemical compounds. In some species, males and females accumulate PCBs at different rates. Clearly, more work on the effects of PCBs on sharks and batoids would provide multiple conservation benefits.

DICHLORO-DIPHENYL-TRICHLOROETHANE (DDT)

DDT, whose production as one of the first synthetic pesticides dates back to the 1940s, is another persistent organochlorine and is similar in structure to PCBs. It was used effectively in insect control in agricultural settings as well as to kill insect vectors of malaria and other diseases, but at the same time had wide-ranging impacts on wildlife, especially birds. DDT is banned in the US and most other countries but its use to control mosquitoes that carry malarial parasites is still allowed in some countries because of its effectiveness and the severity of the problem.

Despite decreased production and use, some of the 1.5 Tg (teragrams; 3.3 billion lb) of DDT used or dumped since it was first produced still exists in the ocean, primarily in sediments and as a result of export to the deep sea. DDT persists longer in seawater than in soil or air.

Thus, it is not surprising that DDT still shows up in shark tissue. Let us consider two cases. In 2013, a 3.7 m (12.1 ft), 600 kg (1320 lb) female Shortfin Mako (*Isurus oxyrinchus*) was caught on hook-and-line (apparently for a TV show) near Huntington Beach, California and was donated to scientists.[10] The liver was analyzed for organic contaminants, and DDT was found to comprise 86% (an estimated 11.4 g, or 0.5 oz) of the total contaminant load. The DDT concentration was nearly 100 times greater than the no-consumption level set for humans by the US EPA. The researchers who measured the DDT as well as PCB levels noted that they were consistent with the shark feeding near a POP-contaminated area (i.e., the Palos Verdes Shelf Superfund site off the coast of Los Angeles County). Among the Mako's stomach contents was the remains of a juvenile California Sea Lion (*Zalophus californianus*) estimated to weigh about 67.6 kg (149 lb) before being consumed.

The second case involves DDT levels in the Greenland Shark (*Somniosus microcephalus*), a top predator found in cold waters of the North Atlantic and Arctic.[11] DDT concentrations in a sample of 15 Greenland Sharks were higher than those of other studies. The authors of the study speculated that the high DDT levels could be attributed to *continuous and chronic exposure* to DDT compounds, but other chemical indicators suggested an old contamination event. Although some of the results were conflicting, recent discoveries that this species is capable of living as long as 400 years could easily mean that the DDT was acquired between the 1940s and 1970s, and that the lower metabolism of this species translated into slow depuration (declines due to metabolism or other loss).

Pharmaceuticals

The last category of organic chemicals we discuss at one time would have been considered a frivolous line of inquiry but now constitutes a rapidly worsening problem. We refer to the exposure to and hazards of human, veterinary, and agricultural pharmaceuticals and personal care products that find their way into marine and coastal environments.

The list of pharmaceuticals in the marine environment is a veritable pharmacy checklist, and includes analgesics, antibiotics, antidepressants, antihistamines, antianxiety drugs, estrogens, contraceptives, bronchodilators, decongestants, nonsteroidal anti-inflammatory drugs (NSAIDs), and so on.

Personal care and cleaning products include ingredients found in shampoos, sunscreens, fragrances, and cosmetics. These pharmaceutical and personal care product chemicals arrive in coastal and ocean waters and sediments through sewage and landfills, and, for pharmaceuticals only, from aquaculture, feedlots, and agriculture.

Once in marine waters or sediments, the ingredients in the pharmaceuticals and personal care products may enter the food chain, either as the parent product or after being degraded.

Initially, it was thought that the diluting effect of the large volumes of water into which the pharmaceutically active compounds and personal care products were discharged would result in negligibly low levels in the water column and sediments and would amount to little biological significance. And initially there was some truth to this, until chemicals from pharmaceuticals and personal care products began to be detected in marine organisms.

More important than the dilution factor are the rates of input, residence time, and toxicity. An important aspect of pharmaceuticals is that they specifically *target* biological functions, as well as possessing a suite of so-called *side effects*, often at low concentrations. Also, long-term exposure to many of the chemicals is likely more harmful than acute impacts. Finally, lipophilic substances may bioconcentrate and biomagnify more so than water soluble ones.

Among marine organisms, the highest concentrations of pharmaceutically active compounds and those derived from personal care products have been found in molluscs, fishes, and mammals.[12] Studies on sharks have been more limited, and these have focused more so on detecting and quantifying a subset of these classes of chemical compounds in sharks and not on their impacts on their health and behavior.

Synthetic polycyclic musks, a group of compounds that are mainstays in the fragrance industry, have been detected in the livers of Scalloped Hammerheads (*Sphyrna lewini*).[13] Parabens, preservatives in pharmaceuticals, food, and personal care products, have been found in Atlantic Sharpnose, Bonnethead (*Sphyrna tiburo*), Shortfin Mako, Spinner, Bull, Blacktip, and Great Hammerhead Sharks (*S. mokarran*), as well as Atlantic Stingray (*Hypanus sabina*),[14] although the highest levels in the study were found in bony fishes.

Finally, researchers[15] examined levels of 10 pharmaceutically active compounds (one synthetic estrogen contraceptive [e.g., Ortho Evra], impotence agent [Viagra], lipid-lowering drug [Lipitor], and eight antidepressants [including, but not all, Paxil, Effexor, Prozac, Zoloft, and Celexa]) in neonate Bull Sharks in pristine and wastewater-impacted systems in Florida. They detected the estrogen and six of the antidepressants in sharks in the wastewater-impacted system, but only one antidepressant in a single shark in the pristine system. The authors concluded that the compounds detected in the Bull Sharks "were generally low in concentration...and not expected to pose acute health risks to these individuals." Whether these Bull Sharks were merely mellower is not known.

Sediments

Pollution by suspended sediment is a growing problem. Fine soil particles (silt and clay) along with organic and inorganic chemicals that originate as runoff from construction sites, logged forests, mining operations, and croplands block sunlight, foul gills of fishes and invertebrates, and can even smother bottom communities (e.g., seagrass beds, oyster reefs, and coral reefs).

Dredging channels to promote port operations enhances the transport of salty oceanic water into estuaries, in the process changing the salinity structure, circulation, flushing, and water residence times. Dredging also resuspends sediments into the water column and may also contaminate the water column with toxic compounds from the substrate as well as adding nutrients and organic matter. Such changes can have serious effects on the health of organisms as well as biological productivity and ecosystem function.

Most studies of the impacts of sediments on marine organisms have focused on channel dredging and beach augmentation (euphemistically better known as *renourishment*). Studies specifically focusing on sharks have been more limited.

Turbidity refers to the relative clarity of water and is one of the factors that influences the distribution of marine organisms. Relatively high turbidity (i.e., murkiness) is a normal persistent or episodic characteristic of ecosystems where elasmobranchs are found (e.g., along the coasts of Virginia, North Carolina, South Carolina, and Georgia in the United States. Additionally, naturally occurring high turbidity occurs seasonally or episodically at river mouths.

Dredging to deepen and widen ship channels or to mine sand for beach augmentation resuspends sediments and can result in large sediment plumes, the duration and extent of which are determined by a number of factors (e.g., type of dredge, sediment characteristics, depth, and current).

According to the US Army Corps of Engineers, more than 400 ports and 25,000 miles of navigation channels in the United States are periodically dredged. Many of these projects are controversial because of their environmental impact and cost. For example, the Port Miami Deep Dredge Project, completed in 2015, deepened the Government Cut seaport channel from 12.8 to 15.8 m (42 to 52 ft) to accommodate larger ships with deeper drafts. In doing so, sensitive coral reefs near the dredging site were damaged.

There are two major questions with respect to sediment pollution and sharks: Does sediment pollution associated with dredging and other projects affect sharks and batoids and, if so, in what ways?

Turbidity is known to influence the presence or absence of sharks. In one study of estuaries in the US state of Georgia,[16] subadult Bonnetheads were found in areas of lower turbidity than juvenile Sandbar Sharks, although other factors (e.g., salinity, depth, dissolved oxygen) also influenced their presence. In the same study, turbidity did not affect the presence of Blacktip and Atlantic Sharpnose Sharks. The study was not designed to assess human impacts and the authors concluded that these results might represent habitat preferences for these species.

In another study, immature Scalloped Hammerheads were shown to prefer habitats with higher turbidity, perhaps because as a mechanism to deter visually orienting predators.[17]

In a third study, turbidity was one of the factors, although not the most important, that predicted the presence or absence of three sharks (Bull, Bonnethead, and Blacktips) along the Gulf of Mexico coastline. Again, this study did not address anthropogenic sedimentation. Studying the direct effects of turbidity is difficult, as there are many covariates that also affect spatiotemporal ecology.

The most comprehensive studies of the impacts of dredging associated with development were conducted by scientists at the Bimini Biological Field Station (BBFS) and others beginning in 1997, when a large resort complex began construction on North Bimini (fig. 12.7). Initial plans called for a 930-room hotel, 3000 m² (32,300 ft²) casino, 18-hole golf course, and two marinas on the small Bahamian island. Dredging began in 1999 and by 2006 about 750,000 m³ (26,500,000 ft³) of sediment had been excavated.

The BBFS was particularly poised to study the impacts of this development because they have been studying the ecology of the system since the early 1990s in order to understand the biology and ecology of Lemon Sharks, which utilize parts of the impacted area as a nursery ground. One study[18] found that the survival rate of juvenile Lemon Sharks in the system and the coverage of seagrass (habitat for their prey) both declined after dredging. Studies of impacts of the development on the biodiversity of the impacted area, as well as impacts on Lemon Sharks, are continuing. We further discuss Lemon Sharks in Bimini below.

Another major case of sediment pollution is that which occurs at the Abbot Point coal loading port in Queensland, Australia, from which as much as 50 million tonnes (110 billion lb) of coal are exported annually. The terminal is 20 km (12 mi) from seagrass beds and is 40 km (24 mi) from the southern section of the Great Barrier Reef, an ecosystem already in poor health as a result of warming oceans, landscape changes causing runoff, predatory starfish, pesticides, and other stressors.

Sediment pollution associated with the port comes from a variety of sources, including dredging, dumping of sediment spoils, and fugitive coal dust (the dust that escapes as coal is moved to the terminal and loaded on to ships). Again, there have been no studies that focus on sharks, but numerous studies have documented the impacts of sediment on coral reef ecosystems, and it is reasonable to hypothesize that these might affect the health, behavior, distribution, and/or abundance of sharks at impacted areas.

Bull Sharks appear to thrive in areas altered by dredging (e.g., in marinas and outfalls from power plants),[19] although the impacts of sediment pollution on their presence has not been examined.

Radioactive Substances

Radioactive substances are produced by the nuclear power industry, the military, electrical utilities, mining industries, and as a result of medical diagnostics and treatment. In the 1960s and 1970s, radioactive waste buried in shallow trenches leached into surrounding soil and groundwater. Modern disposal techniques have been engineered to minimize the potential for leakage. There is also natural background radiation in the ocean.

The meltdowns and use of seawater to cool reactor cores at the Fukushima Daiichi Nuclear Power Plant after the 2011 tsunami led to renewed public interest in radiation pollution. Shark species that foraged near the site of the Fukushima Daiichi plant included Mako, Blue, and Salmon (*Lamna ditropis*) Sharks, high trophic-level predators that might accumulate and transport radioactivity. Pacific Bluefin Tuna (*Thunnus orientalis*) were shown to transport radionuclides from Fukushima across the entire North Pacific Ocean.[20]

Although the amount of radiation, primarily cesium-134 and cesium-137, in the seawater was several orders of magnitude over safe levels after the event, the half-life of the radioactivity was short and the radiation dispersed quickly, minimizing concerns. Cesium is excreted relatively quickly, and biomagnification is minimal, so concerns about the effects of these radionuclides on marine organisms were minimal.

Between 1946 and 1958, the United States detonated 23 nuclear devices on and near the Bikini Atoll in the Marshall Islands in the equatorial Pacific Ocean. Despite rumors of sharks missing their second dorsal fins, recent assessments of the biota of the atoll report a resilient, highly diverse, healthy ecosystem, and no validation of the shark rumor.

Heat

Heat as a form of water pollution is principally produced from the cooling of power plants and industrial processes, although anthropogenic climate change also heats bodies of water. Most threatened are tropical organisms (e.g., corals), which live closer to their upper thermal limit (CTmax), and thus the ecosystems in which the tropical marine organisms live (see Chapter 8).

Studies of the effects of thermal effluent on sharks are lacking. Recall that often the first reaction to environmental stressors is behavioral avoidance, and this likely applies to sharks and batoids in these areas. Notable exceptions occur when there are advantages to warmer waters (e.g., abundant prey or refuge from temperature minima). An example of the latter is that Nurse Sharks (*Ginglymostoma cirratum*) were shown to aggregate at the thermal effluent from the Turkey Point power plant in the US state of Florida.[21] Sharks were also shown to gather at the thermal plume of the Kahe outfall on Oahu, Hawaii.[22]

Plastics

Figure 12.3 shows a Sandbar Shark that we caught on an experimental longline in 2016 in Winyah Bay, South Carolina with a plastic packaging strap fully encircling the shark near the gills. This particular shark likely survived after we removed the strap, since it appeared otherwise healthy, was actively feeding (since it took the bait on our longline), and swam off strongly once released, but many sharks and other marine life do not survive similar entanglements.

Plastic is so commonplace that we need not devote much space to describing its uses. We are in what has been called the *Age of Plastics*. Global production of plastic is more than 335 million metric tons (660 billion lb) per year, and during the 10-year period ending in 2017, an estimated 2.6 billion tonnes (5.7 trillion lb) of plastics were manufactured globally.[23] By 2017, more than 8.3 billion tonnes (18.3 trillion lb) of virgin plastic had been produced, according to one scientific assessment,[24] which also reported that since 2015, 9% of the plastic produced had been recycled, 12% incinerated, and 79% wound up in landfills or the environment.[25] Globally, between 60–80% of all litter is plastic.[26]

How much plastic is in the marine environment? Estimates of the amount of plastic that finds its way into the marine environment annually range from 1.8% to 10% of annual global plastic production. There are two broad categories of plastics in the marine environment, *macroplastics* and *microplastics* (fig. 12.4). The former category includes a wide variety of plastics ranging from beverage bottles and packaging to abandoned commercial fishing gear, some of which may fragment or degrade to smaller pieces. Microplastics are < 5 to 10 mm (0.2 to 0.4 in) in size and may be produced as degradation products of larger plastics in the marine environment or may be manufactured as microplastics (e.g., microbeads in cosmetic and personal care products).

Once in the marine environment, plastics have been shown to cause a suite

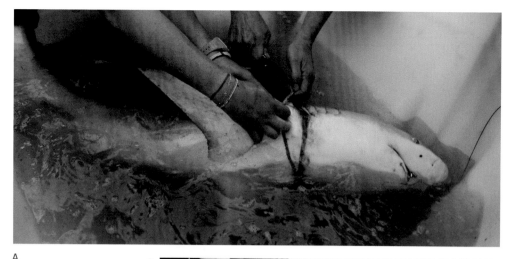

A

B

Figure 12.3. (A) Sandbar Shark with plastic strap encircling body, caught on an experimental longline. (B) Releasing the shark, which swam away strongly. (Courtesy of Caroline Collatos)

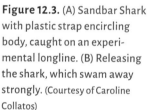

of problems at the organismal and ecosystem level, although much remains to be understood about these impacts. Let us focus on sharks. Broadly, problems are associated with either entanglement or ingestion.

Entanglement of sharks and batoids in plastics, not including bycatch in monitored nets, has been largely reported anecdotally (like our example above), especially from abandoned drift nets.[27] However, the global extent and impact of entanglement on local populations or ecosystems is difficult to assess.

Ingestion of macro- and especially microplastics has been studied more rigorously than entanglement and is a major line of inquiry as we write this book. Direct ingestion of large plastic by sharks is rare. A 2018 review article[28] cites the personal comments of a shark biologist who found that less than 1% of shark stomachs contained ingested plastics, which included PVC fragments, and bottle and pen caps. As of late 2018, sharks found with ingested microplastics include Blackmouth Catshark (*Galeus melastomus*),[29] Blue Sharks,[30] Portuguese Dogfish (*Centroscymnus coelolepis*),[31] Velvet Belly Lantern Shark (*Etmopterus spinax*),[32]

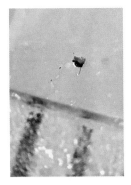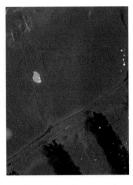

Figure 12.4. Microplastic fibers and fragments found in the gut of Atlantic Sharpnose Sharks. Scale represents 1 mm increments.

Spiny Dogfish (*Squalus acanthias*),[33] Longnose Spurdog (*Squalus blainville*),[34] Brazilian Sharpnose Shark (*Rhizoprionodon lalandii*),[35] and Atlantic Sharpnose Shark.[36] Additionally, Whale Shark (*Rhincodon typus*) skin biopsies showed chemical evidence of microplastic ingestion.[37]

Microplastic toxicity may be due either to adsorption of nonpolar chemical pollutants to the surface of the microplastics (e.g., PCBs, DDE, and DDT), or harmful chemicals in the plastics (e.g., flame retardants and plasticizers). However, studies on the effects of toxic microplastics on sharks and batoids are lacking.

Additionally, although the subject has not yet been thoroughly studied, evidence suggests that plastics are capable of moving up the food chain.

Given the pervasiveness of plastics in the marine environment, the continued growth in their production and use, the absence of meaningful recycling, the continued enumeration of macro- and microplastics in sharks and other marine organisms, it is not farfetched to assert confidently that research in the area will expand considerably.

Noise

Anthropogenic noise in the marine environment is a significant problem to marine mammals, but does background noise from ships and boats, dock and other construction, air guns used in seismic exploration, windmills, and sonar affect sharks?

Unfortunately, studies on the impacts of noise on sharks are lacking. A 2019 study[38] on the behavioral responses of reef sharks to underwater sounds concluded that the behavior of these sharks was altered by relatively low sound levels, but that a *critical need* exists for more studies.

In the absence of such tests, the next best sources of information are comparing what is known about hearing in sharks with the range of marine anthropogenic noises to assess the likelihood of damage, and consulting studies of noise and bony fishes.

Intense sounds have the ability to burst air-filled spaces, like swim bladders in fish, which can lead to hearing impairment or loss. Recall, however, that sharks lack swim bladders. Intense sounds have the potential to damage the hair cells of the lateral line and inner ear, usually causing temporary hearing threshold shifts (i.e., altering the intensity and frequency of sounds detected).[39]

Noise associated with underwater construction could lead to temporary decreases in hearing sensitivity, and high intensity noise when a pile is hit by the hammer could cause tissue damage in the inner ears and other organs of bony fishes and elasmobranchs.[40] Vibrations associated with pile driving and conveyed through the substrate might have greater impacts on benthic sharks and rays. Shipping noise is not considered dangerous to sharks, except perhaps by masking other sounds.[41]

Noises that do not cause short-term or permanent damage might affect sharks and batoids in other ways (e.g., by causing a startle or other avoidance response), affecting foraging or movement, or impacting prey or other organisms in the ecosystem.

One area of particular concern is noise from towed arrays of air guns used in seismic testing for oil and natural gas. These pulsed, high intensity, low frequency sounds are known to impact marine organisms from zooplankton to mammals. Among fishes, sharks and batoids are the most sensitive to low frequency sound and thus might be most susceptible to air gun blasts.

Physical Habitat Destruction

Sharks and batoids are found in a variety of marine ecosystems, including estuaries, salt marshes, intertidal zones (rocky, sandy, muddy), shallow subtidal benthic environments (kelp beds, seagrass communities), deep-sea, open ocean, and tropical ecosystems like coral reefs and mangrove swamps. None of these has escaped anthropogenic impact, and studies to more fully understand how these changes have specifically impacted sharks are lacking. We will examine one of the most biologically productive ecosystems on the planet, mangroves, which play important roles in the life histories of numerous sharks and batoids.

Mangroves

Mangroves are a broad group of tropical, salt-tolerant trees that grow at the water's edge between 32°N and 38°S latitude. Five countries—Indonesia, Australia, Brazil, Nigeria, and Mexico—are home to almost half of the entire global area of mangrove ecosystems.

Stilt roots of mangroves at the water's edge form complex networks, trapping sediment and preventing its transport further seaward. They also provide a myriad of hiding places for a great diversity of marine life, especially the juvenile and larval forms of fish, crustaceans, and mollusks, any of which constitute food for elasmobranchs.

Mangrove communities flourish mainly along the shores of estuaries, along coasts, and at the mouths of large rivers like the Irrawaddy in Myanmar. These locations mean that mangrove communities tend to be at risk from development, especially in countries with few if any land-use controls. Mangrove environments are readily cleared for aquaculture, resort developments, and housing. They are also threatened globally by rapid sea-level rise and locally by agricultural runoff, oil, deforestation for biomass fuel, and gas exploration and production, the latter for example, along the Persian Gulf and the coasts of Australia and Nigeria.

The southwest coast of Florida contains the most extensive mangrove forests in the continental United States, and yet even there, mangrove forests continue to decline because of development.

Aquaculture development along tropical coasts that converts mangroves to fish or shrimp farms directly threatens the survival of mangrove forests. Industrial shrimp farming along the northeast coast of Brazil, for example, puts at risk much of that nation's one million surviving hectares (2.47 million ac) of mangroves, about 7% of the global total.

In addition to serving as habitat for sharks and batoids (e.g., the endangered Smalltooth Sawfish, *Pristis pectinata* [fig. 12.5], and juvenile Giant Shovelnose Ray, *Rhinobatos typus*), mangroves and the waters they fringe serve a suite of critical ecosystem, social, and economic services. Mangrove forests, which are exclusively tropical, cover less than 14 million ha (34.6 million ac) globally, representing from 50% to 65% of historical levels. Additionally, an undetermined

Figure 12.5. Smalltooth Sawfish in shallow water adjacent to mangroves in the Bahamas.

Figure 12.6. Juvenile Lemon Shark on flats adjacent to mangroves in Bimini, Bahamas. (Courtesy of Annie Guttridge)

percentage of remaining mangroves has been degraded by pollution. Losses have slowed in some areas as governments and residents recognize the value of mangroves to commercial, artisanal, and subsistence fishing, and as buffers against storms and tsunami.

A complete list of the sharks and batoids that reside in mangrove ecosystems as neonates, juveniles, and/or adults has not been published, although there are numerous publications on specific species, regional checklists of species, and specific ecosystems.

One particularly well-researched species is the Lemon Shark of Bimini, Bahamas (fig. 12.6). Understanding the biology of the Lemon Shark was a major focus of noted shark biologist Dr. Samuel Gruber of the Bimini Biological Field Station. Research conducted by Gruber and his coworkers over approximately two decades has shown that mangroves and the lagoon fringed by the mangroves are critical to the growth and survival of juvenile Lemon Sharks during their first several years of life, after which they move to more nearshore or coastal habitats. Young Lemon Sharks take advantage of the protection from predation offered by the mangroves (one of their main predators is larger Lemon Sharks), as well as the abundant food supply in the lagoon.

A 2008 study[42] concluded that the development of a resort in Bimini, Bahamas (fig. 12.7; see above) that included removal of mangroves significantly damaged the nursery grounds of Lemon Sharks. By August 2010, more dredging had occurred and about 67 ha (166 ac) representing 39% of the mangrove habitat surrounding the system had been removed[43] (fig. 12.7). This habitat was one of the most important Lemon Shark nurseries in the northwest Bahamas, which also serves to recruit adult Lemon Sharks to Southeastern US habitats.

After the development started, survival rates of Lemon Sharks decreased, growth rates decreased, and sharks remaining in the area were less healthy than comparable sharks in undisturbed areas.

Finally, mangrove and coral reef communities (see Climate Change below) are sometimes cited as examples of *linked habitats* because of their close association. Mangroves are often found landward near coral reefs. Although mangroves require quiet waters containing some sediment in which to proliferate, coral reefs require clear waters free of fine sediment that can smother the corals and kill the reef. The communities are also complimentary in that coral reefs protect mangroves from wave erosion and storm damage, whereas mangroves provide shelter for juvenile stages of numerous reef animals. The removal of mangroves often leads to the degradation or even disappearance of the reef.

Sharks and batoids face similar problems in salt marshes, which typically replace mangroves in protected intertidal areas along temperate coasts and play similar ecological roles as mangroves. Similarly, they are degraded by a suite of physical human activities (e.g., construction of docks, bulkheads, dikes, and so on), as well as long-term impacts associated with their presence (e.g., altering the hydrology of the system leading to changing erosion and deposition of sediments, relocation of channels, different salinity regimes, changing vegetation patterns, and dredging). Boat traffic can lead to erosion of banks. The pace of so-called *land reclamation*, more appropriately draining and filling, as well as removal for dock construction and so on, has slowed, but the process still occurs, albeit on a small scale in the United States. By the early 1990s, an estimated 50% of US salt marshes were lost principally due to draining and filling for agriculture, industry, and residential uses. Over 90% of California's coastal wetlands have been lost. Historical global losses are estimated to be between 25% and 50%. Estimates of remaining salt marshes globally range from 2.2 to 40 million ha (54 to 98 million ac), and a recent effort mapped about 5.5 million ha (13.6 million ac) in 43 countries.[44]

Climate Change and Sharks

There is now complete and unambiguous agreement among scientists that human activity is causing global climate change. Scientific uncertainty remains on issues that include the time frame and severity of climate change impacts, as well as the cost, impact, and effectiveness of mitigation and/or adaptation strategies.

December 2009

May 2010

A

B

Figure 12.7. (A) Photo from North Bimini, Bahamas taken before and after substantial development for the Bimini Bay Resort showing loss of mangroves. (Courtesy of Kristine Stump) (B) Low altitude aerial view of the extent of the mangrove destruction. (Courtesy of Kristine Stump)

What are some of the impacts of climate change on sharks? Let us start with acidification. The scale of impacts of ocean acidification has the potential to be enormous, causing changes in carbonate chemistry that have not been seen in 65 million years, and will affect the vitality and survival of virtually all taxonomic groups. A 2018 paper[45] discusses that the impacts of acidification on sharks may be direct (i.e., affecting the physiology [by acidifying blood and tissues] and behavior of sharks) or indirect (e.g., changing the community structure or prey). In their analysis of other studies, the paper notes that separating acidification from temperature increases may produce conflicting and confounding results.

As of 2018, physiological effects of relevant elevated CO_2 levels had been studied in only four species—Small-spotted Catshark (*Scyliorhinus canicula*), Port Jackson Shark (*Heterodontus portusjacksoni*), Brownbanded Bamboo Shark (*Cephaloscyllium punctatum*), and Epaulette Shark (*Hemiscyllium ocellatum*). These are all benthic species unrepresentative of sharks as a group.

One area of interest is the impacts of acidification on embryonic development. On the one hand, exposure to elevated CO_2 levels in the Port Jackson, Bamboo, and Epaulette Shark did not affect their embryonic development (e.g., their survival, development time, and growth rates). On the other, in the Little Skate (*Leucoraja erinacea*), high CO_2 levels intensified the effects of elevated temperature on embryonic development.

Among the physiological responses to elevated CO_2 levels in Small-spotted Catshark were changes in resting metabolic rate, scope for aerobic activity (an index of maximum aerobic metabolic performance; e.g., faster swimming), and increased bicarbonate and sodium levels. Growth rate was not affected. A suite of similar physiological responses was found in the Epaulette Shark, and included changes in resting oxygen consumption, acid-base balance, hypoxia tolerance, aerobic enzyme activity, and other hematology variables. Because Epaulette Sharks live in shallow reef habitats subject to swings in CO_2 levels, the authors of the study speculated that the species might be capable of tolerating the physiological changes associated with increased CO_2 concentration.

Both studies cited above exposed the sharks to the elevated CO_2 levels for periods of 30 to 60 days. Studies in which neonates were exposed to these conditions for > 200 days demonstrated more significant impacts (e.g., decreased condition, aerobic potential, peroxidative damage in the brain, and others).

Studies on the effects of elevated CO_2 levels on behavior of sharks have shown more direct impact, although results of elevated CO_2 and temperature together were somewhat contradictory. In one study, Port Jackson Sharks maintained in elevated CO_2 levels for > 65 days took about four times longer than those in controls to detect their prey. However, that time was reduced by one-third when the study was conducted in warmer than normal water. A follow-up study validated this finding but, at the same time, reported that elevated CO_2 levels inhibited chemical and visual sensory responses important for foraging. A study on the Dusky Smoothhounds (*Mustelus canis*) showed that high levels of CO_2 were associated with impaired ability to track prey using odors.

In contrast to the few studies of ocean acidification on sharks, there have been a large number of studies on the effects of elevated temperature on this group (e.g., as signals for movements, as well as impacts on metabolism, growth, reproduction, and foraging). However, fewer of these have focused on warming in the range associated with anthropogenic climate change.

Some additional climate change impacts that could affect sharks and rays include changes in precipitation patterns that alter the salinity structure of nearshore and oceanic systems, increased intensity and frequency of tropical storms, rising sea levels that coastal wetland communities may not be able to keep up with, and increased size and severity of dead zones. Some of these impacts may be on a large scale (e.g., changes in ocean circulation that could include a critical slowing of the Gulf Stream).

Moving to ecological impacts, many sharks and rays will likely migrate to higher latitudes or deeper water as a result of temperature increases. It is well-known that temperature (both warm and cold) is a major driver of shark movement.[46] For example, a 2018 study[47] showed the presence of juvenile Bull Sharks in North Carolina estuaries that had not been shown to be frequently used habitat, and correlated their presence with the early arrival of summer temperatures. Another study[48] projected poleward range extensions of 64 km (40 mi) per decade for two shark species near Australia for 2030 and 2070.

In moving to higher latitudes, sharks and rays may encounter systems biotically and/or abiotically foreign or novel to them. These may cause problems like changes in the demographic characteristics of shark populations, including abundance and size structure; changes in trophic structure; changes in behavior; and possibly mortalities, extirpations, and even extinctions for species for which migrations may be difficult or improbable or which may already be depleted or threatened by other stressors. These could include sharks and rays inhabiting coastal ecosystems (e.g., seagrass beds, mangroves, salt marshes, and coral reefs).

Climate change may lead to changes in fundamental large- and small-scale ecological processes (e.g., nutrient cycles), which can alter community structure, food availability, predator-prey dynamics, and so on.

To date, only the New Caledonia Catshark (*Aulohalaelurus kanakorum*) is considered threatened directly by climate change.[49] The list, however, is sure to grow, especially given the bewildering delay in or absence of meaningful action to curb greenhouse gas emissions, especially in the United States and China, as well as the contribution to climate change caused by the growth of population in India and Indonesia.

It is far beyond the scope of this book to attempt a comprehensive summary of additional ways in which climate change have already, and may in the future, impact sharks and rays, especially since they represent a rapidly moving target, with new developments and insights occurring with breakneck rapidity.

Concluding Comments

Sharks have a 400+ million-year evolutionary history, but are they capable of surviving the current human-dominated era, the Anthropocene, and the current sixth great extinction we are currently causing? A 2014 report[50] by 23 shark specialists, led by Nicholas Dulvy, systematically assessed the threat to 1041 sharks, rays, and batoids. In the paper's abstract, they state:

We estimate that one-quarter are threatened according to IUCN Red List criteria due to overfishing (targeted and incidental). Large-bodied, shallow-water species are at greatest risk and five out of the seven most threatened families are rays. Overall chondrichthyan extinction risk is substantially higher than for most other vertebrates, and only one-third of species are considered safe. Population depletion has occurred throughout the world's ice-free waters, but is particularly prevalent in the Indo-Pacific Biodiversity Triangle and Mediterranean Sea. *Improved management of fisheries and trade is urgently needed to avoid extinctions and promote population recovery* (emphasis ours).

In 1992, 1700 scientists, including the majority of living Nobel laureates in the sciences, issued the *World Scientists' Warning to Humanity*[51]:

Human beings and the natural world are on a collision course. Human activities inflict harsh and often irreversible damage on the environment and on critical resources. If not checked, many of our current practices put at serious risk the future that we wish for human society and the plant and animal kingdoms, and may so alter the living world that it will be unable to sustain life in the manner that we know. Fundamental changes are urgent if we are to avoid the collision our present course will bring about.

A human population that is still growing, and which continually practices a resource-consumptive lifestyle, is the single biggest impediment to planetary sustainability. World trade has enormous potential to foster the objectives of development but can also be the source of massive environmental degradation without multinational and international agreements. It remains to be seen if the Paris Agreement or some alternative can stem the seemingly inexorable rise of greenhouse gas emissions. Major failures in the United States are top-down governmental intransigence and the lack of action on the part of the US Environmental Protection Agency to regulate greenhouse gas emissions. The system clearly is broken.

But there are also hopeful signs, the resilience of nature, the human spirit, effective management of fisheries, signs of recovery of some shark and even critically endangered sawfish populations, and you, whom we suspect are reading this book because of your interest and concern for the fate of sharks.

While we all may put at least some of our faith in the ability of larger institutions to begin to turn the supertanker of human insults to the planet in general, and sharks and rays in particular, in the interim or even as an alternative, there are many actions that you can take.

First, you should get out into nature and, conversely, away from the addicting screens of your phone and laptop. Richard Louv, in his groundbreaking book, *Last Child in the Woods*, asserted that a new disorder, Nature Deficit Disorder (NDD), has arisen in an entire generation as a result of replacing outdoor experiences with staring at screens. NDD leads to a lack of interest and concern about environmental issues, and addiction to entertainment and electronics. Sound scholarship demonstrates that time spent on watching screens leads to shorter attention spans, diminished vocabularies, and impaired cognitive skills, and reinforces overconsumption, a key environmental problem. Children and young adults can name more corporate logos than sharks, as well as the trees and birds they see every day. They exhibit brand-loyalty instead of planet allegiance. To value nature, you must first experience it. Seeing a shark swimming in its environment can be a life-transforming experience, one that cannot be replicated on a screen.

Other actions are more intuitive. Actively participate in a movement to make the planet more sustainable.[52] If possible, reduce your ecological footprint by eating responsibly, being a more responsible consumer, and conserving energy. Reduce the amount of, or better yet, eliminate, meat, dairy, and eggs in your diet, and buy local and, if possible, organic vegetables or grow your own. Avoid processed food in wasteful packaging. Educate yourselves about the social and environmental impacts of what you buy—and buy less. Use durable goods instead of disposables, skip the next computer and smartphone upgrades, and use mass transit.

Our last advice: make conservation biology and education a career, or at least a central part of your life. Remember that humans and sharks are both stakeholders in achieving and sustaining a livable planet.

NOTES

1. Retallack, H. et al. 2018. J. Wildlife Dis. 55: 375–386.
2. Borucinska, J.D. and Frasca Jr., S. 2002. J. Fish Dis. 25: 367–370.
3. Flewelling, L.J. et al. 2010. Mar. Biol. 157: 1937–1953.
4. Drymon, J.M. et al. 2013. Fish. Bull. 111: 370–380.
5. Clark, R.B. 2001. Marine Pollution. (5th ed.). Oxford University Press.
6. Matulik, A.G. et al. 2017. Mar. Pollut. Bull. 116: 357–364.
7. Kiszka, J.J. et al. 2015. Deep Sea Res. Pt. I: Oceanogr. Res. Pap. 96: 49–58.
8. See Fig. 4 in Gilbert, J.M. et al. 2015. Mar. Pollut. Bull. 92: 186–194.
9. Steidinger, K.A. and Ingle, R.M. 1972. Env. Let. 3(4): 271–278.
10. Lyons, K. et al. 2015. J. Fish Biol. 87: 200–211.
11. Cotronei, S. et al. 2018. Bull. Env. Contam. Toxicol. 101: 7–13.
12. Cotronei, S. et al. 2018. Bull. Env. Contam. Toxicol. 101: 7–13.
13. Nakata, H. 2005. Env. Sci. Tech. 39: 3430–3434.
14. Xue, X. et al. 2017. Env. Sci. Tech. 51: 780–789.
15. Gelsleichter, J. and Szabo, N.J. 2013. Sci. Tot. Env. 456: 196–201.
16. Belcher, C.N. and Jennings, C.A. 2010. Env. Biol. Fish. 88: 349–359.
17. Yates, P.M. et al. 2015. PLoS One 10(4): p.e0121346.
18. Jennings, D.E. et al. 2008. Env. Biol. Fish. 83: 369–377.
19. Curtis, T.H. et al. 2013. Mar. Coast. Fish. 5: 28–38.
20. Madigan, D.J. et al. 2012. Proc. Natl. Acad. Sci. 109: 9483–9486.
21. Thorhaug, A. et al. 1979. Env. Conserv. 6: 127–137.

22. Coles, S.L. 2018. Mar. Res. Indonesia 19: 52–71.

23. https://www.plasticseurope.org/en/resources/publications/274-plastics-facts-2017. (Accessed 8/9/19).

24. Geyer, R. et al. 2017. Sci. Adv. 3(7): p.e1700782.

25. Although we cite the specific sources of these data, we were informed by this excellent review: Bonanno, G. and Orlando-Bonaca, M. 2018. Env. Sci. Pol. 85: 146–154.

26. Derraik, J.G.B. 2002. Mar. Pollut. Bull. 44: 842–852.

27. Sazima, I. et al. 2002. Mar. Pollut. Bull. 44: 1149–1151 and Wegner, N.C. and Cartamil, D.P. 2012. Mar. Pollut. Bull. 64: 391–394 and Colmenero, A.I. et al. 2017. Mar. Pollut. Bull. 115: 436–438.

28. Thiel, M. et al. 2018. Front. Mar. Sci. 5: 238.

29. Alomar, C. and Deudero, S. 2017. Env. Pollut. 223: 223–229.

30. Bernardini, I. et al. 2018. Mar. Pollut. Bull. 135: 303–310.

31. Cartes, J.E. et al. 2016. Deep Sea Res. Pt. I: Oceanogr. Res. Pap. 109: 123–136.

32. Cartes, J.E. et al. 2016. Deep Sea Res. Pt. I: Oceanogr. Res. Pap. 109: 123–136.

33. Avio, C.G. et al. 2015. Mar. Env. Res. 111: 18–26.

34. Anastasopoulou, A. et al. 2013. Deep Sea Res. Pt. I: Oceanogr. Res. Pap. 74: 11–13.

35. Miranda, D.D.A. and de Carvalho-Souza, G.F. 2016. Mar. Pollut. Bull. 103: 109–114.

36. Elise Pullen, personal communication.

37. Fossi, M.C. et al. 2017. Comp. Biochem. Physiol. C: Toxicol. Pharmacol. 199: 48–58.

38. Chapuis, L. et al. 2019. Sci. Rep.: 6924.

39. Cecilia Krahforst, personal communication.

40. Casper, B.M. et al. 2012. In: Popper, A.N. and Hawkins, A. (eds.). The effects of noise on aquatic life. Springer. 93–97 pp.

41. Casper, B.M. et al. 2012. In: Popper, A.N. and Hawkins, A. (eds.). The effects of noise on aquatic life. Springer. 93–97 pp.

42. Jennings, D.E. et al. 2008. Env. Biol. Fish. 83: 369–377.

43. Jennings, D.E. et al. 2012. J. Coast. Conserv. 16: 405–428.

44. Mcowen, C.J. et al. 2017. Biodiver. Data J.: 5.

45. Rosa, R. et al. 2017. Biol. Let. 13: 20160796.

46. Schlaff, A.M. et al. 2014. Rev. Fish Biol. Fish. 24: 1089–1103.

47. Bangley, C.W. et al. 2018. Sci. Rep. 8: 6018.

48. Robinson, L.M. et al. 2015. Deep Sea Res. II: Topic. Stud. Oceanogr. 113: 225–234.

49. Dulvy, N.K. et al. 2014. Elife 3: e00590.

50. Dulvy, N.K. et al. 2014. Elife 3: e00590.

51. Kendall, H. 1992. http://www.ucsusa.org/about/1992-world-scientists.html. (Accessed 11/11/19).

52. Here are two recommendations: Support Shark Advocates International (for sharks), and Dogwood Alliance (for climate change and forest protection).

Appendix

Conservation: Efforts to Protect Sharks and Rays

The following sharks are considered Critically Endangered or Endangered by the IUCN.

CRITICALLY ENDANGERED SHARKS

Pondicherry Shark, *Carcharhinus hemiodon*
Ganges Shark, *Glyphis gangeticus*
New Guinea River Shark, *Glyphis garricki*
Irawaddy River Shark, *Glyphis siamensis*
Natal Shyshark, *Haploblepharus kistnasamyi*
Daggernose Shark, *Isogomphodon oxyrhynchus*
Striped Smoothhound, *Mustelus fasciatus*
Sawback Angelshark, *Squatina aculeata*
Smoothback Angelshark, *Squatina oculata*
Angelshark, *Squatina squatina*

ENDANGERED SHARKS

Borneo Shark, *Carcharhinus borneensis*
Smoothtooth Blacktip Shark, *Carcharhinus leiodon*
Harrisson's Dogfish, *Centrophorus harrissoni*
Winghead Shark, *Eusphyra blochii*
Speartooth Shark, *Glyphis glyphis*
Whitefin Topeshark, *Hemitriakis leucoperiptera*
Honeycomb Izak, *Holohalaelurus favus*
Whitespotted Izak, *Holohalaelurus punctatus*
Broadfin Shark, *Lamniopsis temminickii*
Narrownose Smoothhound, *Mustelus schmitti*
Whale Shark, *Rhincodon typus*
Scalloped Hammerhead, *Sphyrna lewini*
Great Hammerhead, *Sphyrna mokarran*
Argentine Angelshark, *Squatina argentina*
Taiwan Angelshark, *Squatina formoa*
Hidden Angelshark, *Squatina guggenheim*
Angular Angelshark, *Squatina punctata*
Zebra Shark, *Stegosoma fasciatum*
Sharpfin Houndshark, *Triakis acutipinna*

Protecting Sharks and Rays

In this section, we discuss some of the more prominent measures undertaken by governments and nongovernmental organizations, internationally and in the United States, in the name of shark and ray conservation and management.

International Efforts to Protect Sharks and Rays

Convention on International Trade in Endangered Species of Wild Fauna and Flora (CITES)

CITES is an agreement currently between 183 governments *to ensure that international trade in specimens of wild animals and plants does not threaten their survival.* The following species of elasmobranchs are listed[1]: Basking Shark (*Cetorhinus maximus*), Whale Shark (*Rhincodon typus*), White Shark (*Carcharodon carcharias*), seven species of sawfish, Oceanic Whitetip Shark (*Carcharhinus longimanus*), Porbeagle (*Lamna nasus*), Scalloped, Great, and Smooth Hammerheads (*Sphyrna lewini, S. mokarran,* and *S. zygaena,* respectively), manta and devil rays (Mobulidae), Silky Shark (*C. falciformis*), and threshers (Alopiidae). Sawfish (Pristidae) are listed in CITES Appendix I, *species threatened with extinction. Trade in specimens of these species is permitted only in exceptional circumstances.* All others are in CITES Appendix II, *those not necessarily threatened with extinction, but in which trade must be controlled in order to avoid utilization incompatible with their survival.*

Convention on the Conservation of Migratory Species of Wild Animals (CMS)

Part of the United Nations Environment Programme, CMS works for *conservation and sustainable use of migratory animals and their habitats* . . . through . . . *internationally coordinated conservation measures.* CMS has 127 signatories, and the list does not include the United States.[2] In 2012, the *Memorandum on the Conservation of Migratory Sharks,* the first global document of this type for migratory species of sharks, included a conservation plan.[3]

Currently 29 species of sharks and rays are listed in Annex I of the Memorandum of Understanding. These include most of those listed for CITES above, plus the Longfin (*Isurus paucus*) and Shortfin Mako (*I. oxyrinchus*) and Spiny Dogfish (*Squalus acanthias*).

Global Sharks and Rays Initiative (GSRI)

Begun in 2015 and sponsored by the UN, the GSRI is a collaborative global shark conservation strategy whose goal is to coordinate and catalyze the efforts of global, regional, and national nongovernmental organizations (NGOs). Their plan, documented in *Global Priorities for Conserving Sharks and Rays: a 2015–2025 Strategy,*[4] is that by 2025, *the conservation status of the world's sharks and rays has improved—declines have been halted, extinctions have been prevented, and commitments to their conservation have increased globally.*

International Union for the Conservation of Nature (IUCN) International Plan of Action for Conservation and Management of Sharks (IPOA-Sharks)

The IUCN, founded in 1948, is best known for its *Red List of Threatened Species*,[5] but it is also *the global authority on the status of the natural world and the measures needed to safeguard it.*[6] The IUCN Shark Specialist Group (SSG), one of whose members is co-author Grubbs, was established in 1991 by the IUCN Species Survival Commission to identify and prioritize species at risk, monitor threats, and evaluate conservation action.

In 1999, the SSG produced the Food and Agriculture Organization of the United Nations (FAO) International Plan of Action for the Conservation and Management of Sharks (IPOA-SHARKS).[7] The IPOA-Sharks recommended the following 10 principles:

1 / Ensure that shark catches from directed and nondirected fisheries are sustainable.

2 / Assess threats to shark populations, determine and protect critical habitats, and implement harvesting strategies consistent with the principles of biological sustainability and rational long-term economic use.

3 / Identify and provide special attention, in particular to vulnerable or threatened shark stocks.

4 / Improve and develop frameworks for establishing and coordinating effective consultation involving all stakeholders in research, management, and educational initiatives within and between States.

5 / Minimize the unutilized incidental catches of sharks.

6 / Contribute to the protection of biodiversity and ecosystem structure and function.

7 / Minimize waste and discards from shark catches.

8 / Encourage full use of dead sharks.

9 / Facilitate improved species-specific catch and landings data and monitoring of shark catches.

10 / Facilitate the identification and reporting of species-specific biological and trade data.

International Shark Finning Bans and Policies

In addition to being, or as part of being, signatories to international agreements on shark conservation, numerous countries, as well as other governing bodies (e.g., the Inter-American Tropical Tuna Commission, or IATTC) have enacted full or partial bans on shark finning.[8] In most cases, it remains legal to sell fins from sharks landed with intact fins attached.

To date more than 20 countries have also banned shark fishing within some or all of their territorial waters by designating them as shark sanctuaries.

US Laws

Magnuson-Stevens Fishery Conservation and Management Act

In federal waters of the United States (i.e., from three to 200 mi (4.8–322 km), the primary law governing fisheries management is the Magnuson-Stevens Fishery Conservation and Management Act. From the shore to three miles, management of marine resources is in the jurisdiction of the state, who typically work closely with federal managers.

According to the National Marine Fisheries Service (NMFS), the objectives of the Magnuson-Stevens Act are to prevent overfishing, rebuild overfished stocks, increase long-term economic and social benefits, and insure a safe and sustainable supply of seafood.

The Magnuson-Stevens Act is administered by the NMFS, a part of the National Oceanic and Atmospheric Administration (NOAA). Fisheries in federal waters are managed by one of eight regional councils composed of stakeholders nominated by state governors and representing the public, scientists, commercial and recreational fishers, and the fishing industry.

The major means by which fisheries are regulated are *Fishery Management Plans* (FMPs), which conduct assessments of the health of the *stock* in question, and then establish scientifically sound regulatory measures.

Because many sharks, along with tunas, swordfish, and other billfish, migrate and thus cross numerous jurisdictional lines, they have been managed under the direct auspices of NOAA since 2006, specifically the Consolidated Atlantic Highly Migratory Species Fishery Management Plan.[9]

Two amendments to the Magnuson-Stevens Fisheries Act exclusively concerned sharks: the *Shark Finning Prohibition Act of 2000* which, according to NOAA, *prohibits any person under U.S. jurisdiction from engaging in the finning of sharks, possessing shark fins aboard a fishing vessel without the corresponding carcass, and landing shark fins without the corresponding carcass*; and the *Shark Conservation Act of 2010* which, again according to NOAA, *requires that all sharks in the United States, with one exception, be brought to shore with their fins naturally attached*, closing a loophole in the finning prohibition act.

US Endangered Species Act

Although in danger of being weakened by the federal government as we write, the 1973 US Endangered Species Act still remains one of the most effective tools to protect species in the United States threatened with extinction.

The NMFS administers the program for marine and anadromous (those moving between fresh and salt water) organisms, while the US Fish and Wildlife Service is responsible for terrestrial and freshwater organisms.

Among the 93 Endangered and 73 Threatened fishes listed by 2018 are three species of sharks and five batoids. An organism listed as endangered is in imminent danger of extinction whereas threatened organisms are considered likely to become endangered. The first sharks listed as endangered, in 2014, were two distinct populations of the Scalloped Hammerhead, those in the Eastern Pacific and Eastern Atlantic Oceans. Populations of Scalloped Hammerheads in the Central and Southwest Atlantic and Indo-West Pacific were listed as Threatened. This listing was followed by the addition of the Angel Shark as Endangered and, in early 2018, the Oceanic Whitetip Shark as Threatened. Among batoids, five species of sawfish are listed as Endangered, and these were the first elasmobranchs listed on the US Endangered Species List.

In addition to protecting species, the idea of ecological restoration is also built into the Endangered Species Act, which requires that *the ecosystems upon which [threatened and endangered species] depend* be protected.

If you ever encounter an endangered elasmobranch (e.g., a sawfish) in the wild, you should not approach it or attract it, but you should report it to the National Marine Fisheries Service or, specifically for sawfish, to the International Sawfish Encounter Database (ISED).[10]

Shark Fin Trade Elimination Act of 2017

The *Shark Fin Trade Elimination Act of 2017*, as we write, is still only a bill. The bill makes it illegal *to possess, buy, sell, or transport shark fins or any product containing shark fins, except for certain dogfish fins. A person may possess a shark fin that was lawfully taken consistent with a license or permit under certain circumstances.*

NOTES

1. https://www.cites.org/prog/shark. (Accessed 8/9/19).

2. https://www.cms.int/en/parties-range-states. (Accessed 8/9/19).

3. https://www.cms.int/sites/default/files/document/Sharks_Conservation_Plan_E_0.pdf. (Accessed 8/9/19).

4. Bräutigam, A. et al. Global Priorities for Conserving Sharks and Rays: A 2015–2025 Strategy. Available at: http://fscdn.wcs.org/2016/02/10/1cxcak0agd_GSRI_GlobalPriorities ForConservingSharksAndRays_web_singles.pdf. (Accessed 8/9/19).

5. https://www.iucnredlist.org. (Accessed 8/9/19).

6. https://www.iucn.org/about. (Accessed 8/9/19).

7. https://www.iucnssg.org/ipoa-sharks.html. (Accessed 8/9/19).

8. https://awionline.org/content/international-shark-finning-bans-and-policies.(Accessed 8/9/19).

9. https://www.fisheries.noaa.gov/management-plan/consolidated-atlantic-highly-migra tory-species-management-plan. (Accessed 8/9/19).

10. https://www.floridamuseum.ufl.edu/sawfish/report-encounter/. (Accessed 8/9/19).

Page numbers in *italics* refer to figures.